Weather Radar Information and Distributed Hydrological Modelling

Some Other IAHS Publications

Application of Geographic Information Systems in Hydrology and Water Resources Management. Proceedings of the HydroGIS'96 Conference held at Vienna, April 1996
edited by K. Kovar & H. P. Nachtnebel
Publ. no. 235 (1996) price £58.00, 712 + xii pages
ISBN 0-947571-84-1
With permission this publication was translated (by Liu Dongqing & Liu Yunian) into Chinese and published by: Hohai University Press

Remote Sensing and Geographic Information Systems for Design and Operation of Water Resources Systems. Proceedings of a symposium at the Fifth IAHS Scientific Assembly, Rabat April 1997
edited by Michael F. Baumgartner, Gert A. Schultz & A. Ivan Johnson
Publ. no. 242 (1997) price £38.00, 260 + xii pages
ISBN 1-901502-15-5

Regionalization in Hydrology. Proceedings of a conference held at Braunschweig, March 1997
edited by Bernd Diekkrüger, Mike J. Kirkby & Ulrich Schröder
The German National Committee of the International Hydrological Programme of UNESCO and the Operational Hydrology Programme of the World Meteorological Organization provided financial sponsorship for this publication.
Publ. no. 254 (1997) price £41.50, 260 + xii pages
ISBN 1-901502-80-5

Hydrological Extremes: Understanding, Predicting, Mitigating. Proceedings of Symposium HS1 held during IUGG 99, Birmingham, July 1999
edited by L. Gottschalk, J.-C. Olivry, D. Reed & D. Rosbjerg
The UK Environment Agency provided financial sponsorship for this publication
Publ. no. 255 price £46.50, 314 + xiv pages
ISBN 1-901502-85-6

Interactions between the Cryosphere, Climate and Greenhouse Gases. Proceedings of Symposium HS2 held during IUGG 99, Birmingham, July 1999
edited by Martyn Tranter, Richard Armstrong, Eric Brun, Gerry Jones, Martin Sharp & Mark Williams
Publ. no. 256 (1999) price £42.50, 282 + x pages
ISBN 1-901502-90-2

Integrated Methods in Catchment Hydrology—Tracer, Remote Sensing and New Hydrometric Techniques. Proceedings of Symposium HS4 held during IUGG 99, Birmingham, July 1999
edited by Chris Leibundgut, Jeff McDonnell & Gert Schultz
Publ. no. 258 (1999) price £43.00, 284 + xii pages
ISBN 1-901502-01-5

The Hydrology–Geomorphology Interface: Rainfall, Floods, Sedimentation, Land Use. Proceedings of a conference held at Jerusalem, May 1999
edited by Marwan A. Hassan, Olav Slaymaker & Simon M. Berkowicz
The Hebrew University of Jerusalem provided financial sponsorship for this publication
Publ. no. 261 (2000) price £49.00, 326 + x pages
ISBN 1-901502-16-3

Debris-Covered Glaciers. Proceedings of a workshop held at Seattle, September 2000
edited by M. Nakawo, C. F. Raymond & A. Fountain
The International Arctic Research Center, Fairbanks, Alaska, and the Glaciological Expedition in Nepal provided financial sponsorship for this publication
Publ. no. 264 (2000) price £44.00, 288 + viii pages
ISBN 1-901502-31-7

Remote Sensing and Hydrology 2000. Proceedings of a symposium held at Santa Fe, New Mexico, April 2000
edited by Manfred Owe, Kaye Brubaker, Jerry Ritchie & Albert Rango
Publ. no. 267 (2001) price £80.00, 610 + xiv pages
ISBN 1-901502-46-5

The Extremes of the Extremes: Extraordinary Floods. Proceedings of a symposium held at Reykjavík, July 2000
edited by Árni Snorrason, Helga P. Finnsdóttir & Marshall E. Moss
Publ. no. 271 (2002) price £60.00, 394 + xiv pages
ISBN 1-901502-66-X

FRIEND 2002—Regional Hydrology: Bridging the Gap between Research and Practice. Proceedings of the Fourth FRIEND Conference held at Cape Town, March 2002
edited by Henny A. J. van Lanen & Siegfried Demuth
co-editors E. Servat, R. Ibbitt, C. Prudhomme, M. F. Bari, D. Hughes, M. del C. Llasat & S. H. Mkhandi
Publ. no. 274 (2002) price £73.50, 518 + x pages
ISBN 1-901502-81-3

A Methodology for the Assessment of Surface Resistance and Soil Water Storage Variability at Mesoscale Based on Remote Sensing Measurements: A Case Study with HAPEX-EFEDA Data. A contribution to UNESCO IHP projects H1.1 and H2.1.
by W. G. M. Bastiaanssen, D. H. Hoekman & R. A. Roebeling
Sponsored by the Department of Water Resources of the Agricultural University, Wageningen
Special Publ. no. 2 (1994) price £13.00, 66 + vi pages
ISBN 0-947571-93-0

Remote Sensing Applications to Hydrology compiled on behalf of the IAHS International Committee on Remote Sensing and Data Transmission
edited by Albert Rango & Jerry Ritchie
Hydrological Sciences Journal 41(4) (1996) price £18.00
220 + xii pages (12 papers)

*More details at **www.iahs.info** including abstracts of all papers since 2000.*

Please send orders and enquiries about books to:
Mrs Jill Gash, IAHS Press, Centre for Ecology and Hydrology, Wallingford, Oxfordshire OX10 8BB, UK
tel: +44 1491 692442; fax: +44 1491 692448/692424;
jilly@iahs.demon.co.uk

IAHS members benefit from discounts on book prices: 25% for all members, and a subsidized 80% discount for members in financially disadvantaged countries on titles published before the current year; IAHS membership is free of charge.

All prices include delivery by surface mail. Payment may be made by credit card (VISA, MasterCard, EUROCARD).

*Please address orders and enquiries about **Hydrological Sciences Journal** to:*
Mrs Frances Watkins, IAHS Press, Centre for Ecology and Hydrology, Wallingford, Oxfordshire OX10 8BB, UK
tel: +44 1491 692405; fax: +44 1491 692448/692424;
frances@iahs.demon.co.uk

Weather Radar Information and Distributed Hydrological Modelling

Edited by

YASUTO TACHIKAWA
Disaster Prevention Research Institute, Kyoto University, Uji, Kyoto 611-0011, Japan

BAXTER E. VIEUX
School of Civil Engineering and Environmental Science, University of Oklahoma, 202 West Boyd Street, Norman, Oklahoma 73019, USA

KONSTANTINE P. GEORGAKAKOS
Hydrologic Research Center, 12780 High Bluff Drive, Suite 250, San Diego, California 92130, USA

EIICHI NAKAKITA
Department of Urban and Environmental Engineering, Kyoto University, Sakyo-ku, Kyoto 606-8501, Japan

Proceedings of an international symposium (Symposium HS03) held during IUGG 2003, the XXIII General Assembly of the International Union of Geodesy and Geophysics, at Sapporo, Japan, from 30 June to 11 July 2003. The symposium was organized by the International Commission on Surface Water (ICSW) of the International Association of Hydrological Sciences (IAHS).

IAHS Publication no. 282
in the IAHS Series of Proceedings and Reports

Published by the International Association of Hydrological Sciences 2003
IAHS Press, Centre for Ecology and Hydrology, Wallingford, Oxfordshire OX10 8BB, UK
IAHS Publication no. 282
ISBN 1-901502-37-6

British Library Cataloguing-in-Publication Data.
A catalogue record for this book is available from the British Library.

The papers included in this volume have been peer reviewed and some were extensively revised before publication.

IAHS is indebted to the employers of the Editors for the invaluable support and services provided that enabled them to carry out their task effectively and efficiently.

The conveners of the symposium would like to express their thanks for financial support obtained from a "Leadership for International Scientific Cooperation" Project (PI: Prof. Kaoru Takara, Disaster Prevention Research Institute, Kyoto University) under the Special Coordination Funds for Promoting Science and Technology provided by the Ministry of Education, Culture, Sports, Science and Technology (MEXT) of Japan.

The editors would like to express special thanks to Cate Gardner for coordination and organization of all the publishing processes, to Penny Farnell for her efforts in preparing the final version of this book, and to Toshiharu Kojima for providing the cover illustration.

The papers were checked, formatted and assembled by Penny Farnell at IAHS Press, Wallingford, UK.

Publications in the series of Proceedings and Reports are available *only* from:
IAHS Press, Centre for Ecology and Hydrology, Wallingford, Oxfordshire OX10 8BB, UK
tel.: +44 1491 692442; fax: +44 1491 692448; e-mail: jilly@iahs.demon.co.uk

Printed in The Netherlands by Krips BV, Meppel.

Preface

As radar hydrology is rapidly developing, its combined use with distributed hydrological modelling creates powerful numerical tools for forecasting storms, floods and related natural disasters. Such tools when used with historical observations can also contribute rich information for water resources assessment. In the last 20 years or so, considerable attention has been focused on radar hydrology and distributed hydrological modelling. However, fundamental issues of model development associated with the existence of uncertainties or errors in radar rainfall estimation and prediction, and in distributed model parameters, still remain unresolved. Clarification as to how hydrological prediction accuracy will be improved, or how prediction/estimation uncertainty will be reduced by combining spatially distributed radar rainfall information with a distributed hydrological model, was a basic motivation for organizing the present volume. Scale issues of distributed hydrological modelling also need further study. Suitable model resolution and required forcing resolution might be a function of catchment scale, but the relationship between them is still unknown. In association with the scale issues, assessing in what environment a distributed hydrological model shows better performance than a lumped model, is also an important research topic.

To address these research issues, a symposium entitled "Weather Radar Information and Distributed Hydrological Modelling" was held during IUGG 2003, the XXIII General Assembly of the International Union of Geodesy and Geophysics from 30 June to 11 July 2003 at Sapporo, Japan. The symposium was organized by the International Commission on Surface Water (ICSW) of the International Association of Hydrological Sciences (IAHS).

This volume of proceedings contains 42 papers presented at the symposium, and is organized into four sections:

- Weather Radar Information for Hydrological Prediction
- Assessment of Uncertainty in Distributed Hydrological Modelling
- Assessment of Performance in Distributed Hydrological Modelling
- Application of Distributed Hydrological Models in Watershed Management

The first group of 11 papers addresses radar rainfall estimation and prediction combined with hydrological models and numerical simulation of precipitation, as well as precipitation field analysis related to hydrological predictions. The next group of 10 papers focuses on the uncertainties in radar rainfall estimation/prediction and in parameter distribution in distributed hydrological modelling, and these papers contribute to efforts to assess the utility of a distributed hydrological model. The next 11 papers address the model performance of distributed hydrological models associated with parameter sensitivity as well as scale issues in model resolution and forcing resolution. The final group of 10 papers demonstrates the applications of distributed hydrological models to various areas of many different hydrological environments, and provides examples of distributed hydrological models for watershed management.

The editors hope that this volume contributes to hydrology and water resources management through the development of radar hydrology and distributed hydrological modelling.

Yasuto Tachikawa
Disaster Prevention Research Institute, Kyoto University, Japan

Baxter E. Vieux
School of Civil Engineering and Environmental Science, University of Oklahoma, USA

Konstantine P. Georgakakos
Hydrologic Research Center, San Diego, USA

Eiichi Nakakita
Department of Urban and Environmental Engineering, Kyoto University, Japan

Contents

4 Application of Distributed Hydrological Models in Watershed Management

1 Weather Radar Information for Hydrological Prediction

Effect of radar beam geometry on radar rainfall estimation

SIRILUK CHUMCHEAN
School of Civil and Environmental Engineering, The University of New South Wales, Sydney 2052, New South Wales, Australia
siriluk@civeng.unsw.edu.au

ALAN SEED
Cooperative Research Centre for Catchment Hydrology, Bureau of Meteorology, Melbourne, Victoria, Australia

ASHISH SHARMA
School of Civil and Environmental Engineering, The University of New South Wales, Sydney 2052, New South Wales, Australia

Abstract The frequency, accuracy and resolution of hydrological records is a major limitation in the accurate modelling of hydrological events. The weather radar provides real-time spatially continuous measurements covering a large area at short time intervals. However, considerable uncertainty remains in the procedures used to estimate the rainfall from weather radar observations. This uncertainty may be caused by the variability of raindrop size distribution, the variation of reflectivity with height and with range, and the temporal and spatial resolutions adopted for sampling the radar reflectivity. This paper accounts for the effect of the radar beam geometry as a function of distance. The conical shape of the radar beam causes the observed volume to increase with range from the radar, leading to both bias and an increase in the standard error associated with the measured reflectivity as a function of range. To remove the bias caused by the radar beam spreading, a simple-scaling transformation is proposed. The results show that the transformed reflectivity becomes relatively free from range dependent bias, which leads to more accurate relations with the measured ground rainfall than what is obtained otherwise. A 6-month long rainfall-reflectivity record from the Kurnell radar at Sydney, Australia is used to illustrate the efficiency and applicability of the reflectivity scaling transformation, compared to radar rainfall algorithms used conventionally.

Key words rainfall; scaling; weather radar

INTRODUCTION

Radar rainfall estimation traditionally involves the use of a parametric relation formulated based on measurements of radar reflectivity and ground (point) raingauge rainfall. However, such relations are often uncertain and their use in practical scenarios often leads to significant bias in the estimated rainfalls. Numerous factors are responsible for this uncertainty, including relating of reflectivity measurements that reflect above ground rainfall, to values of rainfall measured at ground level, use of point measurements of ground rainfall as a surrogate of pixel averaged values, the variability of rainfall drop size distribution, the variation of reflectivity with height and with range, the temporal and spatial resolutions adopted for sampling the radar reflectivity, and radar hardware miscalibration and noise. To accommodate for such uncertainties, a probabilistic approach such as the Probability Matching Method (PMM), (Rosenfeld *et al.*, 1993), has been used to eliminate the need for specifying a formal relationship between reflectivity and rainfall. The rationale behind the PMM method is that rainfall and reflectivity values for a specified exceedence probability can be considered to be equal to each other. It is assumed that the cumulative distribution function (CDF) is invariant with range.

A conical shape of the radar beam causes the volume of a radar bin to increase as the square of the distance to the radar. Therefore, the small intense features that are present in a rain field will be averaged out by the measurement process, thereby leading to an underestimation of the probability of high intensity echoes at far range. Assuming that reflectivity's CDF is independent of range results in an uncertain representation of the reflectivity, which leads to uncertainty in the radar estimated rainfall. Hence it is imperative that the measured reflectivity be transformed to a variable that can be considered to be free from range dependent bias before use in estimating radar rainfall. This paper presents two related concepts that attempt to address some of the problems identified above. Firstly, we formulate and evaluate a reflectivity scale transformation function that assumes reflectivity to be a simple-scaling variable; secondly we apply the scale transformation function in two radar rainfall estimation methods and evaluate the effectiveness of using the transformed reflectivity in estimating radar rainfall.

EFFECT OF RADAR BEAM SPREADING IN MEASURED REFLECTIVITY

The volume of a radar bin increases as the square of the range of the bin to the radar. The increased volume means that the small but intense features in a rain field are averaged out as the range from the radar increases. This means that the upper tail of the CDF will be biased at far range. This sensitivity of rainfall statistics to scale is well known and can be modelled using the scaling hypothesis for rainfall (Lovejoy & Schertzer, 1985).

Hourly raingauge rainfall obtained from 6-month data that occurred in Sydney, Australia, during November 2000 to April 2001 and the corresponding 1.5 km CAPPI radar reflectivity data were used to investigate the effect of radar beam spreading in the measured reflectivity. The effect of bright band and a different observation altitude at far range are the other sources of error that cause the range dependent bias in measured reflectivity. In order to avoid the biases caused by these two sources of error, only the measured reflectivity and raingauge data that lie within 100 km from the radar were used in this study. Note that the climatological freezing level of the Sydney area during the study months are above 2.5 km and the height of the base scan beam centre at 100 km from the radar is 1.9 km above the ground, which can be considered to be not overly different from 1.5 km. Therefore, we assume that there is no bias caused by the bright band effect and different observation altitude in the 1.5 km CAPPI data that lie within 100 km from the radar. To avoid the effect of noise and false interpretation caused by hail in the measured radar reflectivity, the reflectivity values that are less

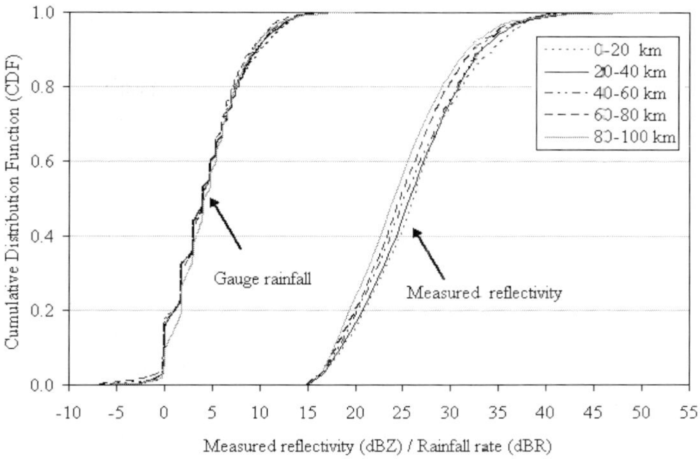

Fig. 1 Measured reflectivity and raingauge rainfall CDF.

than 15 dBZ and greater than 55 dBZ were excluded from the analysis. Raingauge rainfall data used in this study were obtained from tipping bucket raingauges. This type of raingauge records the time of bucket tips, hence they are subject to significant quantization error at low rainfall intensity. Hence, only the rainfall amounts that are greater than the volume of that gauge's tipping bucket were used in this study.

To show that the measured reflectivity CDF is a function of range, hourly raingauge rainfall obtained from the 6-month data and the corresponding 1.5 km CAPPI radar data were used to estimate the CDFs for raingauge rainfall and reflectivity data for each 20-km range interval, as illustrated in Fig. 1. In general, the probability of high reflectivity values decreases as function of range in the radar reflectivity, but this dependence is not observed in the gauge data and must therefore be an artefact.

SIMPLE SCALING HYPOTHESIS FOR HOURLY MEASURED REFLECTIVITY

From Fig. 1 we can see that the measured reflectivity can be considered as a random variable characterized by the same CDF with properly re-scaled parameters. A scale transformation function can be derived if one assumes that measured reflectivity at different ranges are connected through generalized scaling relation. Therefore, in this study the proposed transformation function is derived based on the simple scaling theory of rainfall. The hourly 1.5 km CAPPI data that lie within 100 km from the radar were used in estimating a scaling exponent. The simple scaling theory of time and space properties of rainfall proposed by Menabde *et al.* (1999) has been applied to the measured reflectivity. The reflectivity CDF of the 20-km range interval was selected as a reference reflectivity CDF. We assume that "simple scaling" holds for the measured reflectivity (in dBZ units). This assumption has been verified by estimating the scaling of moments of measured reflectivity at different moment orders (q). Figure 2 shows

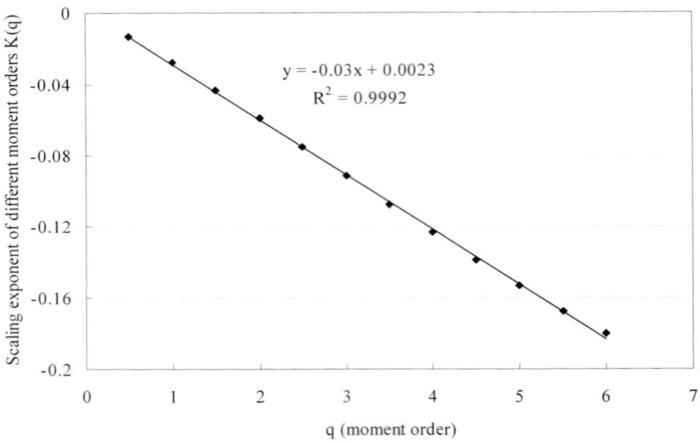

Fig. 2 Scaling exponent for hourly measured reflectivity.

that the dependence on q of the scaling of moments is precisely linear with scaling exponent of 0.03. This confirms the hypothesis about simple scaling in the measured reflectivity, therefore the proposed scale transformation function can be written as:

$$Z_{transformed(dBZ)} = \left(\frac{d}{20}\right)^{0.03} Z_{d(dBZ)} \quad (1)$$

where d (km) is the observation range of the measured reflectivity (beyond 20 km), Z_d (dBZ) is the measured reflectivity at range d. This proposed transformation function was used to transform the measured reflectivity at difference range interval to have same CDF as measured reflectivity at the 20-km range interval. The CDF of transformed reflectivity of each range interval was estimated as illustrated in Fig. 3. It can be seen that applying the scale transformation formula helps to transfer reflectivity's CDF of different range intervals to be close to the 20-km reflectivity's CDF. The mean of the transformed reflectivity of each range interval is corresponding to the mean of raingauge rainfall as shown in Fig. 4. This is a confirmation of the scaling hypothesis of the measured reflectivity, and thereby application of the scale transformation function helps to remove range dependent bias caused by the effect of the radar beam spreading. Radar and gauge data from three events that occurred across the Sydney area were then used to evaluate the effectiveness of using the transformed reflectivity in radar rainfall estimation.

APPLICATION TO KURNELL RADAR DATA

The Kurnell radar transmits radiation with wavelength of 5.3 cm and produces a beam with a 3 dB width of 0.94°. The reflectivity data are in Cartesian grids with 256 km × 256 km extent and a 1 km^2, 10-min resolution. Hourly reflectivity values were obtained by accumulating of 10-min reflectivity data. It should be noted that the hourly reflectivity values used in this study used the method of Fabry *et al.* (1994) to account for the movement of the rainfall field between the instantaneous rainfall

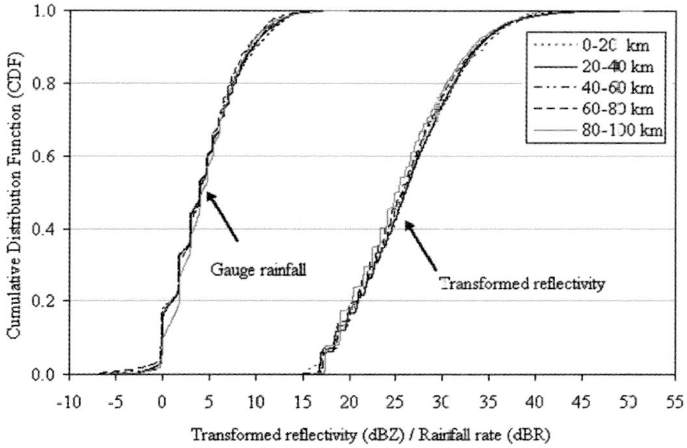

Fig. 3 Transformed reflectivity and raingauge rainfall CDF.

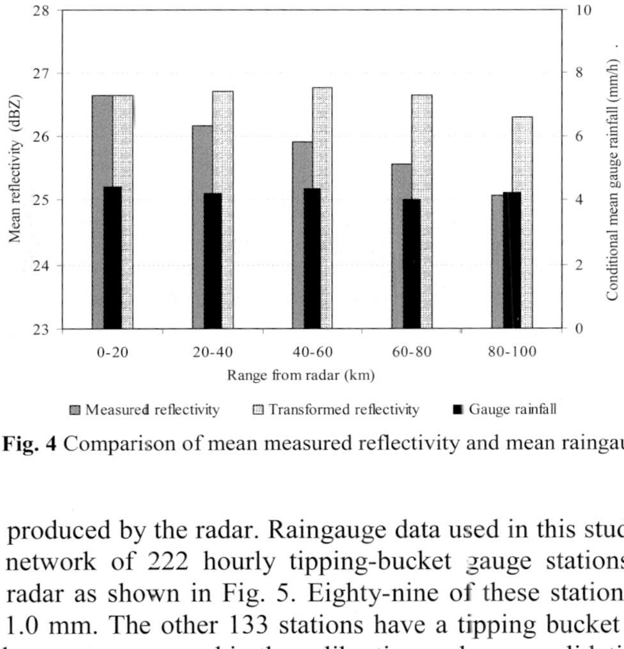

Fig. 4 Comparison of mean measured reflectivity and mean raingauge rainfall.

intensity fields produced by the radar. Raingauge data used in this study were obtained from a dense network of 222 hourly tipping-bucket gauge stations located within 100 km of the radar as shown in Fig. 5. Eighty-nine of these stations have a tipping bucket size of 1.0 mm. The other 133 stations have a tipping bucket size of 0.5 mm. Three case study events were used in the calibration and cross-validation. The first two case study events were widespread rainfalls that pass over Sydney during 30 January 2001–1 February 2001 and 19 April–22 April 2001, respectively. The third case study event was a convective event that occurred during 30 November–1 December 2000.

Altogether, four plausible rainfall calibration strategies were investigated. The methods were: (a) parametric Z–R relationship; (b) parametric Z–R relationship with transformed reflectivity; (c) PMM; and (d) PMM with transformed reflectivity. Note that for the parametric Z–R relationship, the parameter b was fixed to be equal to 1.5 and then calibrated parameter A by minimizing the mean square error (MSE) between

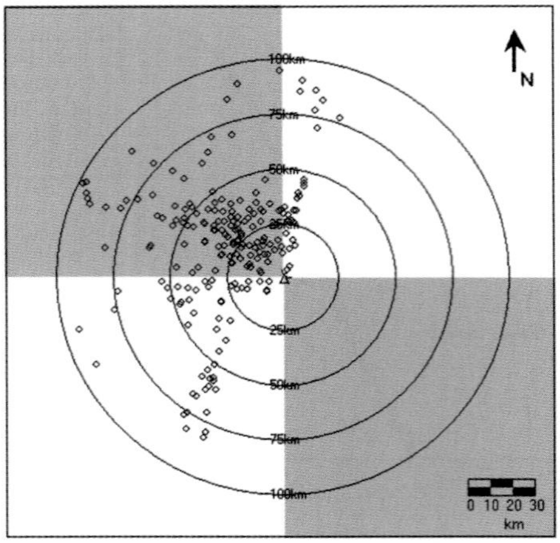

Fig. 5 Raingauge network.

radar and raingauge rainfall. The analysis was performed in an hourly time step by randomly selecting half of the available raingauges for the calibration; the remaining raingauges were used for the cross-validation of the algorithms. In order to investigate the range dependent bias induced by the effect of radar beam spreading, the gauge-radar (*G*/*R*) ratios at the raingauge's locations were estimated. The relative dispersion coefficient of variation about $\overline{G/R}$ (Wilson *et al.*, 1979) of each calibration and cross-validation set was calculated and used to evaluate the effectiveness in reduction of bias in the radar rainfall estimates after applying of scale transformation function in the measured reflectivity. The relative dispersion coefficient represents the ratio (expressed as a percentage) of the standard deviation of (*G*/*R*), where *G* and *R* represent the accumulated and the average (*G*/*R*) ratio over each storm. The accuracy of the radar estimated rainfall was investigated by a comparison of the *R*-square (R^2) statistic obtained from the calibration and cross-validation results. The R^2 is used as a dimensionless measure of model accuracy in the results presented.

RESULTS AND DISCUSSION

To investigate the sensitivity of the model performance obtained to the choice of raingauges network used in calibration and cross-validation, the analyses were repeated 100 times by changing the raingauge combinations used in the calibration and cross-validation. It is noted that the raingauge were selected in a purely random manner. The results presented in this study were the mean values obtained from the 100 re-calibrations. The results presented in Table 1 indicate that the use of transformed reflectivity in estimating radar rainfall can reduce the relative dispersion coefficient of variation about $\overline{G/R}$ approximately by 3% and 2% for calibration and cross-validation, respectively. It is interesting to note that the improvement in terms of relative dispersion is not obviously significant. This is because the random effects dominate, which causes the improvement in the bias to contribute little to the statistic.

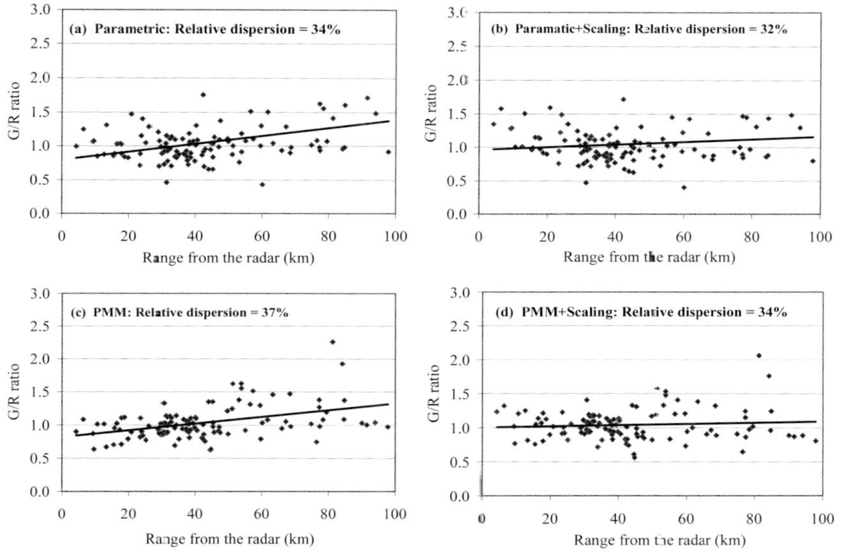

Fig. 6 *G/R* ratio for the validated raingauges of the 19 April 2001 event.

Table 1 Summary of gauge-radar comparisons of the calibration and cross-validation.

Events	Number of gauges	Storm duration (h)	$\overline{R}_g$ (mm)	Relative dispersion about $\overline{G/R}$ (%): Parametric	Parametric + Scaling	PMM	PMM +Scaling
30 Jan 01	105/104	70	6.2/5.9	34/32	31/30	35/33	32/31
19 Apr 01	111/111	82	3.2/3.2	29/28	27/26	31/30	28/28
30 Nov 00	108/107	47	3.5/3.7	45/45	42/44	44/45	41/44

$\overline{R}_g$ is the conditional mean of raingauge rainfall (R_g> raingauge's resolution mm h^{-1}).
(xxx/yyy) = (calibration /cross-validation).

However, it is apparent that slopes of the *G/R* ratios as a function of range are significant flatter when the transformed reflectivity values have been used in both the conventional radar rainfall estimation methods as illustrated in Fig. 6.

The accuracy of radar rainfall was estimated in term of the classical R^2 measures based on 100 sets of gauge combinations. These results are presented in Table 2. The difference in the R^2 results between the calibration and cross-validation of the two widespread events obtained from the parametric method are insignificant, which indicates the stability of the parameters when applied to the widespread rainfall. This optimistic assessment might be influenced by the fact that the validation gauges are placed in the vicinity of the calibration gauges, and thus are not fully independent. However, a reduction in accuracy of radar rainfall when validating for the convective case is evident in Table 2. This is because differences in rainfall drop size distribution are expected to be highly variable in both space and time in the convective rain, leading to a relatively greater level of independence between the calibration and cross validation gauges. The R^2 results of the three case study events indicate that an

Table 2. Sensivity in model performance.

Statistic	R^2				*A*-parameter	
	Parametric	Parametric + Scaling	PMM	PMM + Scaling	Parametric	Parametric + Scaling
30 Jan 01	0.699/0.677	0.703/0.687	0.684/0.671	0.689/0.678	60.11	63.53
19 Apr 01	0.443/0.441	0.450/0.448	0.325/0.211	0.333/0.225	97.99	111.93
30 Nov 00	0.257/0.184	0.260/0.184	0.022/-0.088	0.036/-0.08	275.32	305.44

(xxx/yyy) = (calibration /cross-validation)

increase in accuracy of radar rainfall is insignificant, even if the transformed reflectivity values that can be considered to be independent of range have been used in estimating radar rainfall. This might be because some other sources that cause uncertainty in radar rainfall, such as the variation of rainfall drop size distribution within the event, the reliability of measured reflectivity which is reduced with range due to observation problems, attenuation effect and the error of using point raingauge rainfall in representing mean-areal rainfall of a radar grid size, still have not been taken into account in this study. Although the improvement of the accuracy of radar rainfall estimates in the term of R^2 is insignificant, it still points to the effectiveness of applying scale transformation in the measured reflectivity.

Acknowledgements The authors gratefully acknowledge Mahanakorn University of Technology (Thailand) for funding the first author's PhD studies at the University of New South Wales. We also thank the Australian Bureau of Meteorology and the Sydney Water Corporation for providing the radar and raingauge data used in this study.

REFERENCES

Fabry, F., Bellon, A., Duncan, M. R. & Austin, G. L. (1994) High resolution rainfall measurements by radar for very small basins: the sampling problem re-examined. *J. Hydrol.* **116**, 415–428.

Lovejoy, S. & Schertzer, D. (1985) Generalized scale invariance and fractal model of rain. *Water Resour. Res.* **21**, 1233–1250.

Menabde, M., Seed, A. & Pegram, G. (1999) A simple scaling model for extreme rainfall. *Water Resour. Res.* **35**(1), 335–339.

Rosenfeld, D., Wolff, D. B. & Atlas, D. (1993) General probability-match relations between radar reflectivity and rain rate. *J. Appl. Met.* **32**, 50–72.

Wilson, J. W. & Brandes, E. A. (1979) Radar measurement of rainfall—a summary. *Bull. Am. Met. Soc.* **60**, 1048–1058.

Evaluation of dual polarization radar for rainfall–runoff modelling: a case study in Sydney, Australia

PHILLIP JORDAN, ALAN SEED, PETER MAY & TOM KEENAN
Bureau of Meteorology, PO Box 1289K, Melbourne, Victoria 3001, Australia
p.jordan@bom.gov.au

Abstract Dual polarization radars have advantages that may translate into more accurate measurements of rainfall than are provided by conventional, single polarization radars. An investigation was performed into how much additional benefit will be delivered to practical rainfall–runoff modelling by dual polarization radar over the single polarization radar approach. The benefit of dual polarization is compared against the impact of using calibration raingauge networks for single polarization radar rainfall. The investigation is framed in the context of a case study: a widespread rainfall event of approximately 2 days duration producing a small flood in a partially urbanized basin of 354 km^2 area.

Key words dual polarization radar; single polarization radar; URBS rainfall–runoff model

INTRODUCTION

The application of meteorological radars to quantitative rainfall measurement and hydrological forecasting is increasing. Radar has obvious advantages for rainfall estimation in hydrology: detailed spatial and temporal resolution over an extensive spatial domain, collected by a single remote device, with the ability to make informed nowcasts of future rainfall. However, limitations on the accuracy of conventional single polarization radar in rainfall measurement have been acknowledged for some time. Commonly recognized sources of error include beam blockage, ground detection, hail contamination, variations in raindrop size distribution, attenuation, the vertical profile of reflectivity, wind induced drift, and effects of spatial and temporal sampling (see for example Austin, 1987; Harrold *et al.*, 1974; Jordan *et al.*, 2000).

Over the past decade, significant progress has been made in radar rainfall measurement using dual polarization radar. Signals returned by the horizontal and vertical polarized beams provide additional parameters that can be combined into a multi-parameter estimate of rainfall intensity. Multi-parameter rainfall estimates should: (a) be less susceptible to variations in the power of the radiation emitted by the radar; (b) reduce the impact of variations in rain drop size distribution; (c) be much less affected by attenuation of the rainfall beam as it passes through areas of rainfall; and (d) be much less affected by blocking of the radar beam by the surrounding terrain, which allows the beam to scan at lower elevations (Zrnić & Ryzhkov, 1996).

Hydrologists aim to make the best use of the available radar data while recognizing its shortcomings. Evaluation the accuracy of rainfall–runoff model forecasts made using radar data is a major step in this process (Carpenter *et al.*, 2001)

and there have been many studies that have applied single polarization radar rainfall data to rainfall–runoff models (see Borga, 2002 for a recent list). There are just two case studies in the literature that apply dual polarization radar data to flood forecasting. The papers by Ogden *et al.* (2000) and Yates *et al.* (2000) each considered flooding caused by intense convective thunderstorms on basins of 25 and 130 km^2, respectively. In each case study there was evidence that the dual polarization radar rainfall measurements produce more accurate modelled hydrographs than single polarization radar. However, in each study the hydraulic conductivity parameter, which has significant control on modelled runoff generation, was calibrated to the particular event to improve the rainfall–runoff model results.

Numerous studies (starting with Brandes, 1975) have shown that a network of reporting raingauges can successfully reduce the bias in radar rainfall estimates. Raingauge calibration of a reflectivity based rainfall accumulation field is an alternative or complement to a dual polarization approach. The cost of the calibration network obviously increases with its density.

This paper investigates how much additional benefit will be delivered to rainfall–runoff modelling by dual polarization radar over the conventional single polarization radar approach. The benefit of dual polarization is compared against the impact of using calibration raingauge networks of varying density. The investigation is framed in the context of a case study: a widespread rainfall event of approximately 2 days duration producing a small flood in a partially urbanized basin of 354 km^2 area.

STUDY BASIN AND INSTRUMENTATION

The Georges River basin is located approximately 30 km southwest of Sydney city, Australia. It drains in a roughly south to north direction to Liverpool Weir (33.925°S, 150.925°E). Upper reaches of the Georges basin are forested or partially cleared rural land and the lower reaches are urbanized (approximately 19% of the total basin area) with low-density industrial and residential development. Mean annual rainfall is 870 mm, with flood producing rainfall possible in any month of the year. An automatic water level recorder has been installed at Liverpool Weir since 1968 and the maximum recorded streamflow is 540 m^3 s^{-1}.

The Bureau of Meteorology Research Centre (BMRC) operates a C-Band polarmetric (C-POL) radar (Keenan *et al.*, 2001) that was located at Badgery's Creek (33.894°S, 150.725°E), between 12 and 48 km northwest of the Georges River basin (see Fig. 1). The radar transmits radiation with a wavelength of 5.5 cm and produces a beam with a 3 dB width of 1°. C-POL measures several raw data fields including horizontal reflectivity (Z_h), differential reflectivity (Z_{dr}), differential phase shift (Φ_{dp}) and zero lag cross-correlation coefficient (ρ_{hv}). C-POL was operated in volume-scan mode, collecting data from 15 elevation angles every 10 min. The data resolution was 1.0° in azimuth and 300 m in range.

Raingauge data were obtained from 128 tipping bucket rainfall stations that are located within 105 km of C-POL. Tipping bucket sizes vary across the network between 0.2, 0.5 and 1.0 mm. The spatial distribution of raingauges is non-uniform, with none over the Pacific Ocean (east of the radar) and none in parts of the Blue Mountains (to the west). Therefore, the effective area of coverage of the gauge

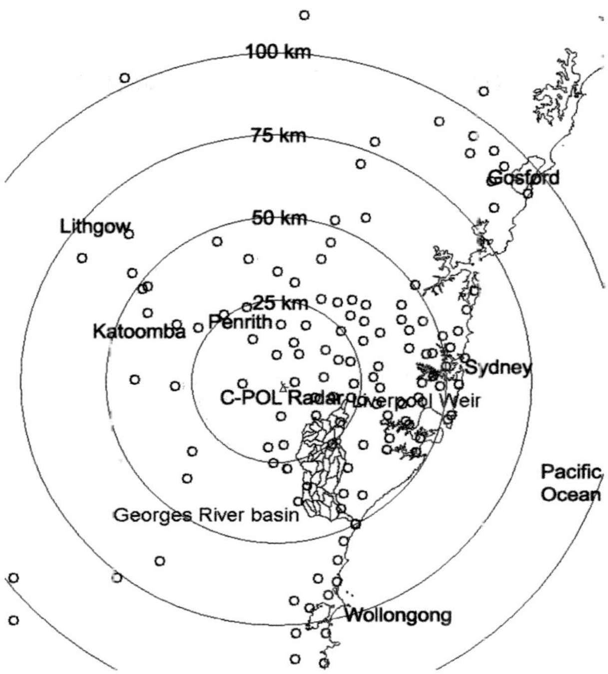

Fig. 1 Map showing area surrounding the C-POL radar, including the outline of the Georges River basin, the sub-areas of its URBS model, and the 128 raingauge network (small circles).

network is approximately 19 000 km^2 (see Fig. 1). This represents a raingauge network density of about 1 per 150 km^2. Less dense raingauge networks were simulated by randomly selecting gauges from the full network.

The Australian Bureau of Meteorology provides a flood warning service for the Georges River. Flood forecasts are made using the URBS rainfall–runoff routing model. The URBS model (Carroll, 1996) subdivides the Georges River basin along ridges in the topography into 47 sub-areas of similar area (as shown in Fig. 1), each represented by a rainfall-input node. Non-linear storages are used to represent the sub-area routing response, with the storage–discharge equation for each subarea given by:

$$S_{sa} = \beta\sqrt{A}Q^m \qquad (1)$$

where S_{sa} is the sub-area storage volume, β is the sub-area lag parameter, A is the area of the sub-area, Q is the outflow from sub-area and m is the sub-area nonlinearity parameter.

URBS connects the sub-area storages using channel segments, with flows routed using the linear storage–discharge equation:

$$S_{ch} = \alpha L[xQ_u + (1-x)Q_d] \qquad (2)$$

where S_{ch} is the channel storage volume, α is the channel lag parameter (the inverse of average flood wave speed), L is the length of reach, Q_u is the inflow to the upstream

Table 1 Calibrated parameter values for the Georges River basin URBS model.

Parameter name	Value
Model time step	60 min
Channel lag parameter, α	0.2448
Channel translation parameter, x	0.3
Sub-area lag parameter, β	3.473
Sub-area nonlinearity parameter, m	0.8
Impervious area	50% of urban area
Volumetric runoff coefficient (Impervious areas)	1.0
Volumetric runoff coefficient (Pervious areas)	0.2237

end and Q_d is the outflow from the downstream end of the reach and x is the translation parameter. A simple initial loss-proportional loss model is used to determine runoff generation from pervious areas. The semi-distributed nature of URBS allows for the spatial variability in the rainfall pattern to be utilised in prediction of flood hydrographs.

The URBS model for the Georges River basin was calibrated to five small flood events that occurred between November 2000 and March 2001. The floods varied in duration between 1 and 5 days and had peak flows of between 29 and 88 $m^3 s^{-1}$. Model parameters were calibrated using data from the raingauge network (with a density of 1 per 300 km^2) and the streamflow gauge at Liverpool Weir. The calibrated parameter values, shown in Table 1 were obtained using the shuffled complex evolution optimization algorithm (Duan *et al.*, 1992). The initial loss from pervious areas was calibrated independently for each of the five calibration events.

MODELLING APPROACH

A widespread rainfall system passed over the domain of C-POL from 19 April 2001 06:00 UTC to 22 April 2001 15:00 UTC. This system produced rainfall accumulations over the Georges basin of between 40 and 70 mm, with the most common rainfall intensities being less than 5 mm h^{-1}. A small flood event occurred in the Georges River, with a peak flow of 22.5 $m^3 s^{-1}$ recorded at 21 April 2001 23:00 UTC. The total recorded runoff is 2150 Ml, which represents average runoff from the basin of 6.08 mm.

Six different algorithms were used to estimate rainfall intensity fields, as listed in Table 2. The first algorithm is a conventional $R(Z_{hS})$ relationship for convective precipitation where Z_{hS} is the reflectivity at horizontal polarization that was determined after carrying out quality control using the single polarization data only and attenuation correction using the Hitschfeld & Bordan (1954) method. The second algorithm, $R(Z_{hD})$, uses the same rainfall intensity estimation algorithm but it uses Z_{hD}, which is the reflectivity at horizontal polarization that was determined using the dual polarization radar data. In the $R(Z_{hD})$ algorithm, a threshold was applied on the ρ_{hv} field of 0.95 to identify probable areas of ground clutter, anomalous propagation and non-liquid precipitation and then the self-consistent rain profiling method (Bringi *et al.*, 2001) was used to perform attenuation correction on the reflectivity data. The four

Table 2 Equations for radar rainfall estimation algorithms.

Algorithm	Equation
$R(Z_{hS})$	$R = 0.015Z_{hS}^{0.734}$
$R(Z_{hD})$	$R = 0.015Z_{hD}^{0.734}$
$R(A_h)$	$R = \begin{cases} 243.2A_h^{0.79} & Z_{hD} > 35\,\text{dBZ and } A_h > 0\,\text{dBZkm}^{-1} \\ 0.015Z_{hD}^{0.734} & \text{otherwise} \end{cases}$
$R(A_h, Z_{dr})$	$R = \begin{cases} 677.1A_h^{0.96}Z_{dr}^{-2.08} & Z_{dr} > 1\text{dB}, Z_{hD} > 35\,\text{dBZ and } A_h > 0\,\text{dBZkm}^{-1} \\ 243.2A_h^{0.79} & Z_{dr} \le 1\text{dB}, Z_{hD} > 35\,\text{dBZ and } A_h > 0\,\text{dBZkm}^{-1} \\ 0.015Z_{hD}^{0.734} & \text{otherwise} \end{cases}$
$R(K_{dp})$	$R = \begin{cases} 32.4K_{dp}^{0.83} & Z_{hD} > 35\,\text{dBZ and } K_{dp} > 0.5\,°\text{km}^{-1} \\ 0.015Z_{hD}^{0.734} & \text{otherwise} \end{cases}$
$R(K_{dp}, Z_{dr})$	$R = \begin{cases} 46.8K_{dp}^{0.88}Z_{dr}^{-1.31} & Z_{dr} > 1\text{dB}, Z_{hD} > 35\,\text{dBZ and } K_{dp} > 0.5\,°\text{km}^{-1} \\ 32.4K_{dp}^{0.83} & Z_{dr} \le 1\text{dB}, Z_{hD} > 35\,\text{dBZ and } K_{dp} > 0.5\,°\text{km}^{-1} \\ 0.015Z_{hD}^{0.734} & \text{otherwise} \end{cases}$

phased-based algorithms, which are appropriate for estimation of rainfall intensity at the C-Band, were the same as those given in Keenan *et al.* (2001). The third and fourth algorithms, $R(A_h)$ and $R(A_h, Z_{dr})$, used the estimates of attenuation of horizontal reflectivity, A_h, produced by the self-consistent rain profiling method. The fifth and sixth algorithms, $R(K_{dp})$ and $R(K_{dp}, Z_{dr})$, used the specific differential phase, K_{dp}, which is the first derivative of the differential phase shift field with respect to range.

Accuracy of the radar estimated accumulation fields might be improved by calibrating the fields using a raingauge network. Real-time calibration of the rainfall

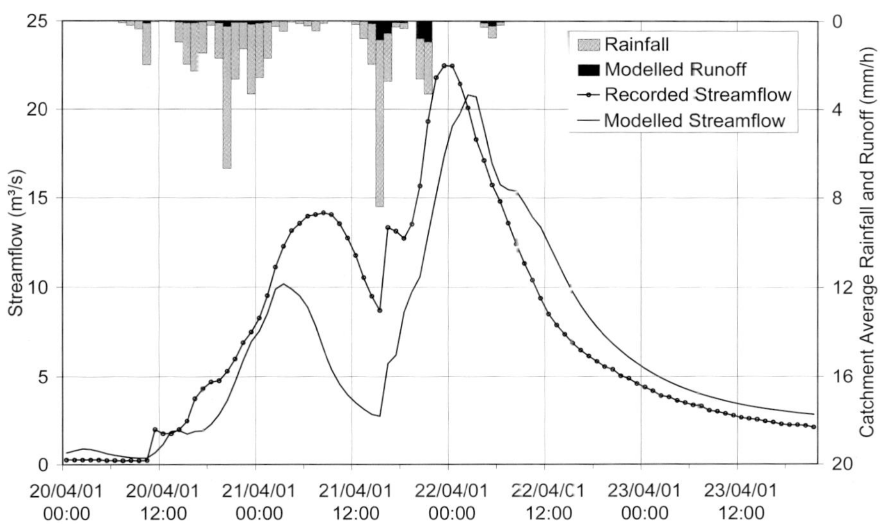

Fig. 2 Recorded and modelled streamflow hydrographs obtained using $R(Z_{hS})$ radar rainfall algorithm.

fields was simulated by calculating a calibration factor as the average ratio of the raingauge to radar estimated rainfall accumulations for the hour. Only locations where the radar and raingauge accumulations exceeded the greater of 1 mm or two bucket tips were included in the calculation of the calibration factor. The hourly radar accumulation field was then multiplied by calibration factor. Raingauge calibrated rainfall fields were produced for the $R(Z_{hS})$ algorithm rainfall fields using network densities of 150, 300, 600, 1200 and 2400 km^2 per gauge.

Rainfall estimated using each of the six radar algorithms was used as input to the URBS model for the Georges basin. The URBS model was also run with the single polarization rainfall input, calibrated with each of the five different gauge network densities. The quality of each modelled hydrograph was determined by computing the sum of square errors and coefficient of efficiency (CoE) (Nash & Sutcliffe, 1970) with the recorded hydrograph. URBS model parameter values were set to those determined from the calibration events, except for initial loss from pervious areas, which was calibrated individually for each of the rainfall input fields. However, results were not sensitive to the initial loss parameter, as more than 90% of the runoff in this small flood event was generated from the impervious areas.

RESULTS AND DISCUSSION

Figure 2 shows a typical modelled flood hydrograph produced by the URBS model, in this case with rainfall input from the $R(Z_{hS})$ algorithm. The fit to the observed hydrograph is good, with the total runoff, peak flow and hydrograph shape being well matched. The modelled flood hydrographs produced from all six radar measurement algorithms produce good predictions of the recorded hydrograph, as evidenced by the CoE values given in Table 3. Two of the dual polarization algorithms, $R(A_h)$ and $R(A_h, Z_{dr})$, produce better predictions of the recorded hydrograph than the prediction based upon the single polarization algorithm, $R(Z_{hS})$. This confirms that the rainfall measurement algorithms based upon attenuation fields, as estimated from differential phase shift data, produce more accurate flood hydrographs than those based on conventional reflectivity measurements.

Table 4 shows that calibration with the raingauge network improves the accuracy of the flood hydrographs predicted using single polarization radar data. The CoE generally improves as the network density increases, peaking for a raingauge density

Table 3 Summary statistics for the comparison between the recorded and modelled hydrographs for the different radar rainfall estimation algorithms.

Rainfall algorithm	Basin average rainfall (mm)	Modelled runoff (mm)	Sum of square streamflow errors ($m^6 s^{-2}$)	Coefficient of efficiency
$R(Z_{hS})$	56.12	5.39	1057.4	0.7380
$R(Z_{hD})$	57.05	5.49	1267.5	0.6859
$R(A_h)$	68.11	5.38	936.0	0.7681
$R(A_h, Z_{dr})$	61.16	5.28	993.9	0.7537
$R(K_{dp})$	67.60	5.23	1045.9	0.7409
$R(K_{dp}, Z_{dr})$	65.37	5.20	1090.2	0.7299

Table 4 Summary statistics for the comparison between the recorded and modelled hydrographs using the $R(Z_{hS})$ algorithm with different densities of raingauge calibration networks.

Average raingauge density (km^2 per gauge)	Number of calibration raingauges	Basin average rainfall (mm)	Modelled runoff (mm)	Sum of square streamflow errors (m^6s^{-2})	Coefficient of efficiency
Uncalibrated	*0*	*56.12*	*5.39*	*1057.44*	*0.7380*
2400	8	59.94	5.49	887.84	0.7800
1200	16	62.31	5.44	865.02	0.7857
600	32	61.85	5.52	774.93	0.8080
300	64	60.17	5.44	749.18	0.8144
150	128	61.44	5.60	848.12	0.7899

of 1 per 300 km^2. In this case study, even sparse calibration raingauge networks (2400 km^2 per gauge) were able to improve the predictions from the $R(Z_{hS})$ fields by more than any of the dual polarization algorithms on their own. Calibration was so effective in this case study because the rainfall was widespread enough to obtain an accurate estimate of the radar rainfall bias from the raingauge networks. Calibration of the best performing dual polarization algorithm field, $R(A_{h,} Z_{dr})$, with the 300 km^2 per gauge network produced a less accurate modelled flood hydrograph (CoE = 0.7585) than the single polarization radar data calibrated with the same network.

The results of this case study are conditioned upon the use of raingauge rainfall in calibration of the parameters of the rainfall–runoff model (Troutman, 1983). It is possible that, if radar rainfall were used to calibrate the URBS model, parameter values would adjust themselves so that predictions from the model using radar data would be further improved (Borga, 2002).

Dual polarization rainfall estimation algorithms have their greatest advantage over single polarization algorithms at higher rainfall intensities, typically greater than 10 mm h^{-1}.

It should also be noted that rainfall intensities in this widespread rainfall event were relatively low and that, under these conditions, improvements from dual polarization are likely to be relatively modest. Likewise, raingauge calibration is likely to be most effective in a widespread case, such as this, with a large number of radar and raingauge comparisons. It is likely that convective events, with higher rainfall intensities, will demonstrate the advantages of dual polarization more clearly. Future analysis of other flood events, both convective and widespread, should clarify these issues.

CONCLUSIONS

A case study was analysed of a widespread rainfall event of approximately 2 days duration, producing a small flood in a partially urbanized basin of 354 km^2 area. A semi-distributed rainfall–runoff routing model was used to predict flood flows from six different radar rainfall measurement algorithms and these were compared to the observed flood hydrograph.

Dual polarization radar rainfall measurement algorithms based upon attenuation fields, as estimated from differential phase shift data, produce more accurate flood hydrographs than those based on conventional reflectivity based algorithms. However, for this widespread rainfall event, calibration of the reflectivity based rainfall accumulation fields with an hourly reporting raingauge network at a density of 1 per 2400 km^2 produced a more accurate flood hydrograph than the best of the dual polarization algorithms. Accuracy of the modelled hydrograph improved with calibration network density, peaking at 1 raingauge per 300 km^2.

It is very likely that dual polarization algorithms will demonstrate clearer advantages in convective rainfall events, where rainfall intensities are greater and raingauge calibration is less effective at removing bias from the reflectivity based rainfall estimates.

Acknowledgements The authors thank Mr Ken Glasson of BMRC for operating and maintaining the C-POL radar during its operation at Badgery's Creek. We also acknowledge Mr Michael Whimpey of BMRC for his assistance in preparation of the C-POL radar data. The authors thank Sydney Water for providing most of the raingauge data used in the research.

REFERENCES

Austin, P. M. (1987) Relation between measured radar reflectivity and surface rainfall. *Mon. Weath. Rev.* **115**, 1053–1070.

Borga, M. (2002) Accuracy of radar rainfall estimates for streamflow simulation. *J. Hydrol.* **267**, 26–39.

Brandes, E. A. (1975) Optimising rainfall estimates with the aid of radar. *J. Appl. Met.* **14**, 1339–1345.

Bringi, V. N., Keenan, T. D. & Chandrasekar, V. (2001) Correcting C-Band radar reflectivity and differential reflectivity data for rain attenuation: a self-consistent method with constraints. *IEEE Trans. Geosci. Remote Sens.* **39**(9), 1906–1915.

Carpenter, T. M., Georgakakos, K. P. & Sperfslagea, J. A. (2001) On the parametric and NEXRAD-radar sensitivities of a distributed hydrologic model suitable for operational use. *J. Hydrol.* **253**, 169–193.

Carroll, D. G. (1996) *URBS-CM, A Catchment Management and Flood Forecasting Rainfall Runoff Routing Model.* Gutteridge Haskin and Davey, Brisbane, Australia.

Duan, Q., Sorooshian, S. & Gupta, V. (1992) Effective and efficient global optimisation for conceptual rainfall runoff models. *Water Resour. Res.* **28**(4), 1015–1031.

Harrold, T. W., English, E. J. & Nicholass, C. A. (1974) The accuracy of radar-derived rainfall measurements in hill terrain. *Quart. J. Roy. Met. Soc.* **100**, 331–350.

Hitshfeld, W. & Bordan, J. (1954) Errors inherent in the radar measurement of rainfall at attenuating wavelengths. *J. Appl. Met.* **11**, 58–67.

Keenan, T., Nan, Z., Lei, F., Bringi, V., Nystuen, J. & Whimpey, M. (2001) A comparison of radar rainfall estimators during the South China Sea Monsoon Experiment (SCSMEX) (30th Int. Conf. Radar Meteorology, Munich, Germany), 603–605. Am. Met. Soc.

Jordan, P. W., Seed, A. W. & Austin, G. L. (2000) Sampling errors in radar estimates of rainfall. *J. Geophys. Res.* **105**(D2), 2247–2257.

Nash, J. E. & Sutcliffe, J. V. (1970) River flow forecasting through conceptual models 1. A discussion of principles. *J. Hydrol.* **10**, 282–290.

Ogden, F. L., Sharif, H. O., Senarath, S. U. S., Smith, J. A., Baeck, M. L. & Richardson, J. R. (2000) Hydrologic analysis of the Fort Collins, Colorado, flash flood of 1997. *J. Hydrol.* **228**, 82–100.

Troutman, B. M. (1983) Runoff prediction errors and bias in parameter estimation induced by spatial variability of precipitation. *Water Resour. Res.* **19**, 791–810.

Yates, D. N., Warner, T. T. & Leavesley, G. H. (2000) Prediction of a flash flood in complex terrain. Part II: A comparison of flood discharge simulations using rainfall input from radar, a dynamic model and an automated algorithmic system. *J. Appl. Met.* **39**, 815–825.

Zrnić, D. S. & Ryzhkov, A. (1996) Advantages of rain measurements using specific differential phase. *J. Atmos. Oceanic Tech.* **13**, 454–464.

Radar raingauge with volume-scanning function

TAKEO KINOSITA
Suimon Kankyo, FT Building 6, 10 Nihonbasi-Hisamatsucho, Chuo-ku, Tokyo 103-0005, Japan
s.k.kinosita@mx9.ttcn.ne.jp

MINORU IRISAWA
Tone Dam Control Office, Ministry of Land, Infrastructure and Transport, 593 Moto-Soja, Maebasi, Gunma Prefecture 371-0846, Japan

Abstract The volume-scanning radar has been operated for 10 years at the top of Mount Akagi in Japan, monitoring rain clouds three-dimensionally every 15 min with 20 kinds of angles in a 200 km radius. The authors attempted to develop the availability of the weather radar for operational use, especially for operation of dam reservoirs. The main topics of this paper are to present parameter development and selection of altitude for rainfall intensity estimation, variable *B* and β method, bright band elimination, and vertical profile observations of rain clouds for monitoring heavy rainfall for disaster prevention. The authors developed the variable *B* and β method by using the volume-scanning mode, then implemented the method successfully for rainfall observation by weather radar.

Key words bright band; radar; Japan; variable *B* and β method; vertical profile of rain cloud; volume-scanning radar

INTRODUCTION

Radar is, needless to say, useful to observe rainfall instantaneously in a wide area. The authors have worked for weather radar development to utilize it as a raingauge for 40 years (Kinosita, 1984). In Japan, 26 sets of radar raingauges are now operating for monitoring rainfall from a hydrological viewpoint. At the early stage of the radar development, it could only show rain clouds on the cathode-ray tube qualitatively. Later it became possible to measure rainfall quantitatively by digital technology. Radar is generally operating as PPI (plane position indication) mode, which displays rain clouds on the tube like a rainfall map. Volume-scanning radar, called three-dimensional radar, was introduced for the advanced use of weather radar for hydrological purposes about 10 years ago at the top of Mount Akagi, which is located 100 km northwest of Tokyo, Japan. It has 20 kinds of beam angles and observes rainfall within a 200 km radius (see Fig. 1). Scanning almost the whole space of the atmosphere in the radius, it can get the radar reflectivity at any altitude with horizontally homogeneous conditions. By using the observed data, the accuracy of rainfall estimation will be much improved (Nakagawa *et al.*, 1998). On the other hand, PPI radar can scan the atmosphere on a flat conic plane due to a slight beam angle. The beam is low near the radar site and high at places far from the radar site, which means the height of the observed rainfall is different depending on the location. In this paper, parameter development and selection of the observation altitude for rainfall intensity estimation by using the volume-scanning mode are described, and the variable *B* and β method is presented for better estimation of precipitation.

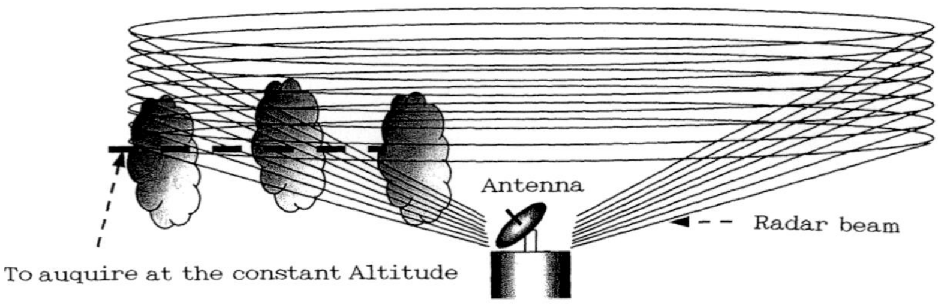

Fig. 1 Explanation of the volume-scanning mode.

APPLICATION

Parameter development

Conversion parameters B and β are indispensable to obtain rainfall intensity R from radar reflectivity Z. The relation is formulated as $Z = BR^{\beta}$ where B and β are dependent upon atmospheric and cloud-physical conditions. In most cases of parameter development, large numbers of observed R and Z pairs are collected and the regression line is empirically obtained. In contrast, in this paper, B and β are determined by limited numbers of pairs of R and Z at the most suitable sites. Two criteria were proposed to select preferable pairs of R and Z. First, radar data where the radar reflectivity is frequently missing are eliminated to identify the parameters. Second, rainfall data on ground sites at the radar shadows made by mountain topography are not used. These processes make the reasonable relations between R and Z, i.e. reasonable B and β.

Selection of altitude

As volume-scanning radar is operating three-dimensionally, any altitude can be selected for observation of rain clouds. For monitoring of rainfall on the ground, the radar beam should be as low as possible, but should not touch the ground. Even the ground clutter is removed by the MTI method. From the viewpoint of structures of rain clouds, a certain altitude may be the best height for Z measurement. It seems preferable to select the middle layer of rain clouds, because the atmosphere is perhaps well mixed and the raindrop distribution is uniform there. However, such layer changes depend on precipitation events. Thereby, from the topographical aspect, the Z measurement height is decided at 3000 m above mean sea level. The reason is that the radar site is located at 1600 m and mountains of 2000 to 2500 m height are located to the north and west of the radar site.

Variable B and β method

The conversion parameters B and β are usually given as fixed parameters, while atmospheric conditions change rapidly. The parameters should change due to changes

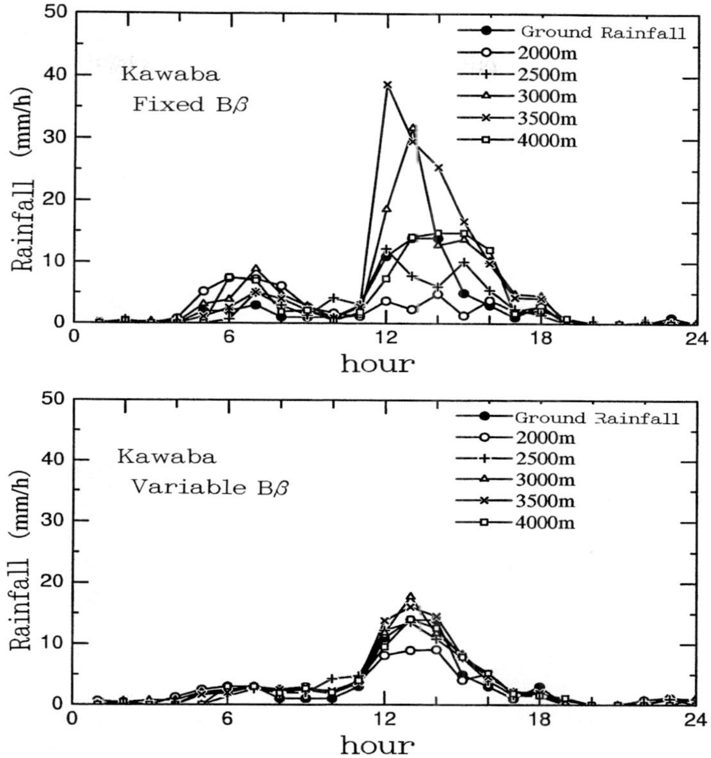

Fig. 2 Comparison of estimated precipitation used by fixed and variable B and β on 22 September 1996. Radar reflectivity at altitudes of 2000, 2500, 3000, 3500 and 4000 m above mean sea level were used.

in atmospheric conditions. The volume-scanning radar observes the reflectivity at various altitudes. Here, the parameters determined by the regression equation of R and Z were revised every hour by using telemetric rainfall data and the radar reflectivity at the adequate altitude. In Fig. 2, the upper figure shows the rainfall calculated by the fixed values $B = 200$ and $\beta = 1.6$. Dependent upon the altitude, converted rainfalls are different and also they do not coincide with ground rainfall. The lower figure is the estimated precipitation using the variable B and β. They show good agreement with the ground gauged rainfall.

Bright band

A bright band is sometimes found in radar observation. It looks like a doughnut on a PPI image. Erroneous reflectivity emerges in the layer of atmosphere just below 0°C, where liquid water attaches to ice particles in the layer. When a bright band appears, radar observation becomes erroneous in PPI mode. While in the volume-scanning mode, the problem of a bright band is avoidable. If a bright band appears on the

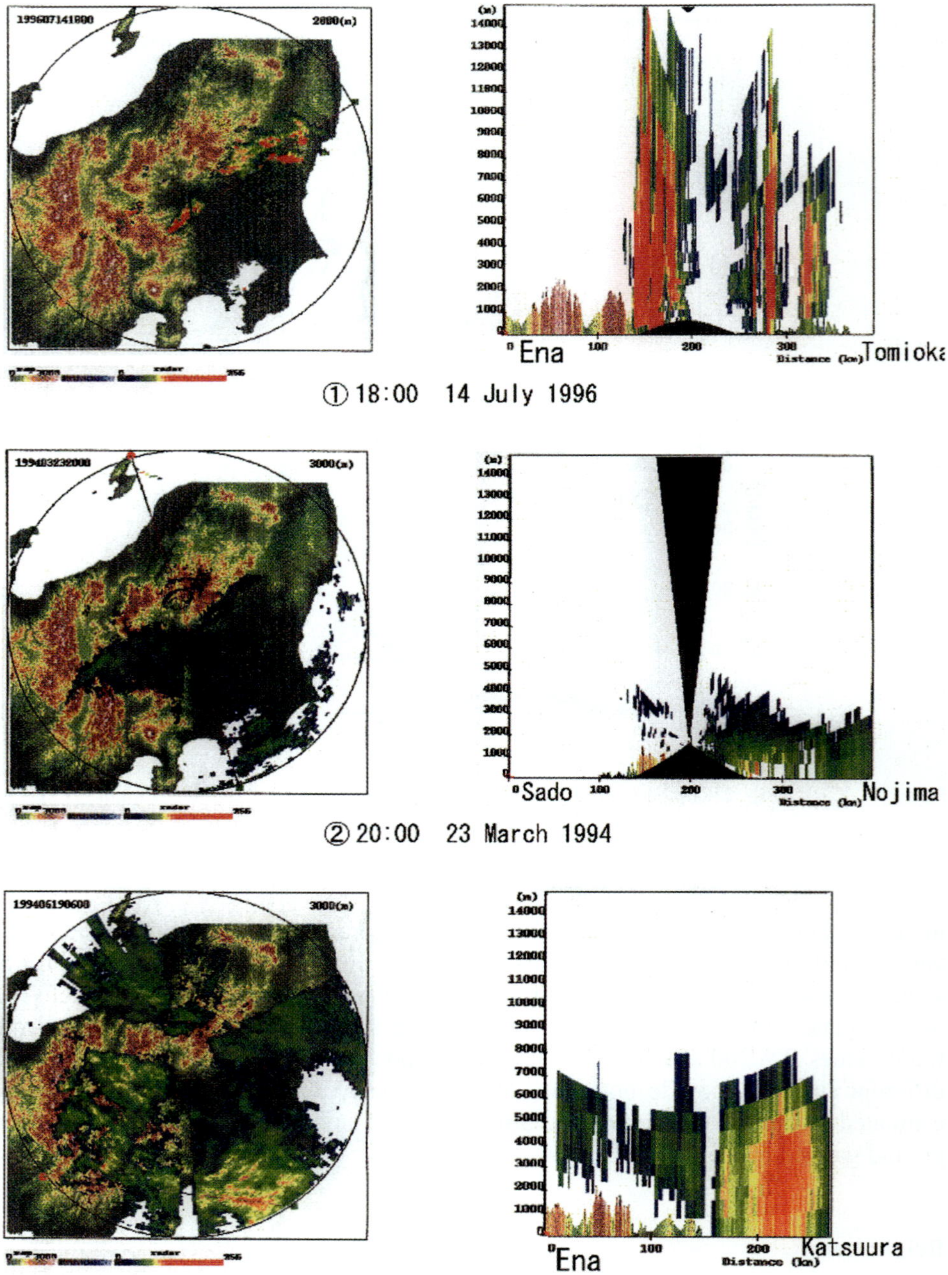

Fig. 3 Vertical profiles of rain clouds. The left images show plane indications of rain clouds at the altitude indicated at the upper right corners of the images. The right images show vertical profiles at the cross-sections indicated by a straight line between red and green points on the left image. In the right images, the abscissa axis distance is in km, and the ordinate is the altitude above mean sea level in m.

observation height, the variable B and β method automatically adjusts the parameters in accordance with the bright band; also Z values at other heights are used for avoiding the use of the bright band level reflectivity.

Vertical profile

The most beneficial volume-scanning function is to make images of a vertical profile. It is well known that RHI (range height indication) mode is available to show a vertical profile in radar. This profile is only made in the cross-section including the radar site, while volume-scanning radar can show any vertical profile in the observing area. Examples are shown in Fig. 3, a tall thundercloud whose top reaches 14 000 m higher than the tropopause, a low cloud cluster in spring, and a high rain cloud in a Baiu front, respectively. The images are useful to predict the movement of rainfall field for reservoir operation at least qualitatively (Nakagawa *et al.*, 1997).

CONCLUSION

The authors reported an advanced use of volume-scanning radar operated at Mount Akagi in Japan. Homogeneous reflectivity can be obtained by this mode, therefore the conversion of rainfall from the reflectivity is easy and accurate. By using the property, the variable B and β method was applied to obtain more reliable rainfall estimation in a real-time fashion for operational hydrology. The vertical profile of rain clouds is also drawn for monitoring of heavy rainfall.

REFERENCES

Kinosita, T. (1984) Development of operational use of radar raingauges. *J. Hydrol. Coast. Environ. Engng JSCE* **351**(II-2), 1–15 (in Japanese).

Nakagawa, K., Nakakita, E., Sato, T. & Ikebuchi, S. (1997) Parameters of rain drop size distribution and their vertical profile depending on rainfall type. *Ann. J. Hydraul. Engng JSCE* **41**, 141–146.

Nakagawa, K., Nakakita, E., Sato, T. & Ikebuchi, S. (1998) Analysis on vertical profile of DSD and building up an algorithm for estimating ground rainfall using radar. *Ann. J. Hydraul. Engng JSCE* **42**, 7–12.

Weather Radar Information and Distributed Hydrological Modelling (Proceedings of symposium HS03 held during IUGG2003 at Sapporo, July 2003). IAHS Publ. no. 282, 2003.

Evaluation of the applicability of radar rainfall information to operational hydrology

TADASHI MATSUURA, KAZUHIKO FUKAMI & JUNICHI YOSHITANI

Hydraulic Engineering Research Group, Public Works Research Institute, 1-6, Minamihara, Tsukuba, Ibaraki 305-8516, Japan

tmatuura@pwri.go.jp

Abstract This paper proposes a framework for evaluating the usefulness of radar rainfall information by different usage, and discusses the prospects and limitations of the further development of the radar system. The authors evaluate the usefulness of radar rainfall from the viewpoint of the user, not the radar developer or researcher, and they implement evaluations for each purpose. It was found that the radar raingauge is highly effective in estimating rainfall averaged over a river basin, which is important for flood forecasting and reservoir management; and a rainfall observation system composed of combination of ground raingauges and radar raingauges should be established for further development of the radar system.

Key words calibration; evaluation method; flood forecast; radar rainfall

INTRODUCTION

It has been more than 50 years since the concept of precipitation observation with weather radar was established, and more than 30 years since the radar system was fully studied and operated in Japan. However, it is still difficult to answer the questions: how accurately does the radar raingauge observe and how much can we expect its accuracy to improve? This is because the true values to test the rainfall accuracy observed with radar cannot be obtained. We usually define rainfall depth measured with the ground raingauge as rainfall. However, the ground raingauges only provide the amount of rain that pours on the raingauge of only about 20 cm in diameter, and does not widely measure distributed rainfall in the atmosphere. The true value to be compared cannot be known. Therefore, the accuracy of radar is always evaluated with an assumption or guess, and is different depending on various conditions or case studies, and thus quantitative evaluation of observation accuracy is difficult.

In this paper, rainfall observation accuracy of the radar raingauge was attempted from the viewpoint of radar data users. Because the required performance and accuracy of radar observation varies according to the purposes, such as weather forecast, flood forecast, sediment disaster warning, traffic management, and reservoir management, it is necessary to estimate accuracy and capability for each purpose to decide whether radar information is practically effective or not. For example, point rainfall information is required for sediment disaster warning and traffic management, while areal averaged rainfall in a large area in a river basin is important for flood forecasting and reservoir management. At first, an appropriate evaluation method for observation accuracy according to each usage is proposed. Secondly, the authors demonstrate case studies to show the advantages of the use of different evaluation methods according to usage and to evaluate the usefulness of the radar rainfall system from multi-functional aspects. The authors also evaluate the effectiveness of the calibration of radar rainfall with ground rainfall data for various operational applications.

CALIBRATION OF RADAR DATA

Calibration methods such as the uniform correction method, Ninomiya & Akiyama's method (1978), the range weight method, RWM (Brandes, 1975), the dynamic window method, DWM (Yamaguchi *et al.*, 1993a) and the kriging method (Yoshino *et al.*, 1988), have been proposed. The RWM and DWM methods are used in Japan.

The Japan Meteorological Agency (JMA) has worked to develop a practical application of radar observation and has provided Radar–AMeDAS rainfall, which is generated from the combination of weather radar and AMeDAS ground rainfall information with RWM (JMA, 1995). Values of calibration coefficient f for each radar mesh are calculated by giving range weight W, which takes a smaller value as ground raingauges are far from the radar mesh.

The Ministry of Construction (MOC) (Ministry of Land, Infrastructure and Transport, MLIT from 2001) had not provided the calibrated radar rainfall, because the radar system has been under development and the incomplete calibration system may produce erroneous rainfall intensity, while from 2001, MLIT began to distribute the calibrated radar data by using DWM, a method to decide the most appropriate sampling area by balancing variation of estimation accuracy. The calibration is carried out by using the averaged inverse number of spatial fluctuation coefficient (relation of variation between observed rainfall and radar rainfall) as the weight factor (Yamaguchi *et al.*, 1993a).

Traditionally, radar rainfall data have been calibrated to make the radar rainfall closer to the point ground rainfall observation. However, the DWM calibration method proposes to improve the accuracy of areal rainfall by combining both the advantages of point observation accuracy with ground raingauges and spatio-temporal observation ability with a radar raingauge. The calibration method is more appropriate when accurate areal rainfall is needed, such as flood forecasting, than when point rainfall observation accuracy is required.

EVALUATION OF HARDWARE ERROR ON RADAR RAINGAUGE SYSTEM

Before observation accuracy with radar raingauges is discussed, the error in the hardware itself has to be understood quantitatively. Ishizaki *et al.* (1986) analysed electrical and mechanical errors of the MOC radar raingauge hardware, focusing on a specific resolution of a radar instrument, a range of electrical loss and noise, etc. The results showed that, except for the attenuation due to the water film formed on the radome, a radar raingauge error of the hardware was about 1.0 dB as the amount of electrical wave fluctuation, which corresponds to less than about 15% of rainfall observation error when the radar constant $\beta = 1.6$ was used.

A ground raingauge also has various measurement errors. A point rainfall observation error of a ground raingauge properly placed on the flatland with no shielding, according to the installation standard, can be within 3% of verification tolerance, but when it is strongly affected by wind, the observation accuracy changes to a large extent. Koschmieder (1934) compared values of a raingauge under the ground not affected by wind and a gauge installed without any guard 1.1 m above the ground. The difference between the two observed values was large when wind velocity v was strong. According to the research, the observed value of the underground raingauge was about 1.4 times larger than the value of the other raingauge when v = 8 m s^{-1}, twice when v = 12 m s^{-1}, and about three times when v = 15 m s^{-1} (Kawabata *et al.*, 1972).

ACCURACY FROM THE VIEW OF POINT RAINFALL OBSERVATION

Nakao *et al.* (2001) analysed the MOC radar rainfall data all over Japan from June to October 1999, and calculated the correlation and the total rainfall ratio (TRR) between ground point rainfall and radar rainfall right above the ground stations. According to the analysis, the correlation coefficient r of hourly rainfall was higher than 0.6 at most of the stations and higher than 0.8 at many stations. In addition, r showed a tendency to become higher as rainfall intensity was stronger. Although the result gives an indication of current radar observation accuracy, raindrop drift effect by wind in the atmosphere should be considered to evaluate the point rainfall accuracy.

Nakao (1999) analysed radar data observed at the MOC's Takasuzu radar rain-gauge during the rainstorm in August 1998. He proved that the correlation value r between ground rainfall and the mesh radar rainfall right above the ground station was 0.71, while the correlation between the ground rainfall and the mesh radar rainfall on the windward side always showed higher values (r = 0.85–0.98). Ishizaki *et al.* (1986) examined correlation and average deviation between ground gauged data and radar data right above the ground station and eight neighbourhood meshes by using the high density ground raingauge observation in the Kurihashi area with 24 ground raingauges installed in an area of 15 km^2. In the research, it was shown that neighbourhood meshes had high correlation, but they did not correspond to wind direction and velocity in all cases.

The feature of sediment disaster caused by heavy rainfall is that there is almost no time left to evacuate once sediment disaster occurs. It is important to issue an effective warning and order for evacuation at an appropriate time because people have difficulties with evacuation for this sort of disaster. Local governments set up in guidelines of rainfall amount for each stream to issue evacuation warning and order against debris flow disaster based on the MOC guideline (1984). In practice, the warning and evacuation order are announced when the snake line (time series of rainfall intensity–rainfall depth relation) reaches WL (warning line) and EL (evacuation line), but there is some difficulty in giving the precise rainfall because of the sparse density of ground raingauge networks in cases of small-scale debris flow disaster.

The introduction of weather radar was expected to resolve the problem, but there is a difficulty in that the spatial resolution of radar information (several km) is too coarse for the scale of sediment disaster (250–500 m). Yamaguchi *et al.* (1993b) examined the radar data when rainstorm caused debris flow disaster in the southern part of Kyoto in 1986. They showed that the radar estimated rainfall intensity was an averaged value at 3 km resolution, which could not observe high regional variation of precipitation in a mountainous area; thus, the radar estimated rainfall could be underestimated in small-scale rainfall variation.

ACCURACY FROM THE VIEW OF POINT RAINFALL OBSERVATION WITH CALIBRATION

Yamaguchi *et al.* (1993b) examined the advantage of the calibration with DWM for the aforementioned debris flow disaster in 1986. They reported that the calibration improved the entire observation accuracy of rainfall distributions, but in the sediment disaster area the cumulative rainfall changed too little before and after the calibration. The ground raingauge density was too low to improve the underestimation of rainfall.

Nakao *et al.* (2001) applied RWM and DWM to MOC's radar data and examined the improvement of point rainfall observation accuracy by the calibration method. Both of the methods improved TRR for ground rainfall up to about 1.0, and there was a small difference in the calibration performance between the two methods. The research also showed that real-time calibration had better accuracy than extrapolated time calibration.

To use rainfall forecast based on radar rainfall observation for issuing sediment disaster warning, Yamakoshi *et al.* (2001) developed a new method to produce the snake line by calculating effective rainfall from short-term precipitation forecast data (1 h 10 min time resolution and 2.5 km spatial resolution) predicted by the Japan Weather Association. Hara *et al.* (2001) and Watari *et al.* (2002) improved the efficiency of the method to avoid underestimation by applying the maximum radar rainfall from about nine neighbouring radar meshes, including the mesh right above the study area as a forecast value.

ACCURACY FROM THE VIEW OF AVERAGED AREAL RAINFALL OBSERVATION

It is essential for flood forecasting and reservoir management to obtain averaged rainfall over river basins. However, there is a problem that areal rainfall cannot be gauged to use as a true value for the evaluation of radar rainfall.

Hashimoto (1977) discussed the accuracy and reliability of areal rainfall estimated from ground point rainfall data by using a sample design method (SDM). In the discussion, he reviewed the extant studies and research results with the thinning-out method (TOM) from within and outside of countries and found that the relation between the raingauge control area and the estimated error was different from basin to basin. He also illustrated the relation among the basin area, the number of rainfall observation stations and the average relative error (Hashimoto & Satou, 1974). It shows that in a 1000 km^2 basin, for instance, the relative estimation error of the basin average rainfall is less than about 15% with eight rainfall observation stations, and less than 10% with 20 stations.

Ishizaki *et al*. (1986) applied TOM and SDM for the high-density ground rainfall observation data in the Kurihashi area. They showed that, for the rainfall with more than certain rainfall intensity, the estimated error for representing average rainfall in a range of 15 km^2 by one point raingauge station was 40–60% and by six stations was about 10%. Also, with SDM, the estimated relative error was 6–8% and about 13% in thunderstorm with 24 observation stations at risk rate $2\alpha = 50\%$. Ishizaki *et al.* (1986) investigated the basin average rainfall estimation error by radar rainfall data with TOM at 900 km^2 in the Kurihashi and Shimokubo-dam areas in a mountainous district. The result revealed that the areal rainfall by radar was good.

ACCURACY FROM THE VIEW OF AREAL RAINFALL OBSERVATION WITH CALIBRATION

Matsuura *et al*. (2001) clarified the improvement effect of the areal rainfall observation accuracy by DWM with rainfall in the Syounai-gawa River basin (A = 1010 km^2) in August 1996. The result showed that TRR of ground areal rainfall and radar areal rainfall was 0.7 and correlation r was 0.95 before the radar data were calibrated, while TRR was improved to nearly 1.0 after the calibration. The paper also focused on rainfall averaged over the river basin in the 2000 Tokai heavy rainfall disaster. The rainfall amount averaged over the river basin using all ground rainfall data from JMA AMeDAS, MOC telemeter and Aichi prefecture was estimated, and the estimated value was compared against the previous result obtained by Hashimoto (1974). The ground rainfall averaged over the river basin was estimated with less than 10% error. First, Matsuura *et al*. (2001) compared the Thiessen and kriging methods for obtaining the averaged rainfall over the river basin using 20 stations and found little difference between these two methods. Then, they compared the areal rainfall amounts (up to 100–600 km^2) calculated using 20 ground raingauge data with the ones using MOC's radar data, which were calibrated with MOC telemeter rainfall data at 12 points in the river basin. The result was that the correlation r was larger than 0.95 in all averaged cases and TRR was 1.00 ± 0.06. Also it was shown that the difference between the values of cumulative rainfall during the flood period was quite small.

CONCLUSION

Availability and limitations of radar rainfall estimation for disaster prevention purposes have been discussed from the viewpoint of users. It was found that the radar raingauge was effective to estimate rainfall averaged over the river basin for flood forecast. On the other hand, point rainfall estimations in each mountain stream need to be improved in accuracy. As radar observation is the practical method to obtain rainfall where no ground raingauges are installed, it is necessary to establish and maintain the rainfall observation system composed of the combination of ground raingauge and radar raingauges. In addition, the data dissemination system and data format should be established to meet the various requirements of users.

Acknowledgement The authors would like to express our sincere thanks to Dr Fumio Yoshino for his valuable advice to achieve this study.

REFERENCES

Brandes, E. A. (1975) Optimizing rainfall estimates with the aid of radar, *J. Appl. Met.* **14**, 1339–1345.

Hashimoto, T. & Satou, I. (1974) Accuracy of spatial averaged rainfall and the number of rainfall observatories, *Civil Engng J.* **16-12,** 35–41 (in Japanese).

Hashimoto, T. (1977) Study on the estimation of the accuracy and reliability of areal rainfalls by sample design method, Report no. 149 of Public Works Research Instn, Ministry of Construction (in Japanese).

Hara, Y., Itou, A., Ono, H., Takagi, H., Tatara, H. & Yoshida, N. (2001) Applicability of every 10-minute short-term precipitation forecast (part 3), In: *Erosion Control Engineering* (Proc. Annual Conf.), 196–197 (in Japanese).

Ishizaki, K., Nakao, H., Ohkura, H., Yoshino, F. & Kozeki, D. (1986) A study on the characteristics of rain fall observation based on the radar raingauge. Report no. 167 of Public Works Research Instn., Ministry of Construction (in Japanese).

Japan Meteorological Agency (1995) Analysis process and accuracy of the estimated amount of rainfall based on radar and AMeDAS observations, *Weath. Serv. Bull.* **62**(6), 279–339 (in Japanese).

Koschmieder, H. (1934) Method and results of definite rain measurements, *Mon. Weath. Rev.* **62**, 5–7.

Kawabata, *et al.* (1972) Hydrometeorology. *Chijin-Syokan*, **35** (in Japanese).

Matsuura, T., Fukami, K., Kaneki, M. & Yoshitani, J. (2001) Accuracy evaluation of September 2000 Tokai storm radar observation by areal-averaging. In: *Hydrol. Appl. of Weather Radar* (Proc. Fifth Int. Symp.), 39–44.

Ministry of Construction (1984) Rainfall setting guideline for warning and evacuation order against debris flow(scheme) (in Japanese).

Nakao, T. (1999) Observation of 1998.8. heavy rain by Mt. Takasuzu radar raingauge. In: *Japan Society of Hydrology & Water Resources* (Proc. Annual Conf.), 194–195 (in Japanese).

Nakao, S., Nakao, T., Gosyu, H. & Ishikawa, S. (2001) Accuracy verification of radar raingauge network system in MLIT of Japan. In: *Hydrol. Appl. of Weather Radar–Radar Hydrology* (Proc. Fifth Int. Symp.), 51–56.

Ninomiya, K., & Akiyama, T. (1978) Objective analysis of heavy rainfalls based on radar and gauge measurements. *J. Met. Soc. Japan* **56**, 206–210.

Watari, M., Yamakoshi, T. & Tsunakawa, H. (2002) Applicability of MOC every 10-minute short-term 1 km mesh precipitation forecast for warning and evacuation system preparing for sediment disaster. In: *Erosion Control Engineering* (Proc. Ann. Conf.), 294–295 (in Japanese).

Yoshino, F., Mizuno, M. & Tamamoto, S. (1988) Application of kriging method to calibration of radar raingauge. In: *Japan Society of Hydrology & Water Resources* (Proc. Ann. Conf.), 194–197 (in Japanese).

Yamaguchi, T., Kitagawa, A., Onodera N., Takamizu, K. & Yuki, H. (1993a) Accuracy improvement of radar raingauge. *Mem. River Inform. Res.* **1**, 91–104 (in Japanese).

Yamaguchi, T., Kitagawa, A., Irasawa, M., Sugiura, Y. & Takemori, S. (1993b) Sediment-related disasters forecasting using radar rainfall. *Mem. River Inform. Res.* **1**, 71–79 (in Japanese)

Yamakoshi, T., Yamada, T., Nakano, M. & Watari, M. (2001) Applicability of every 10-minute short-term precipitation forecast for warning and evacuation system preparing for sediment disaster. *Civil Engng J.* **43**(11), 26–31 (in Japanese).

Weather Radar Information and Distributed Hydrological Modelling (Proceedings of symposium HS03 held during IUGG2003 at Sapporo, July 2003). IAHS Publ. no. 282, 2003.

Weather radar distributed hydrological modelling: a case study in the Indus Basin

SHAUKAT ALI AWAN
Flood Forecasting Division, 46-Jail Road, Lahore 54000, Pakistan
nffb@paknet4.ptc.pk

Abstract Radar rainfall estimation through echo reflectivity detects heavy precipitation areas. The intensity of rainfall depicted quantitatively through radar echo variations with pseudo colouring technique and CAPPI (Constant Altitude Plain Position Indicator) representation can pinpoint the areas of intense rainfall. The weather radar equipment plays an extremely vital role for flood monitoring/forecasting and its assessment for subsequent mitigation measures applied to the Pakistan river basins on a real-time basis. An application of radar rainfall estimation and runoff prediction, which was studied in the Jhelum River basin, a tributary of the Indus River in Pakistan, is presented here.
Key words flash flood prediction; radar rainfall estimation

INTRODUCTION

Rainfall measurement in drainage basins, particularly in the mountainous region, poses a great problem, though some mitigation of the situation is made by the installation of manual and telemetric rainfall stations. Still, uncertainty persists in areal precipitation measurement because the ground gauged rainfall stations give only the point rainfall estimation. To solve this problem and predict river runoff, a simple two-dimensional hydrological model with quantitative precipitation measurement radar (QPM radar) is utilized. It covers spatio-temporal rainfall distributions for river runoff estimations, particularly with regard to integrated streamflow in a main river channel. The system was applied to the Jhelum River basin, a tributary of the River Indus in Pakistan, for the flood season from 15 June to 15 October 2002.

Figure 1 shows the location of the ground gauged telemetric rainfall stations and the study catchment area. The telemetric stations reported almost no significant rainfall, therefore it was difficult to estimate the inflow into the Mangla Dam due to rainfall during this period in the upper drainage basin. Table 1 shows the hourly ground gauged rainfall intensity and inflow to the Mangla Dam.

The C-band QPM radar during the period showed a sequential growth and decay of the clouds, which was a clue for prediction of the increase of inflow into the reservoir. To utilize the suitability of QPM radar for reservoir management, a two-dimensional hydrological model has been developed and applied to the Jhelum River basin.

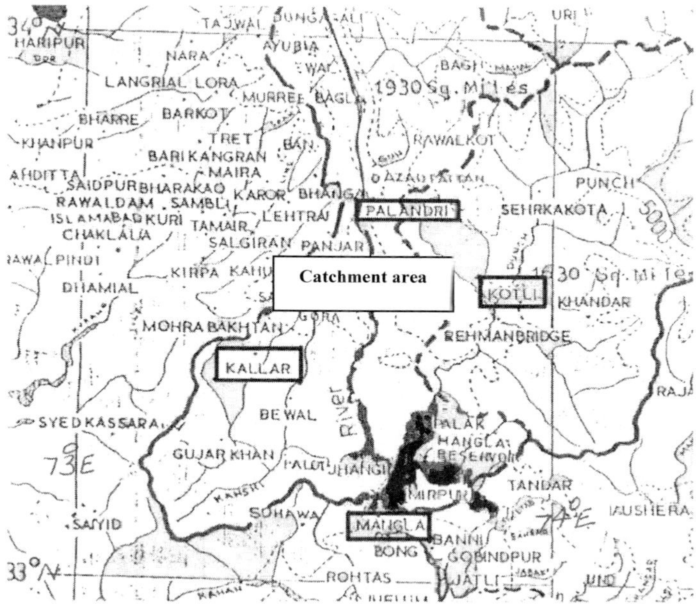

Fig. 1 Location of telemetric station and study area.

Table 1 Ground gauged telemetric rainfall and inflow on 26 August 2002.

Time (h GMT)	Telemetric rainfall (mm h^{-1}):			Inflow (m^3)
	Kotli	Kallar	Plandari	Mangla Dam
00:00	0.0	0.0	0.0	753.0
01:00	0.0	0.0	0.0	755.6
02:00	14.5	6.3	0.0	713.0
03:00	26.2	10.7	0.0	906.0
04:00	1.3	2.5	2.5	906.0
05:00	0.5	0.0	0.0	3197.9

FLOOD OF AUGUST 2002

The C-band QPF image obtained on 26 August 2002 at 04:00 PST (Pakistan Standard Time) showed hardly any rainfall activity over the upper catchment of the Jhelum River. Between 04:30 and 05:00 h (Figs 2 and 3) the rainfall at the rate of 2.5 mm h^{-1} started over the reservoir, while the telemetric rainfall measuring stations located at Plandri (PLI), Kotli (KTI) and Mangla (MAG) indicated no rainfall in 2 h from 04:00 to 06:00 h. The whole catchment area (250 km^2) of the reservoir was covered with rainfall intensity up to 5 mm h^{-1} until this time while no ground verification could be made as the telemetric station reported no rainfall. The heavy rain band approached after 07:00 h (Fig. 4). The rainfall considerably decreased over the reservoir at 08:30 h PST, as indicated through radar echoes (Fig. 5). The rainfall generated runoff into the Mangla Dam, which is the second largest reservoir in the country. The inflow jumped from 755.6 m^3 at 06:00 h to 3197.9 m^3 at 10:00 h (PST).

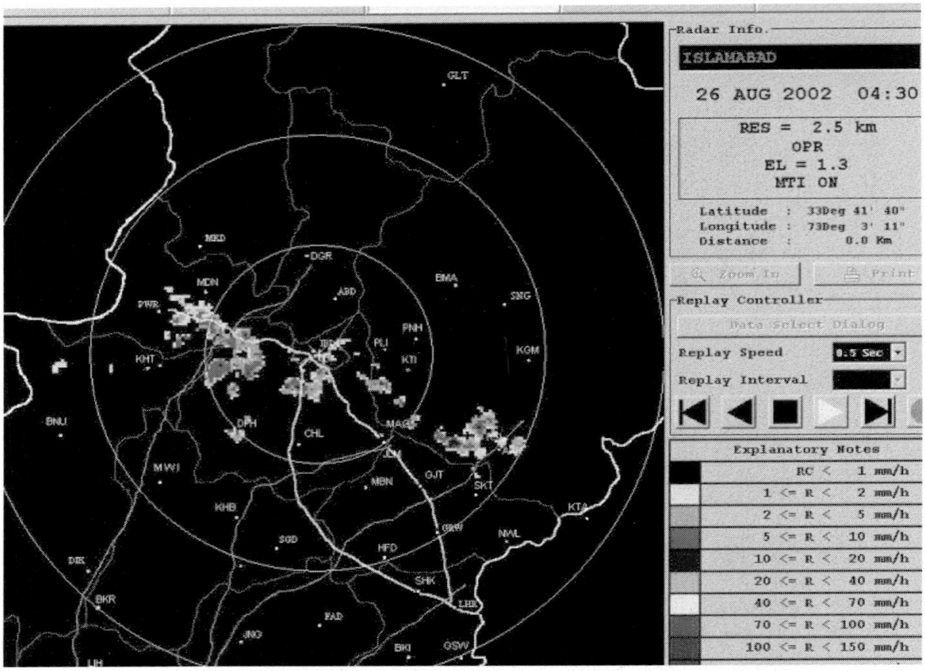

Fig. 2 C-band radar image showing amount of precipitation at 04:30 h on 26 August 2002.

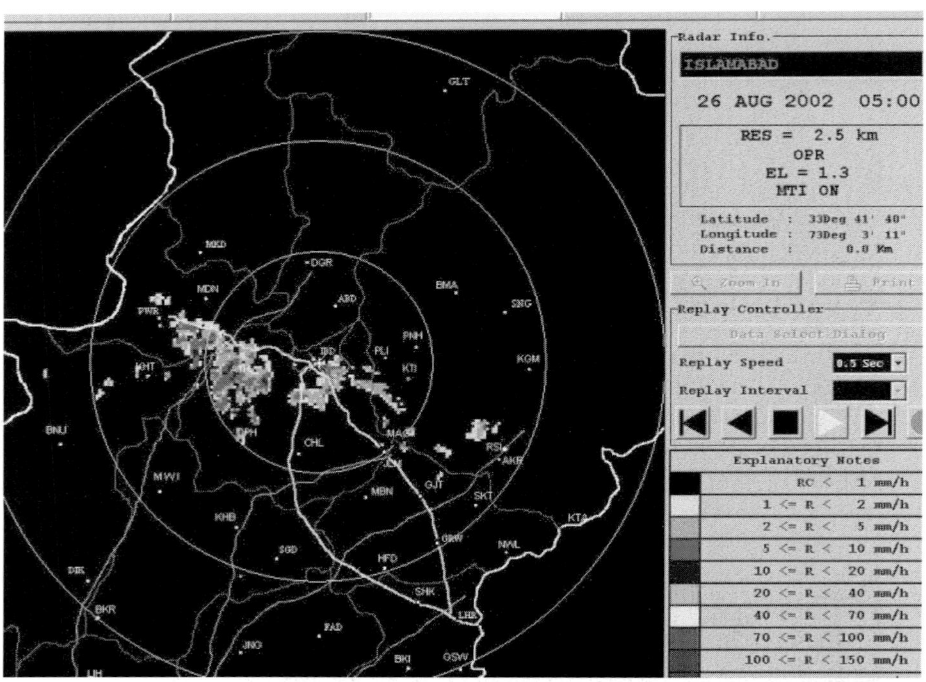

Fig. 3 C-band radar image showing amount of precipitation at 05:00 h on 26 August 2002.

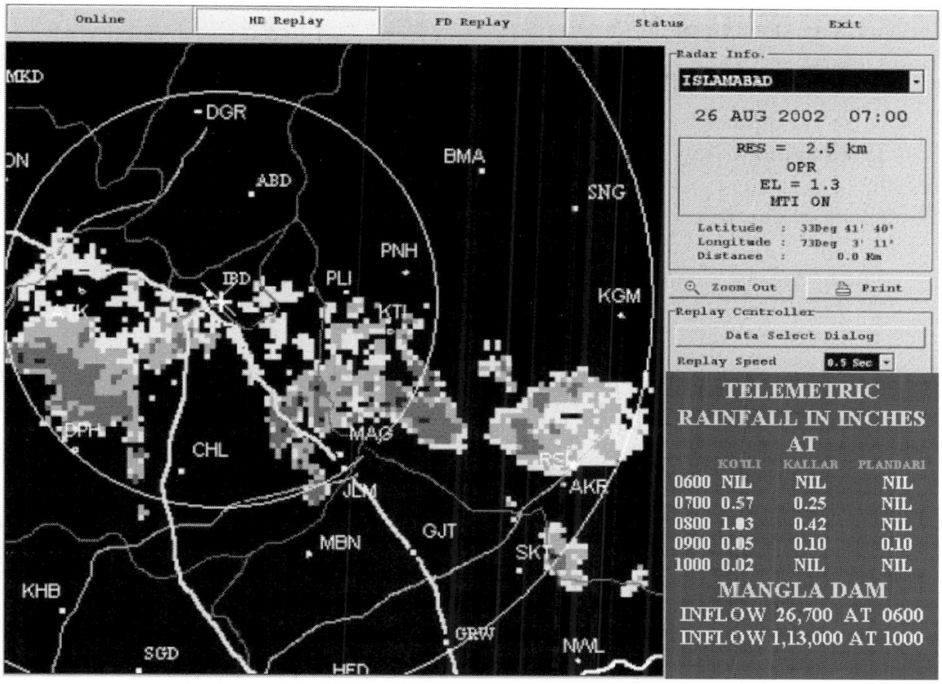

Fig. 4 C-band radar image showing amount of precipitation at 07:00 h on 26 August 2002.

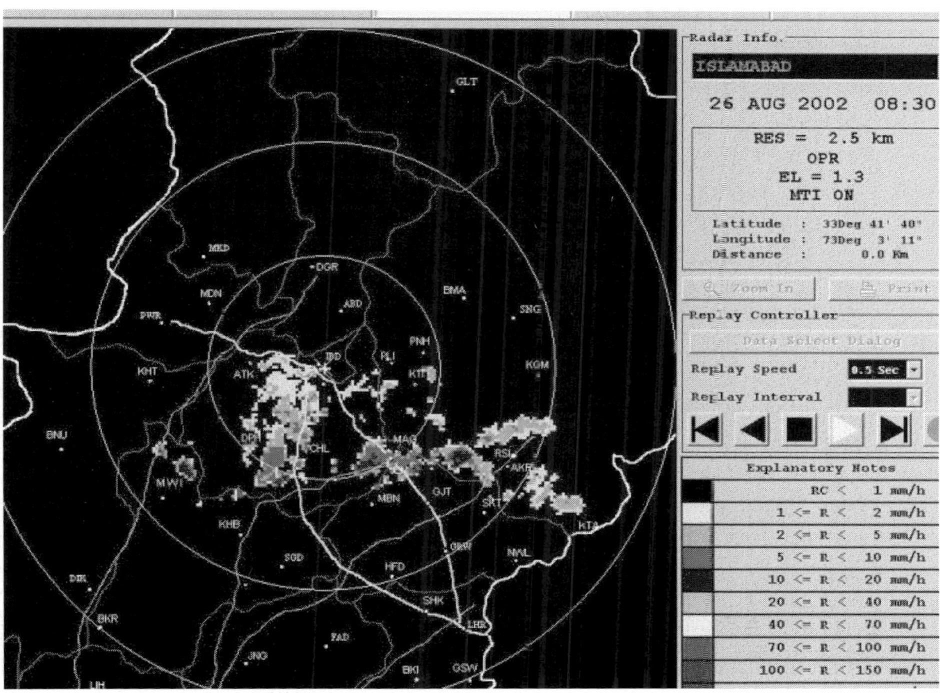

Fig. 5 C-band radar image showing amount of precipitation at 08:30 h on 26 August 2002.

RESULTS

Though there was a sudden increase of inflow into the Mangla Dam from 755.6 m^3 to 3197.9 m^3 in almost 3 h, the inflow could be effectively predicted in the distributed hydrological modelling. From the viewpoint of the reservoir operation, the prediction resulted in effective reservoir operation to treat additional inflow without causing any flood peak or danger of spillover.

Radar precipitation for winter hydrological modelling

S. R. FASSNACHT

Department of Forest, Rangeland, and Watershed Sciences, Colorado State University, Fort Collins, Colorado 80523-1470, USA

srf@cnr.colostate.edu

E. D. SOULIS & N. KOUWEN

Department of Civil Engineering, University of Waterloo, Waterloo, Ontario N2L 3G1, Canada

Abstract Interpolated precipitation gauge measurements and weather radar snowfall estimates were used as input to a physically-based hydrological model (WATCLASS). The gauge measurements were corrected for wind undercatch according to WMO standards. Post-processing of the radar data was undertaken to consider underestimation due to the use of winter radar coefficients for liquid precipitation. Streamflow was simulated for the Upper Grand River basin in central southwestern Ontario, Canada for the five winters from 1993 to 1997. For each year, except 1995, the radar data provided precipitation estimates that were better, in terms of simulated runoff volumes, than those provided by the gauge data. Along with runoff volumes, peak streamflows were more closely estimated from radar precipitation than gauge precipitation. Gauge estimates consistently yielded lower than observed peak streamflows.

Key words precipitation gauges; snow hydrological modelling; weather radar; winter hydrology

INTRODUCTION

For hydrological modelling, weather radar can provide the spatial distribution of rainfall, whereas the gridding of point gauge rainfall produces questionable areal estimates, especially where gauge data are sparse. For colder climates, gauge snowfall estimates require correction for wind undercatch and other errors (Goodison *et al.*, 1998) and the use of weather radar for snowfall estimation has focused primarily on short time intervals, such as hourly and sub-hourly, to single storm events lasting up to two days. Improvements in radar estimates have emphasized the use of ground-based measurements as truth.

Numerous *Z–R* relationships have been developed for snowfall (Fujiyoshi *et al.* (1990) provides a good summary of previous *Z–R* relationship determinations). Most are based on individual storm events and involve comparisons to precipitation gauges. The analysis by Sekhon & Srivastava (1970) developed the *Z–R* relationship that is widely used for the cold weather season precipitation estimation. Recently the advanced capabilities of weather radar have been explored for complex snowfall identifications. For example, Ryzhkov & Zrnic (1998) used polarimetric radar to distinguish between rain and snow, and Matrosov (1998) showed that the difference between the reflectivities at the dual wavelengths yielded a snowflake size estimate that could be combined with the longer wavelength reflectivity to approximate a snowfall rate.

The use of weather radar to estimate snowfall across watersheds for modelling purposes has been limited to work by Houck *et al.* (1995) who recommended using weather radar snowfall estimates as input to a GIS hydrological analysis, and Fassnacht *et al.* (1998) who used weather radar for modelling the winter season on the Grand River basin in southern Ontario.

This paper investigates the use of weather radar *vs* gridded gauge data as the precipitation inputs to a hydrological model. Peak streamflows and cumulative runoff volumes are compared for five winters across a 3520 km^2 watershed in southern Ontario, Canada.

STUDY AREA

The Upper Grand River watershed is located in central southwestern Ontario and has a climate that is moist with cold winters (a Koppen Dbf climate, as per Gough *et al.*, 2002). Synoptically, it is located near the moving boundary between continental polar air originating in northern Canada, maritime polar air from the Pacific modified by the Rockies, maritime tropical air which forms over the Gulf of Mexico, and continental tropical air from the south central area of the North American continent (Gough *et al.*, 2002). The polar air flows in from the northwest and is dominant in the winter. Winter temperatures are warmed due to the Great Lakes, and snow belts extend from Lake Huron into the northern portions of the basin.

The Meteorological Service of Canada (MSC) radar in King City, north of Toronto, covers the basin. The installation currently operates in a Doppler mode, but the imagery used in this paper are 2 × 2 km hourly precipitation accumulation maps derived from a composite of C-band (5.2 cm) conventional scans of 10 min Constant Altitude Plan Position Indicator images. For the warm season, the *Z–R* coefficients developed for southern Ontario by Richards & Crozier (1983) are used, and for the cold season, the Sekhon & Srivastava (1970) *Z–R* coefficients are used.

The watershed is mostly covered by crops and low vegetation (59%), with some regions of wetland (18%), mixed deciduous–coniferous forest (14%), small bare (8%) and impervious (1%) areas. The terrain in this portion of southern Ontario is glacial material, composed primarily of clayey till. Streamflow was monitored at eight hydrometric stations across the study basin for the study period, i.e. the winters of 1993–1997.

METHODOLOGY

Hydrological modelling

The WATCLASS physically-based distributed hydrological model was used to simulate streamflows. This is a linkage of the University of Waterloo WATFLOOD hydrological model and the Environment Canada CLASS land surface scheme (Soulis *et al.*, 2000). The vertical water and energy budgets near the land surface are computed using CLASS, while the lateral water budget and streamflow routing is performed using WATFLOOD. The Upper Grand River basin was modelled using 10 × 10 km grid elements, for which the slope and flow direction have been determined by Tao & Kouwen (1989).

Gauge undercatch correction

Precipitation was measured using Nipher-shielded gauges. Gauge undercatch was estimated using wind speed data (Goodison *et al.*, 1998) and snowfall was corrected to the DFIR reference. The corrected gauge data were further corrected to the bush setting (Yang *et al.*, 1993). The gauge data were corrected for wetting loss by adding 0.15 mm to each precipitation event.

Meteorological data gridding

The WATCLASS model requires air temperature, precipitation, barometric pressure, specific humidity, wind speed, shortwave and longwave radiation data. Gauge data were gridded using the inverse squared distance interpolation (Tao & Kouwen, 1989). Longwave radiation data were computed from the air temperature, specific humidity, and an estimate of the cloud cover. The daytime cloud cover estimates were derived from a ratio of measured to computed shortwave radiation (Fassnacht *et al.*, 2001a).

Radar data adjustment

For the study area, Fassnacht *et al.* (2001b) identified a local scaling issue, that caused overestimation of precipitation during periods of strong anomalous propagation. Reprocessing of the data to the lowest possible increment minimized the forced over-scaling problem.

Radar reflectivity from melted or partially melted hydrometeors yield precipitation underestimates when snowfall Z–R coefficients are used for rainfall, freezing rain, etc. The probability of snow *vs* air temperature curve derived by Auer (1974) was used to estimate the quantity of mixed precipitation and the proportion of rain was adjusted based on the rain Z–R coefficients (Fassnacht *et al.*, 2001b).

RESULTS

Areal estimation of snowfall is a problem related to gauge density across a watershed and to the technique used to grid the point data. The use of weather radar can overcome the gridding problem. For example, the 10 km resolution gridded gauge precipitation, using the inverse squared distance approach, and the 10 km radar precipitation for 26 January 1996 from 12:00 to 13:00 h EST are presented in Fig. 1(a) and (b). From the various gauges across central southwestern Ontario, precipitation was only observed at the north central gauge, around which the gridded precipitation is

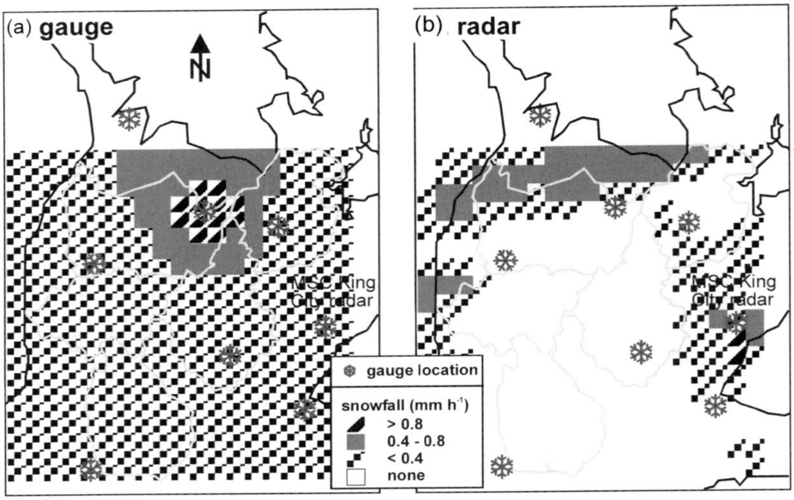

Fig. 1 Hourly precipitation at a 10 km resolution across the five central southwestern study basins on 26 January 1996 at 13:00 EST derived from: (a) gridded gauge data, and (b) radar data.

concentrated. No precipitation was observed at this location according to the radar. Precipitation across the remainder of the domain is a result of gridding (Fig. 1(a)). From the radar estimates, precipitation was only observed over the Toronto station in the eastern portion of the domain, just below the MSC King City radar (Fig. 1(b)). However, the radar precipitation near Toronto has often been observed to be urban clutter.

On 28 January 1996 from 02:00 to 03:00 h EST, no precipitation was observed at any of the gauges. Thus, the gridding of the gauge data yielded no precipitation (Fig. 2(a)). The left-middle gauge was not recording and the right-middle gauge did not measure precipitation. However, radar observations indicated the presence of precipitation, but not near the recording gauges (Fig. 2(b)). The discrepancy between the spatial gauge and radar precipitation is due primarily to the gridding of the point data.

Streamflows were simulated at eight stations within the basin for the winters of 1993–1997 using the gridded gauge and adjusted radar precipitation data. These model results were compared to the observed streamflows that were obtained from the HYDAT CD-ROM (Environment Canada, 1997).

A double mass curve of cumulative modelled *vs* observed streamflow at the mouth of the basin for the winter of 1993 illustrated that the adjusted radar provided a better precipitation estimation for streamflow modelling than did the gridded corrected gauge data (Fig. 3). This is consistent for both peak flow and runoff volume at all hydrometric stations for 1993, as illustrated by the residual plots Fig. 4(a) and (b), respectively. The plots illustrate the difference between modelled and observed peak flows or runoff volumes using radar precipitation *vs* using gauge precipitation at the eight stations for the five study years. A better radar than gauge precipitation estimate for modelling yields a data point closer to the x-axis (the shaded area) and a better gauge estimate yields a data point closer to the y-axis (unshaded area).

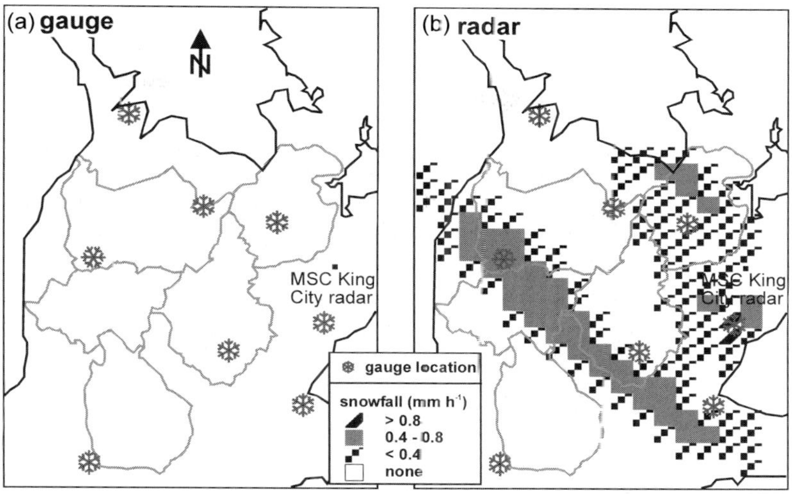

Fig. 2 Hourly precipitation on 28 January 1996 at 03:00 EST derived from (a) gridded gauge data, and (b) radar data.

There is little annual net difference in peak runoff for 1994, 1995, and 1997, typically with varying extents of over- and underestimation from modelling (Fig. 4(a)). Peak flows are more appropriately modelled with gridded gauge data than with radar data for 1996. Comparatively, radar was slightly better overall in terms of absolute error. However, 80% of the modelled peak flows using gauge precipitation were underestimated with an average peak flow underestimation of 23.1 $m^3 s^{-1}$. Using the radar precipitation, the modelled peak flows were almost evenly distributed in number between underestimates (47.5%) and overestimates (52.5%) with an average overestimation of 14.1 $m^3 s^{-1}$.

The peak streamflows from radar data were overestimated for 1995, and removal of the 1995 results yielded an average peak flow overestimation of 1.66 $m^3 s^{-1}$ for the radar precipitation versus an underestimation of 36.7 $m^3 s^{-1}$ for the gauge precipitation.

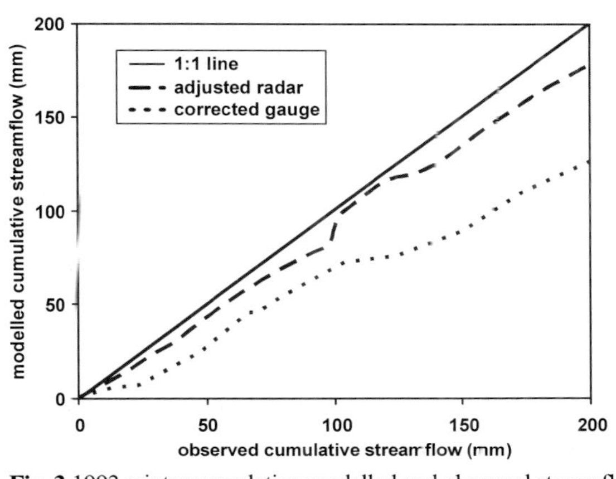

Fig. 3 1993 winter cumulative modelled and observed streamflow at the mouth of the basin.

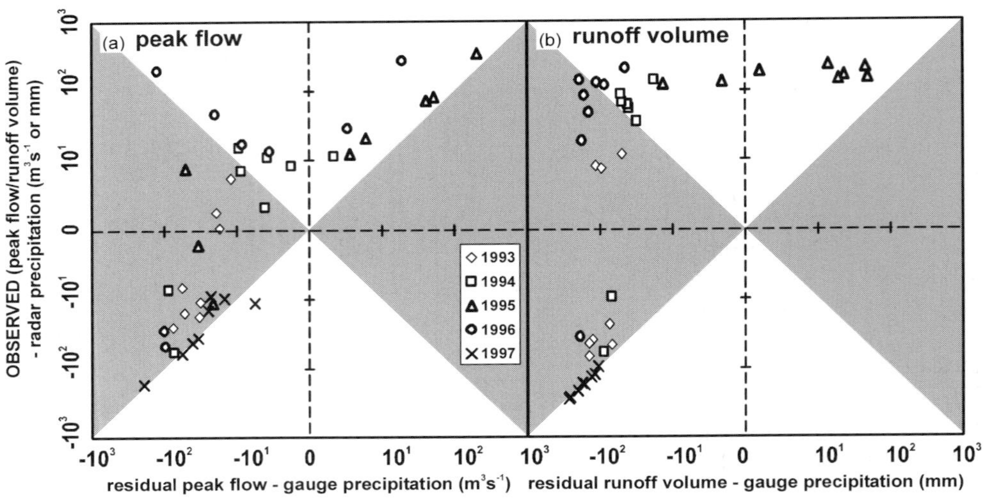

Fig. 4 Residual of modelled minus observed (a) peak flow and (b) runoff volume from snowmelt streamflow for radar precipitation *vs* gauge precipitation. Each symbol denotes one winter's peak flow at one hydrometric station.

The discrepancy for 1995 was more obvious from the simulated runoff volumes using radar precipitation (Fig. 4(b)). Whereas the 1996 peak flows were better estimated from gauge precipitation, the 1996 runoff volumes were better estimated from radar precipitation. The results from 1994 and 1997 were similar. However, for all years except 1995, the gauge precipitation yielded underestimates of simulated runoff volume. The gauge precipitation yielded on average 91 mm less simulated runoff and the radar yielded 20.1 mm more simulated runoff. Ignoring the 1995 data yielded an average underestimation of 41.7% from the gauge precipitation and an underestimation of 0.3% from the radar precipitation.

DISCUSSION

The radar precipitation yielded better simulated peak streamflows and runoff volumes for four of the five study years. The gridded gauge precipitation yielded underestimates for most sub-watersheds for most years. These differences occurred in part due to the gridding of the precipitation data. An inverse weighted distance scheme was used to grid the gauge precipitation data, as per Tao & Kouwen (1989). However, they modelled summer storm events and there are more rainfall gauges than snowfall gauges across this and most study areas. The use of another distance-type interpolation scheme, such as kriging, should not improve the gridded gauge underestimation. Similarly, the use of hypsometric interpolation would be limited by the lack of gauge data, as well as the limited topographical variation across the region.

The gauge data have been corrected to consider undercatch due to wind. Without this correction, the gridded gauge data yielded larger underestimates of streamflow. The radar data have been corrected due to a local scaling issue, which reduced much of the radar overestimation, and resulted in underestimation of streamflow. The data were adjusted to consider underestimation by using winter *Z–R* coefficients for liquid precipitation. This adjustment improved the radar precipitation estimation.

The radar precipitation overestimation for 1995 was a problem, as illustrated by Fassnacht *et al.* (2001b). The 1997 precipitation was underestimated by the radar. Similar patterns were observed during the 1995 and 1997 winters for the state of Michigan from the National Weather Service NEXRAD system (Hollingsworth, personal communication, 1999). These variations may be due to differences in synoptic storm types. These years must be further investigated to remove seasonal systematic biases, when and where they exist.

Due to systematic seasonal differences in radar precipitation estimation and the lack of station data, it is recommended that a combination of the gauge and radar data be investigated for estimating winter precipitation. This combination of gauge and radar data has been performed for rainfall with different degrees of success by Wilson (1970), Brandes (1975) and others.

Acknowledgments This research was supported in part by the Meteorological Service of Canada and the MSC-CRYSYS project. The radar images were provided by MSC-King City Radar, and meteorological data were provided by MSC-Downsview. A discussion with Jane Hollingsworth of the National Weather Service Indiana Office in 1999 provided information into the radar problems with snowfall estimation during the 1995 and 1997 winters across the state of Michigan.

REFERENCES

Auer, A. H., Jr. (1974) The rain versus snow threshold temperatures. *Weatherwise* **27**, 67.

Brandes, E. A. (1975) Optimizing rainfall estimates with the aid of radar. *J. Appl. Met.* **14**, 1339–1345.

Environment Canada (1997) 1996 HYDAT Hydrometric Data CD-ROM. Water Survey of Canada, Ottawa, Canada.

Fassnacht, S. R., Soulis, E. D., Snelgrove, K. R. & Kouwen, N. (1998) Application of weather radar to model the snow hydrology of southern Ontario (Proc. 55th Annual Eastern Snow Conference), 115–123. Jackson, New Hampshire, USA.

Fassnacht, S. R., Snelgrove, K. R. & Soulis E. D. (2001a) Daytime incoming longwave radiation approximation for physical hydrological modelling. In: *Soil-Vegetation-Atmosphere Transfer Schemes and Large-Scale Hydrological Models* (ed. by A. J. Dolman, A. J. Hall, M. L. Kavvas, T. Oki & J. W. Pomeroy) (Proc. Maastricht, Symp., July 2001), 279–286. IAHS Publ. no. 270.

Fassnacht, S. R., Kouwen, N. & Soulis, E. D. (2001b) Surface temperature adjustments to improve weather radar representation of multi-temporal winter precipitation accumulations. *J. Hydrol.* **253**(1–4), 148–168.

Fujiyoshi, Y., Endoh, T. Yamada, T. Tsuboki, K. Tachibana, Y. & Wakahama, G. (1990) Determination of a Z–R relationship for snowfall using a radar and high sensitivity snow gauges. *J. Appl. Met.* **29**(2), 147–152.

Goodison, B. E., Louie, P. Y. T. & Yang, D. (1998) WMO Solid Precipitation Measurement Intercomparison Final Report. WMO Instruments and Observing Methods Report No. 67, WMO/TD No. 872. World Meteorological Oranization, Geneva, Switzerland.

Gough, W. A., Lillyman, C., Karagatzides, J. & Tsuji, L. J. S. (2002) Determining the validity of using summer monitoring to estimate annual deposition of acidic pollutants in Southern Ontario, Canada. *Water, Air and Soil Pollut.* **137**, 305–316.

Houck, R. E., Waldstreicher, J., Hassett, J. M. & Blottman, P. F. (1995) Preliminary investigation of WSD-88D data for winter hydrometeorological events in upstate New York (Proc. 52nd Annual Eastern Snow Conference), 39–50. Toronto, Ontario, Canada.

Matrosov, S. Y. (1998) A dual-wavelength radar method to measure snowfall rate. *J. Appl. Met.* **5**, 165–166.

Richards, W. G. & Crozier, C. L. (1983) Precipitation measurement with a C-band weather radar in southern Ontario. *Atmos. Ocean*, **21**, 125–137.

Ryzhkov, A. V. & Zrnic, D. S. (1998) Discrimination between rain and snow with a polarimetric radar. *J. Appl. Met.* **37**(10), 1228–1240.

Sekhon, R. S. & Srivastava, R. C. (1970) Snow-size spectra and radar reflectance. *J. Atmos. Sci.* **27**, 299–307.

Soulis, E. D., Snelgrove, K. R., Kouwen, N. & Verseghy, D. L. (2000) Toward closing the vertical water balance in Canadian atmospheric models: coupling of the land surface scheme CLASS with the distributed hydrological model WATFLOOD. *Atmos. Ocean*, **38**(1), 251–269.

Tao, T. & Kouwen, N. (1989) Remote sensing and fully distributed modelling for flood forecasting. *J. Water Resour. Plan. Manage.* **115**(6), 809–823.

Wilson, J. W. (1970) Integration of radar and raingauge data for the improved rainfall measurement. *J. Appl. Met.* **9**, 489–497.

Yang, D., Metcalfe, J. R., Goodison, B. E. & Mekis, E. (1993) True snowfall: an evaluation of the Double Fence Intercomparison Reference gauge (Proc. 50th Eastern Snow Conference), 105–111. Quebec City, Canada.

Numerical study of rainfall–topography relationships in mountainous regions of Japan using a mesoscale meteorological model

YOSHIHARU SUZUKI
Department of Architecture and Civil Engineering, Utsunomiya University, 7-1-2 Yoto, Utsunomiya 321-8585, Japan
ysuzuki@cc.utsunomiya-u.ac.jp

EIICHI NAKAKITA
Department of Urban and Environmental Engineering, Kyoto University, Sakyo-ku, Kyoto 606-8501, Japan

SHUICHI IKEBUCHI
Disaster Prevention Research Institute, Kyoto University, Gokasho, Uji 611-0011, Japan

Abstract Rainfall–topography relationships are investigated through numerical simulations using the mesoscale atmospheric model MM5. For a quantitative evaluation of topographic effects on rainfall distribution, a rainfall–elevation relationship, the Dependence Line on Topographic Elevation (DLTE) of rainfall distribution, was examined in four mountainous regions in Japan. The results confirmed that DLTE is one of the fundamental properties of rainfall distributions. The conditions required for the formation of DLTE were investigated as well as the fluctuation properties around DLTE in each region. The effects of wind on rainfall distribution were also examined to understand the mechanism of rainfall–topography relationships. The results show that there is a definite correlation between the strength of horizontal wind and the slope of DLTE, although the strength of the correlation differs from region to region.

Key words mesoscale meteorological model, MM5; numerical simulation; rainfall distribution; rainfall–elevation relationships; topographic effect

INTRODUCTION

For hydrological applications and water resources management, it has been required to understand the spatial and temporal structures of rainfall distribution related to topographic properties in a watershed. For understanding the mechanism of rainfall–topography relationships and developing a stochastic model of rainfall distribution, the rainfall–topography relationships are analysed through numerical simulations using the mesoscale meteorological model MM5.

The fluctuations of rainfall distribution depend on atmospheric conditions and topographic features such as moisture distribution, wind speed, wind direction, topographic elevation, slope orientation, barrier characteristics and scale of mountains. This study focuses on topographic elevation as one of the most significant factors because it has predominant effects on rainfall distribution on the spatial and temporal scale of meso β or meso α.

A great number of investigations on rainfall–elevation relationships have been conducted from the beginning of the twentieth century (e.g. Lee, 1911; Alter, 1919; Henry, 1919; Varney, 1920). According to these authors, it is already evident world-wide that rainfall amounts generally increase with topographic elevation. However, the quantitative nature of rainfall–elevation relationships has not yet been determined, and therefore analyses using weather radar were required for understanding the physical mechanism and for developing a rainfall model that well describes the stochastic properties. Recent studies investigate the relations between rainfall and topography using a numerical meteorological model (e.g. Colton, 1976; Alpert, 1986; Barros & Lettenmaier, 1994; Oishi *et al.*, 1996). The main factors that determine the amount and the distribution of rainfall were made clear to some extent through the simulations under various conditions of the height of a mountain, the slope of a mountain, the speed of synoptic wind, etc. These investigations, however, have not yet led us to a comprehensive discussion with consideration to the universality or the singularity of actual rainfall–topography relationships.

The authors investigated rainfall–elevation relationships through the analysis of the data from a weather radar in a mountainous region in Japan, which has an area of 240 × 240 km (Suzuki *et al.*, 2002). As a result, it was found that a definite linear relationship between stratified (piecewise) sampling averages of accumulated rainfall amount and topographic elevation was represented by a semi-logarithmic relation (i.e. rainfall–elevation relationships are exponential as a whole). This relationship is called the Dependence Line on Topographic Elevation (DLTE) of rainfall distribution. Figure 1 shows an example of rainfall–elevation plots and the linear relation of DLTE obtained from a weather radar observation in the Kinki region of Japan. The correlation between the stratified sampling averages represented by black circles and topographic elevation is more than 0.9. The topographic properties in the study region and the properties of rainfall events are well reflected in the value of the slope of DLTE. In other words, the linear relation of DLTE represents the fundamental and average properties of rainfall–elevation relationships in a study region. However, it has not been clarified whether the linear relation of DLTE is a universal property. In this paper, the characteristics of DLTE obtained through numerical simulations using a physically-based numerical model MM5 are investigated.

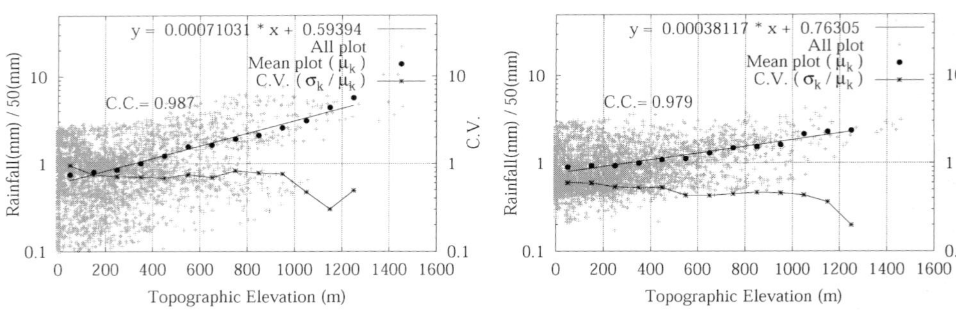

(a) 1988.8.10 (convective rainfall) (b) 1989.9.13 (stratiform rainfall)

Fig. 1 Rainfall–elevation relationships and the linear relation of DLTE obtained by the Miyama radar raingauge observation in the Kinki region of Japan.

DESCRIPTION OF SIMULATIONS IN THIS STUDY

MM5, the fifth-generation NCAR/Penn State mesoscale model, has a multiple-nest capability, non-hydrostatic dynamics and a four-dimensional data assimilation capability. The nesting method simulates several domains running at the same time with two-way interaction. In this study, three domains with the grid scales of 27 km, 9 km and 3 km (Domain 1, Domain 2 and Domain 3, respectively) were set in all simulations. Initial and boundary conditions were produced by interpolating the six hourly GPV data, including geopotential height, temperature, wind speed and humidity, which were provided by the Japan Meteorological Agency.

Simulations were carried out in the four mountainous regions of southern Kyushu, southern Kinki, Shikoku and Chubu in Japan. By comparison of simulation results among the regions, rainfall–topography relationships were investigated. For each region, 15 atmospheric simulations were conducted. The simulation period of each case was for 2 days.

DLTE OF RAINFALL DISTRIBUTION FROM NUMERICAL SIMULATIONS

Figure 2 represents the simulation results of 2-days-accumulated rainfall amount and rainfall–elevation relationships in each region. The figures show a linear relation is formed in every case between the stratified sampling averages of accumulated rainfall and topographic elevation. The linear relations of DLTE are found not only in observed rainfall but also in simulated rainfall fields. This means the DLTE is one of the universal properties which could be measured in various regions. This result leads us to the thought that DLTE is not one of the statistical properties of rainfall distribution, but one of the properties based on a physical mechanism of rainfall.

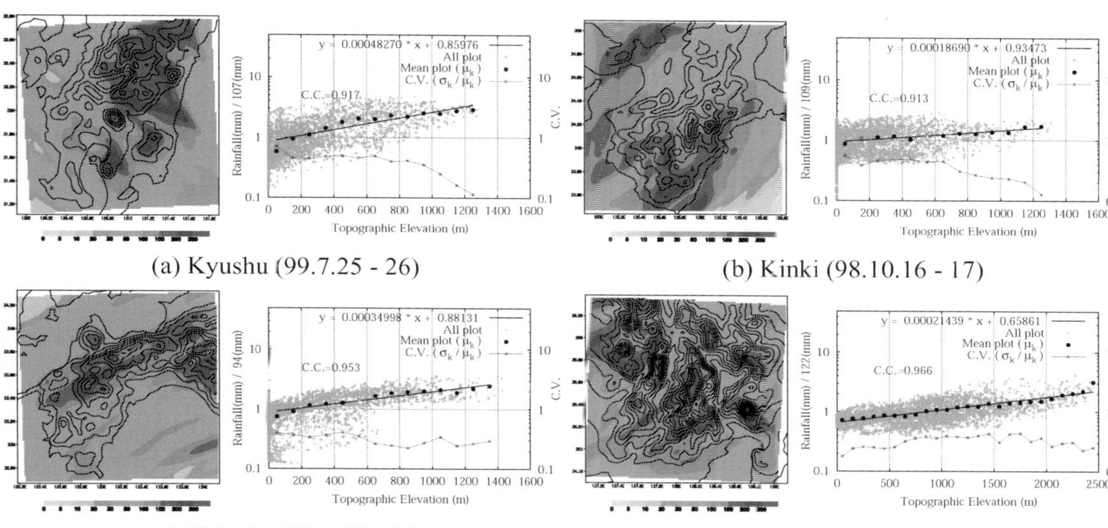

(a) Kyushu (99.7.25 - 26) (b) Kinki (98.10.16 - 17)

(c) Shikoku (99.6.18 - 19) (d) Chubu (98.6.21 - 22)

Fig. 2 Simulation results of 2-days-accumulated rainfall amount (mm) (left figures), and rainfall–elevation relationships (right figures).

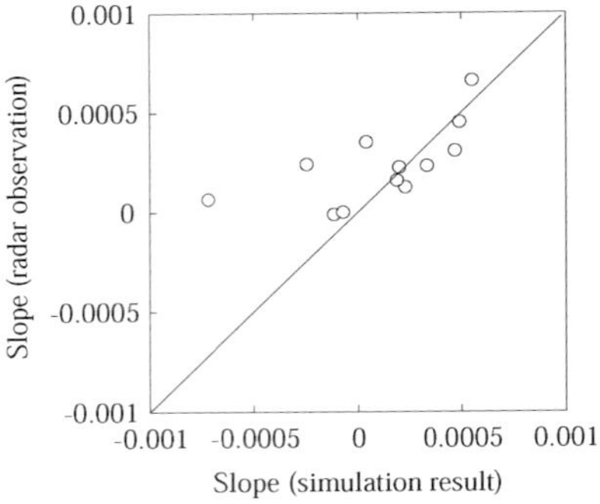

Fig. 3 Comparison between the slope of DLTE from the numerical simulations and that from the observations by Miyama radar raingauge in the Kinki region of Japan.

Furthermore, the slopes of DLTE from the simulation results were compared with those obtained from a weather radar in the Kinki region of Japan. The result is shown in Fig. 3. The slope of DLTE represents the degree of topographic effects on rainfall distribution. The figure shows that there is a strong correlation between the slopes from the numerical simulations and those from the radar observations, except for a few cases. Rainfall–elevation relationships shown by DLTE are reproduced with sufficient accuracy by a physically-based numerical model, even if the reproducibility of the position or the amount of rainfall is not so high. This also proves that DLTE is one of the universal properties of rainfall distribution.

The authors have found, through the analysis with a weather radar, that one of the conditions required for the formation of DLTE is a regional average rainfall (RAR) over 50–100 mm (Suzuki *et al.*, 2002). Here, RAR represents the scale of rainfall accumulation. In this study, the conditions in each region were investigated through the numerical simulations. The root mean square error (*RMSE*) of the stratified sampling averages against DLTE was used as the criterion for judging the formation of DLTE. The authors decide the DLTE is formed when the *RMSE* is less than 0.1. The *RMSE* was computed for 15 cases of the simulation results in each region, and the relation between RAR and the percentage of the cases which meet the conditions was investigated. Figure 4 shows that DLTE is formed on more than 90% cases when RAR

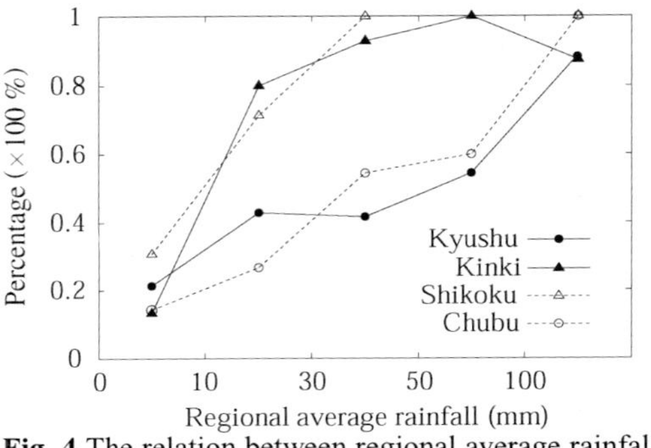

Fig. 4 The relation between regional average rainfall (RAR) and the percentage of the cases where *RMSE* of the stratified sampling averages against DLTE is less than 0.1.

is more than 30 mm in the Kinki and the Shikoku regions and more than 100 mm in the Kyusyu and the Chubu regions. DLTE is formed in most of the cases where RAR is over 100 mm in every region, which corresponds to the author's previous analysis using radar observation.

EFFECTS OF WIND ON RAINFALL–TOPOGRAPHY RELATIONSHIPS

Various physical values such as wind speed, humidity and temperature, are obtained through numerical simulations. Among them, wind speed and wind direction have great impacts on rainfall distribution because they determine the strength and direction of ascending flows which could develop precipitation. The effects of wind on rainfall distribution are examined for understanding the mechanism of rainfall–topography relationships.

Relation between wind speed and rainfall distribution

The simulated results in Kyushu region in 13–14 June 1998 (case-1) and in 6–7 October 1998 (case-2) were analysed. In case-1 the slope of DLTE is relatively large and in case-2 small among the simulation results. Figure 5 represents the simulation

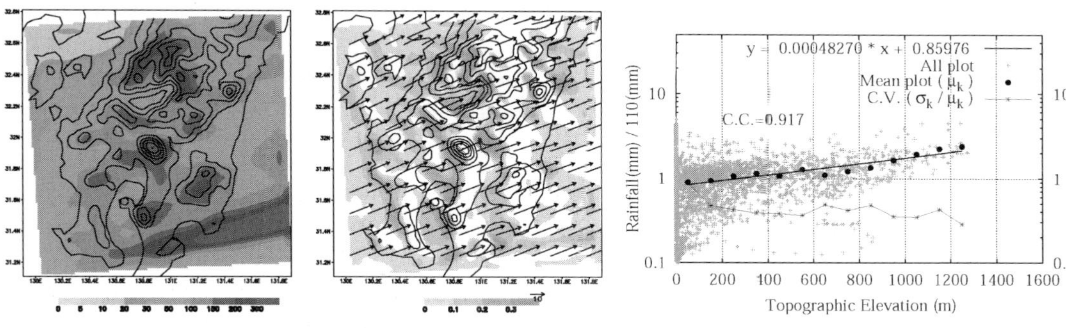

(a) Case 1 (98.6.13–14)

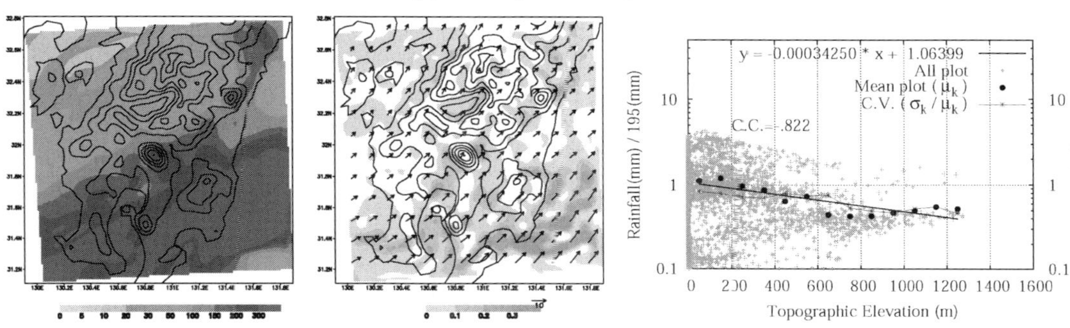

(b) Case 2 (98.10.6–7)

Fig. 5 Simulation results of 2-days-accumulated rainfall amount (mm) (left figures), two-days-averaged speed of horizontal wind (arrows) and vertical wind (contours) (m) (centre figures), and rainfall–elevation relationships (right figures) in the Kyushu region.

results of 2-days-accumulated rainfall, 2-days-averaged speed of horizontal and vertical wind fields, and rainfall–elevation relationships. The wind field data on an isobaric surface of 800 hPa (i.e. an altitude of about 2000 m) was used in this investigation.

In case 1, horizontal wind is strong as a whole and strong ascending flows occur on the windward side of mountains as shown in the centre figures. In case 2, horizontal wind is weak and ascending flows occur on the southeast sea far from mountains. In case 1, the rainfall distribution shown in the left figure has peaks on the top of mountains, and in case 2 much rainfall occurred on the southeast sea. The area with much rainfall corresponds well to the area with strong ascending flows, which occur above the mountains when horizontal wind is strong. This means that flows are forced to ascend along the slope of mountains when horizontal wind is strong and the height of mountains is high enough. The ascending flows then become convection that brings much rainfall. That is the reason why the slope of DLTE becomes large when horizontal wind is strong, and there is a strong correlation between the strength of horizontal wind and the slope of DLTE.

Relation between horizontal wind speed and the slope of DLTE

The relation between regional averaged speed of horizontal wind (*H*) and the slope of DLTE (*S*) were examined in each region. The value of *H* was obtained by averaging wind spatially in a study region and then averaging temporally during 2 days. Figure 6 shows that there is a definite correlation between *H* and *S* and the strength of the correlation differs from region to region. Correlation coefficient (*CC*) is high in the

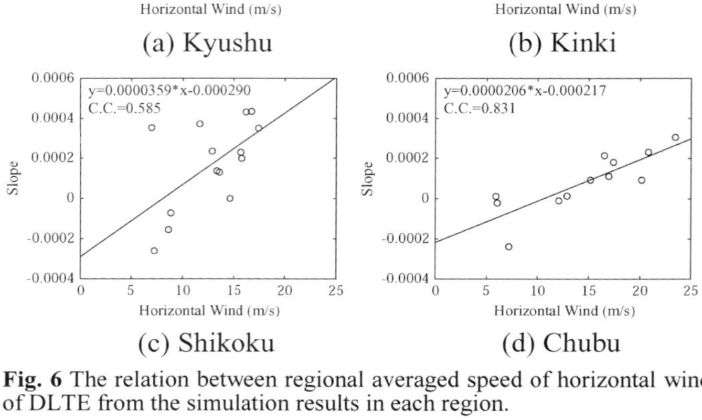

Fig. 6 The relation between regional averaged speed of horizontal wind and the slope of DLTE from the simulation results in each region.

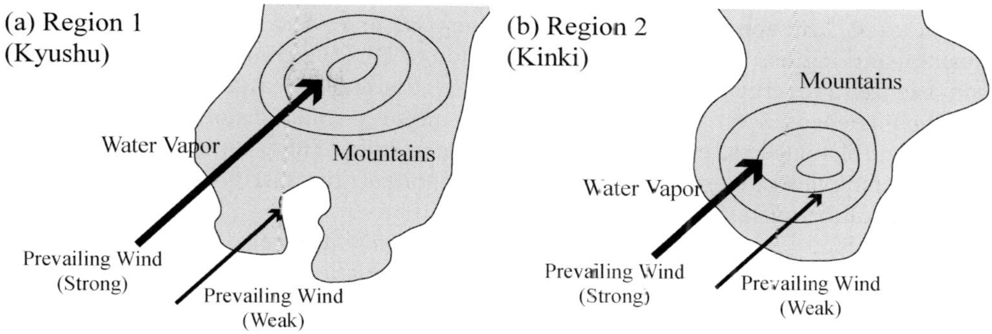

Fig. 7 Schematic of the transportation of water vapour in (a) Region 1 (Kyushu) and (b) Region 2 (Kinki).

Kyushu and the Chubu region (Region 1), and is not high in the Kinki and the Shikoku region (Region 2). This means that meteorological factors other than horizontal wind have relatively more effect on rainfall–elevation relationships in Region 2 than those in Region 1.

In Region 1, as represented in Fig. 2(a)–(d), mountains are located mainly far from the coast in a southwesterly wind direction, which is the prevailing wind in Japan. The amount of water vapour transported from the sea to the mountains varies with the strength of horizontal wind as a whole. This means that the degree of topographic effects on rainfall distribution depends on the strength of horizontal wind in each region, and that the spatial fluctuations of rainfall–elevation relationships become large. That is why the correlation between H and S is high, and much accumulation is required for the emergence of topographic effects in Region 1. On the other hand, water vapour is easy to transport from the sea to the mountains in Region 2 because most of the mountains are located near the seaside, as represented in Fig. 2(b) and (c). Thus, the atmosphere becomes saturated by even a few ascending flows forced by mountains because enough water vapour is supplied, even if horizontal wind is weak. The reason for the low correlation between H and S is that the degree of topographic effects is large according to the atmospheric conditions regardless of the strength of horizontal wind. Much accumulation is not, then, required for the emergence of topographic effects because the spatial fluctuations of water vapour in mountainous regions become relatively small as a whole. Figure 7 represents the of the transportation of water vapour in the regions.

The authors also investigated the effects of the direction of wind through numerical simulations in the four regions. A correlation was not found between the direction of wind and the degree of topographic effects.

SUMMARY AND CONCLUSIONS

This study analyses rainfall–topography relationships through numerical simulations using the mesoscale meteorological model MM5. For a quantitative estimation of topographic effects on rainfall distribution, rainfall–elevation relationships and the

Dependence Line on Topographic Elevation (DLTE) were investigated in four mountainous regions of Japan. The linear relation of DLTE represents the fundamental and average properties of rainfall–elevation relationships in a study region. When attention is focused on mesoscale (especially meso γ-scale) disturbances, rainfall–elevation relationships exhibit complicated behaviour. Thus, an analysis from a macroscopic viewpoint such as DLTE is important for clarifying the universal properties of the relationships.

The effects of wind on rainfall distribution were also investigated for understanding the mechanism of rainfall–topography relationships. The results showed that there is a definite correlation between the strength of horizontal wind and the slope of DLTE, although the strength of the correlation differs from region to region. This means that the strength of topographic effects depends on the distance from the coast to the mountains in each region because the amount of water vapour transported from thesea to the mountains fluctuates with the distance and the strength of horizontal wind. Therefore, the distance from the coast to the mountains plays a significant role in meso β-scale rainfall–topography relationships.

This study conducted diagnostic simulations under some actual conditions. The authors, in future, will conduct virtual simulations including various conditions of topography and the atmosphere for further determining the mechanism of rainfall–topography relationships.

REFERENCES

Alpert, P. (1986) mesoscale indexing of the distribution of orographic precipitation over high mountains. *J. Clim. Appl. Met.* **25**, 532–545.

Alter, J. C. (1919) Normal precipitation in Utah. *Mon. Weath. Rev.* **47**, 633–636.

Barros, A. P. & Lettenmaier, D. P. (1994) Dynamic modeling of orographically induced precipitation. *Rev. Geophys.* **32**(3), 265–284.

Colton, D. E. (1976) Numerical simulation of the orographically induced precipitation distribution for use in hydrologic analysis. *J. Appl. Met.* **15**(12), 1241–1251.

Henry, A. J. (1919) Increase of precipitation with altitude. *Mon. Weath. Rev.* **47**, 33–41.

Lee, C. H. (1911) Precipitation and altitude in the Sierra. *Mon. Weath. Rev.* **39**, 1092–1099.

Nakakita, E., Suzuki, Y. & Ikebuchi, S. (2001) Hierarchical time-scale structure in the dependence of rainfall distribution on topography. *J. Hydrosci. Hydraul. Engng* **19**(2), 1–10.

Oishi, S., Kitani, Y., Nakakita, E. & Ikebuchi, S. (1996) Numerical approach on the effect of mountainous topography to development of severe rainfall (in Japanese with an English abstract). *Ann. Disaster Prevention Research Institute of Kyoto University* **39**(B-2), 1–20.

Suzuki, Y., Nakakita, E. & Ikebuchi, S. (2002) Study on the dependence properties of rainfall distributions on topographic elevations. *J. Hydrosci. Hydraul. Engng* **20**(1), 1–11.

Varney, B. M. (1920) Monthly variations of the precipitation-altitude relation in the central Sierra Nevada of California. *Mon. Weath. Rev.* **48**, 648–650.

Simulation of precipitation caused by a Baiu front: an evaluation study with radar data

SOICHIRO SUGIMOTO & HIROMARU HIRAKUCHI

Central Research Institute of Electric Power Industry, 1646 Abiko, Abiko-shi, Chiba 270-1194, Japan

soichiro@criepi.denken.or.jp

Abstract A numerical simulation of 24-h duration down to the 3-km horizontal resolution was performed with the PSU/NCAR mesoscale model Version 5 (MM5) to evaluate the ability of a mesoscale model to forecast heavy precipitation caused by a Baiu front. Evaluation of the simulated results using both the ECMWF analyses and the radar-based precipitation from the Radar-AMeDAS indicates that the simulation successfully reproduces most principal features of the event, including the synoptic atmospheric circulations, the frontal passage, and the distribution of precipitation. One of the shortcomings of the simulation is that the model tends to underestimate precipitation totals in some heavy rainfall areas. Comparison with winds retrieved from single-Doppler radar data indicates that the model cannot capture a terrain-induced convergence zone. Despite the several deficiencies, the simulated results are encouraging and data assimilation is needed for further improvements.

Key words Baiu front; heavy rainfall; mesoscale meteorological model, MM5; precipitation forecasting; radar

INTRODUCTION

Heavy precipitation and flooding are undoubtedly some of the most significant weather events. Precipitation is an especially difficult weather element to forecast. Considerable studies based on simulations of mesoscale models have provided useful insights on the physical mechanisms and important factors for the formation of heavy precipitation. Recent remarkable improvements of a mesoscale model allow its potential application for short-term (or medium-range) precipitation forecasting to be wider range.

A Baiu front is one of the most significant circulation systems for the hydrological cycle in the East Asia monsoon region. During the Baiu season, the quasi-stationary Baiu front extends from southern China to Japan, and very intense and concentrated rainfall events are often associated with organized mesoscale convective systems, which are embedded within and propagated along the Baiu front (Chen *et al.*, 1988). The objective of this study is to investigate the capability of a mesoscale model to reproduce the spatial variability of precipitation and the associated fields along the Baiu front.

MODEL DESCRIPTION

The numerical model chosen for this study is the nonhydrostatic version of the Pennsylvania State University—National Center for Atmospheric Research Mesoscale

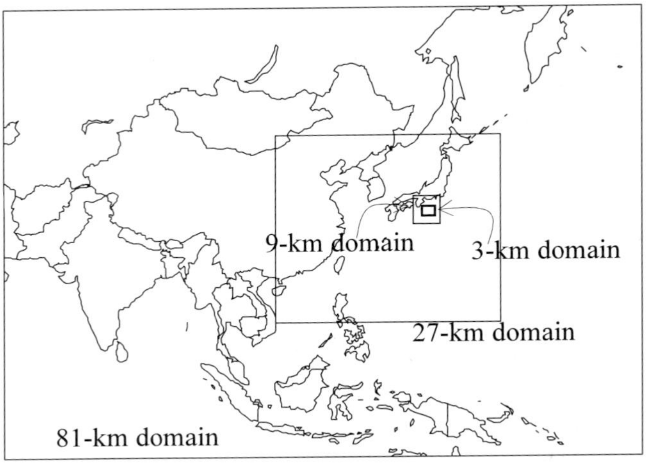

Fig. 1 Geographical location of domains for the MM5 simulation.

Model version 5.3.4 (hereafter referred to as MM5), described in detail by Dudhia (1993). The model configurations and parameterization schemes used in this study are as follows: the coarsest grid domain has a horizontal grid spacing of 81 km, and nested domains have 27, 9 and 3 km, respectively. The model contains terrain-following full-sigma levels in the vertical. The 31 sigma levels are set unevenly as 1.0, 0.99, 0.98, 0.96 …, 0.13, 0.09, 0.05, and 0.0. The model top is located at the 50 hPa level. The geographic location for four grids can be seen in Fig. 1. The number of grid points in each of four domains is 163 × 121, 172 × 151, 55 × 61, and 85 × 64, respectively. The outermost domain is designed to simulate the synoptic-scale environment. The size of the outermost domain is chosen sufficiently large to minimize the influence of the lateral boundary conditions into the MM5 interior.

MM5 includes parameterizations of dynamical and microphysical processes that are important for mesoscale systems. The Dudhia explicit moisture scheme is used on all the four domains. This scheme accounts for the resolvable scale convection with explicit treatment of cloud water, rainwater, snow, and ice (Dudhia, 1993). The Grell cumulus parameterization is applied, except for the two innermost domains where convective processes could be resolved explicitly. The model uses the Blackadar's high-resolution planetary boundary layer parameterization (Zhang & Anthes, 1982) to calculate vertical fluxes of heat, moisture, and momentum at each vertical layer.

The model is initialized using archived analyses from the European Centre for Medium-Range Weather Forecasts (ECMWF) and Reynolds sea-surface temperature (SST) from the National Oceanic and Atmospheric Administration (NOAA), from which first-guess fields are generated on the coarsest model grid by interpolation at 6-h intervals (00:00, 06:00, 12:00 and 18:00 UTC). The ECMWF datasets are the spectral analyses with triangular truncation at wavenumber 106 (T106) with a horizontal resolution of approximately 1.125°. The time-dependent lateral boundary conditions for the 81-km domain are provided by repeating the above procedure at 6-h intervals and linear interpolation in time using a 16-point, two-dimensional parabolic function.

The two-way interactive nested grid system is adopted between the 27-km and 81-km domains. The 3-km and 9-km domains, however, receive information from the next coarser grid, but do not feed information back to the larger domain (that is called the one-way interface). The model simulation starts at 12:00 UTC on 27 June 2000 and runs for 24 h. A time step of 135 s is used in the coarsest grid domain.

SYNOPTIC-SCALE CIRCULATION

The event studied here occurred in East Asia on 27–28 June 2000. Figure 2 shows the Geostationary Meteorological Satellite (GMS) Imagery at 00:00 UTC 28 June. The GMS imagery indicates that a cloud band ahead of the Baiu front stretches nearly continuously from the Yangtze River to Japan. The 500-hPa ECMWF analysis at 00:00 UTC, shown in Fig. 3(a), reveals that the contour line at 5840 m is highlighted to indicate the approximate southern boundary of the Baiu front. This boundary is related to the region of transition zone between the subtropical high and the westerly trough. The southerly low-level jet (LLJ) is also found around the Yangtze River basin. This flow is important for intensification of the mesoscale convective systems (e.g. Chen *et al.*, 1998). The 850-hPa relative humidity from the ECMWF analysis suggests that a west–east oriented high humidity belt coincides with the cloud band (not shown). The belt-like region is primarily determined by the northwest Pacific subtropical high, but the orientation of the belt seems to be difficult to forecast because of some oscillations during the Baiu season.

Here, we evaluate the ability of MM5 to forecast the synoptic atmospheric circulations using the forecasts for the 81-km domain and the ECMWF analyses. That is, the ECMWF analyses are used for initial and boundary conditions as well as the evaluation of the simulation. Since the simulation on inner grids is driven by larger-

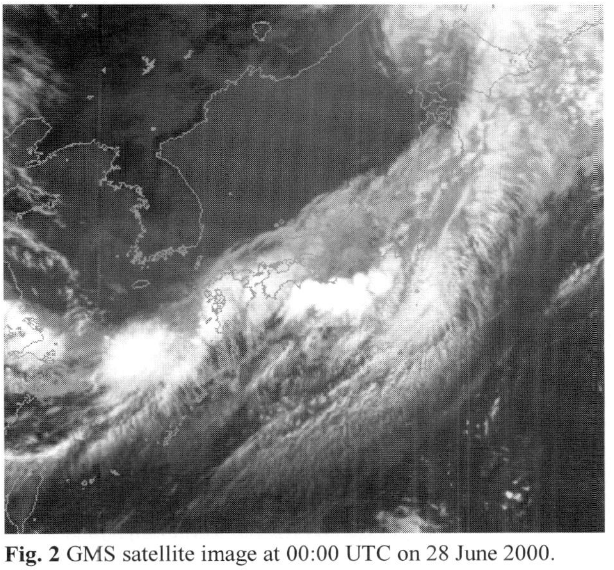

Fig. 2 GMS satellite image at 00:00 UTC on 28 June 2000.

scale atmospheric background, the model has to simulate reasonably well atmospheric circulations to reproduce regional features of the rainbelt and local rainfall maxima (Liu *et al.*, 1996). Figure 3(b) shows the 500-hPa geopotential heights and winds on the 81-km domain from the simulation at t = 12 h. Comparisons of the simulated results with the analyses demonstrate that the model can capture the synoptic important features of the event. The simulated pattern of the atmospheric circulations at other levels agrees as well with the pattern of the analyses (not shown). Figure 3(b) indicates, however, that MM5 apparently simulates the earlier propagation of the Baiu front (contour line at 5840 m) better than the analysis. This is more evident in the simulation at t = 18 h. Such a bias in the simulation would significantly affect distribution of precipitation.

PRECIPITATION FORECASTS

The model ability to forecast precipitation is verified using radar-based rainfall data from the Radar-AMeDAS. The radar-based data are produced by calibration of data

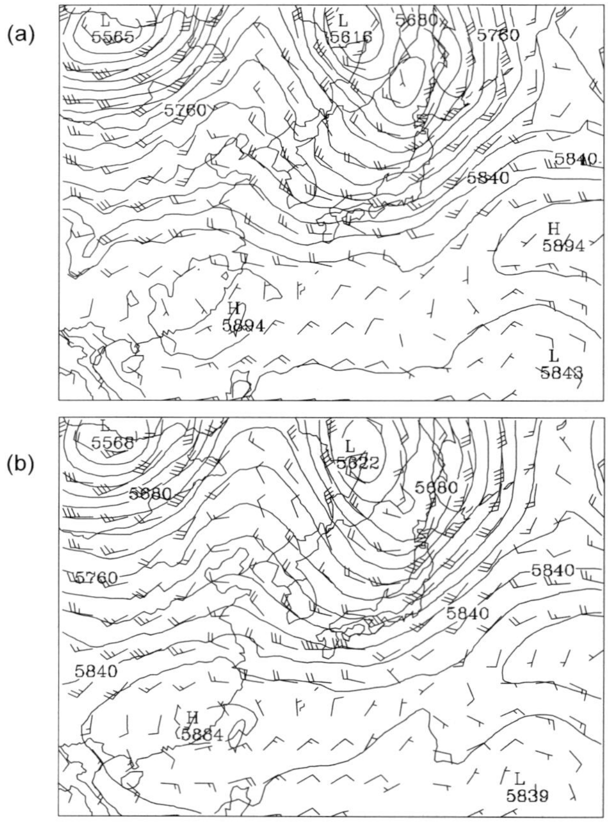

Fig. 3 The 500-hPa geopotential heights and winds for 00:00 UTC on 28 June 2000 from (a) the ECMWF analysis and (b) the MM5 simulation (t = 12 h). Heights are contoured every 20 m. Full wind barb represents 5 m s^{-1}.

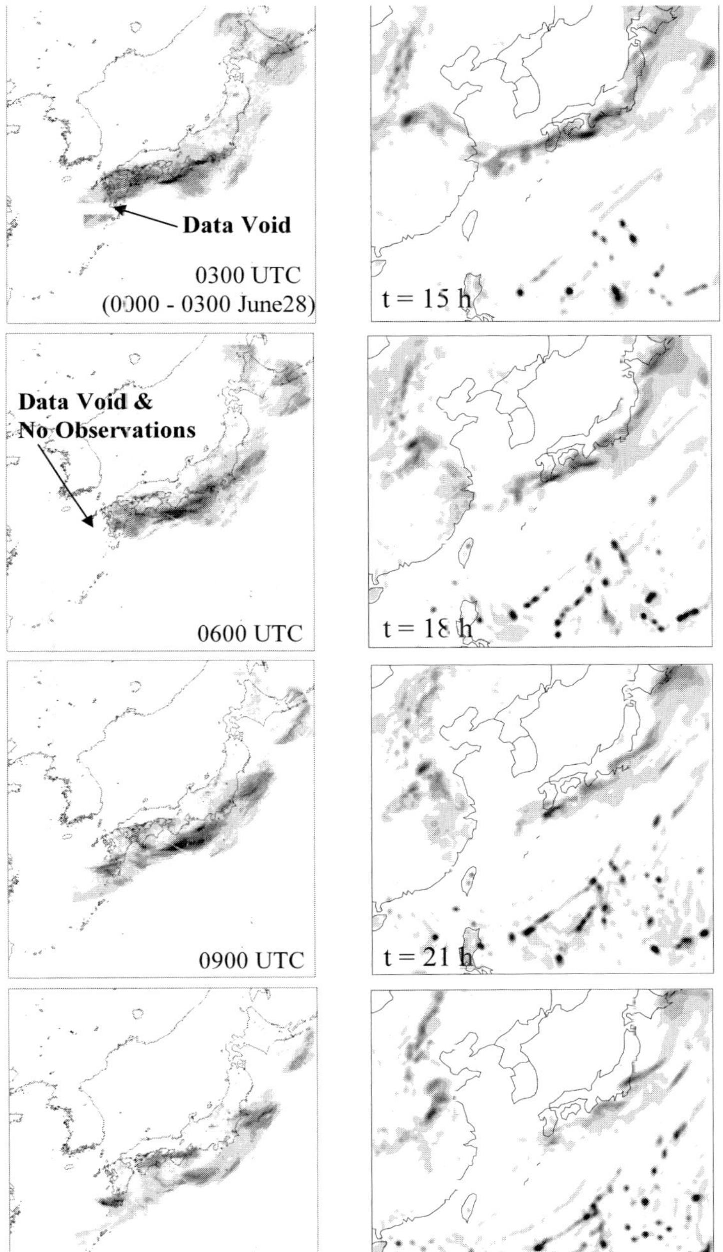

Fig. 4 The radar-based (left) and the simulated (right) sequential distributions of 3-h accumulated precipitation (mm) from 03:00 to 12:00 UTC on 28 June 2000 at intervals of 3 h.

from operational C-band radars around Japan using ground gauged rainfall data collected by the automated meteorological acquisition system (AMeDAS). Both the Radar-AMeDAS and the AMeDAS data are originally produced by the Japan Meteorological Agency. The resolution of radar-based data is about 5 km. Because of the difficulty to forecast, we focus on evaluating forecasts of the 3-h accumulated precipitation.

The left-hand column of Fig. 4 depicts sequential distributions of the 3-h accumulated precipitation yielded by the Radar-AMeDAS data from 03:00 to 12:00 UTC on 28 June at intervals of 3 h. The major aspects of the precipitation to be considered include:

(a) A west–east oriented rainbelt along the Baiu front is prominent, and several precipitation centres are found in the rainbelt.
(b) The Baiu front propagates southward, while mesoscale convective systems embedded move eastward.

The distributions of the 3-h accumulated precipitation simulated on the 27-km domain during the same period (15, 18, 21, and 24 h) are shown in the right-hand column of Fig. 4. Despite the relatively coarse horizontal resolution, the MM5 simulation can provide a general description of the mesoscale features listed above. The approximate position and orientation of the precipitation maxima are realistically simulated. The major deficiencies of the model are its underestimation of the precipitation totals in some strong precipitation centres and to fail in reproducing some rainfall areas during the period of t = 18–24 h, mainly due to the model forecast of the earlier frontal passage.

Figure 5 shows the distributions of the 3-h accumulated precipitation simulated on the 81-km, 27-km, and 9-km domains at 00:00 UTC on 28 June (t = 12 h). The majority of the model simulated precipitation is generated by the convective scheme rather than by the grid-scale explicit scheme. Quantitative comparisons between the simulations and the observations indicate that several improvements are found in the result on the 27-km domain compared with the one on the 81-km domain, and that the simulation on the 9-km domain shows better performance than the simulation on the 27-km domain. However, there are not large discrepancies between the precipitation on the 9-km domain and the one on the 3-km domain (not shown). This suggests that simply decreasing the grid size of the model will not always yield a better representation of convective activities.

The simulated winds on the 3-km domain are shown in Fig. 6 together with the retrieved winds from Doppler radar data using the EVVP method (Sugimoto, 2002). The region displayed in Fig. 6 is very orographic, and the convergence zone (circled), which was persistent for several hours, was responsible for the formation of a terrain-induced and localized rainband in the meso-β scale. Unfortunately, the simulated winds cannot resolve the convergence zone. The model terrain could be smoothed so that the interaction of the front and the terrain-induced flow cannot be represented.

The understanding of the statistical model behaviour for precipitation forecasting needs further studies on the effects of a microphysical parameterization, a scheme for smoothing the terrain, and others. In effect, precipitation and winds estimated from radar data are valuable information for an evaluation of the model. Moreover, it must be recognized that the resolution of the ECMWF analyses in this study is too coarse to

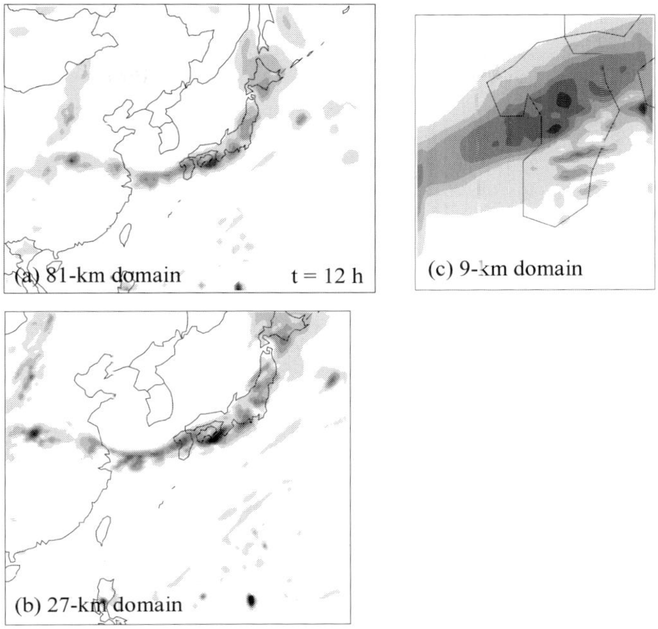

Fig. 5 Model-accumulated precipitation for 21:00 UTC on 27 June to 00:00 UTC on 28 June 2000 (t =12 h) on: (a) the 81-km domain, (b) the 27-km domain, and (c) the 9-km domain.

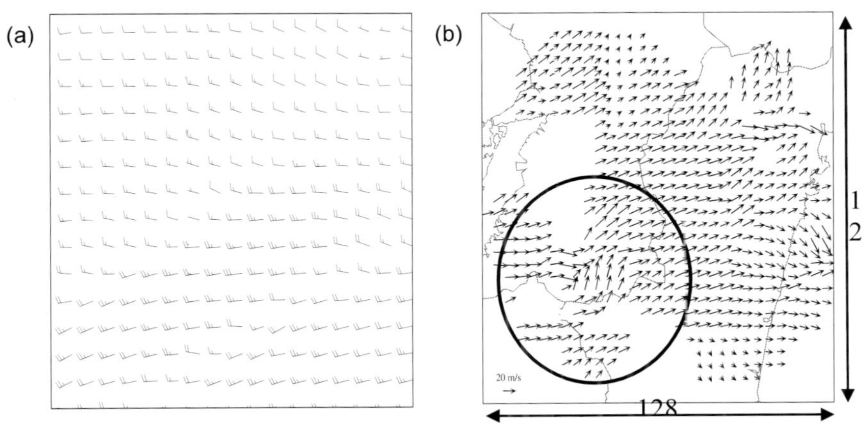

Fig. 6 Horizontal winds for the height of 3 km at 0200 UTC June 28 2000, derived by (a) the MM5 simulation and by (b) the EVVP method. Full wind barb represents 10 m s^{-1}. The displayed region for the simulation corresponds with the analysis area of the EVVP method.

resolve the variability of convections in the meso-β scale. For circumventing the insufficient resolution of the initial and boundary conditions, mesoscale data assimilation should be required. Therefore, Doppler radar data is expected to provide a potentially valuable source of upper-air data suitable for four-dimensional data assimilation.

REFERENCES

Chen, S.-J., Kuo, Y.-H., Wang, W., Tao, Z.-Y. & Cui, B. (1998) A modeling case study of heavy rainstorms along the Mei-Yu front. *Mon. Weath. Rev.* **126**, 2330–2351.

Dudhia, J. (1993) A nonhydrostatic version of the Penn State—NCAR mesoscale model: validation tests and simulation of an Atlantic cyclone and cold front. *Mon. Weath. Rev.* **121**, 1493–1513.

Liu, Y., Avissar, R. & Giogi, F. (1996) Simulation with the regional climate model RegCM2 of extremely anomalous precipitation during the 1991 East Asian flood: an evaluation study. *J. Geophys. Res.* **101**, 26199–26215.

Sugimoto, S. (2002) Kinematic properties in convective storms derived from Doppler radar observations in mountainous regions of Japan. In: *Mesoscale Convective Systems and Heavy Rainfall/Snowfall in East Asia* (Proc. Int. Conf.), 497–502. Japan Science and Technology Corporation & Chinese Academy of Meteorological Sciences, China.

Zhang, D. & Anthes, R. A. (1982) A high-resolution model of the planetary boundary layer—Sensitivity tests and comparisons with SESAME-79 data. *J. Appl. Met.* **21**, 1594–1609.

Weather Radar Information and Distributed Hydrological Modelling (Proceedings of symposium HS03 held during IUGG2003 at Sapporo, July 2003). IAHS Publ. no. 282, 2003.

Rainstorm characteristics in temperate areas derived from multiple time series of weather radar images

GABRIELLE J. M. DE LANNOY, INGE L. M. DE JONGH, NIKO E. C. VERHOEST & FRANCOIS P. DE TROCH

Laboratory of Hydrology and Water Management, Ghent University, Coupure links 653, B-9000 Ghent, Belgium

inge.dejongh@rug.ac.be

Abstract To define the spatial and temporal characteristics of rainstorms, 20 time series of weather radar images over northwestern Europe were analysed. The velocity of rainfall events and the direction of movement were determined for 1632 radar images using the cross-correlation technique. A technique was developed to discriminate single rainstorms within a radar image. After calculating the area and perimeter of these rainstorms, different methods to define the dimensions of rainstorms were proposed. Statistical calculations revealed that the dimensions of rainstorms in the direction of the rainfall movement were significantly different from the dimensions in the direction perpendicular to the movement. Rainstorms are composed of several rain cells which are clustered together. In order to describe the occurrence of cell clusters, a new technique was used for localizing cluster centres within a rainstorm. The positions of the cluster centres along and across the rainfall movement follow a Poisson distribution.

Key words cross-correlation; rain cell clusters; rainfall movement; weather radar

INTRODUCTION

Several studies over the last decades have shown that hydrological models give better results when the spatial variability of the rainfall input is taken into account (e.g. Krajewski *et al.*, 1991; Obled *et al.*, 1994; Willems & Berlamont, 1999). When performing a long-term hydrological simulation using uniform rainfall over the basin, the peak discharges will be overestimated since in reality the largest rainfall intensities will not be found over the whole basin at the same time. As a consequence a rainfall generator, which preserves the spatial organization of rainfall patterns, is necessary for simulating long time series of spatial rainfall. Weather radar images are a perfect source for obtaining the amount of spatial rainfall information that is needed for building a spatial rainfall generator, considering their spatial and temporal resolution. The objective of this study is to have a more accurate description of some rainfall characteristics at mesoscale. Not only the velocity of rainfall events and the direction in which they move are studied thoroughly, but also the dimensions of rainstorms and the localization of clusters within a rainstorm are considered. Therefore a clear definition of a rainstorm has to be formulated first. The statistical distributions of the rainfall characteristics that are determined in this study can directly be used as an input for a stochastic spatial rainfall generator (e.g. Willems, 2001).

DATA

The data used for this study are pCAPPI (pseudo Constant Altitude Plan Position) radar images provided by the KNMI (Royal Meteorological Institute of The Netherlands). The Netherlands, the western part of Germany and the northern part of Belgium are covered with these images. The radar images are parallel to the Greenwich meridian, with a 200 × 200 pixel grid of pixel size 2.4 × 2.4 km^2. Twenty time series of radar images, with a 15-minutes time lag between two consecutive images, a minimum of 45 images and a maximum of 96 images per series, were selected within the period 1998–2000. To include possible seasonal variation, rainfall events were chosen throughout the year and only days with a high total amount of rainfall were selected. All selected rainfall events are parts of frontal weather systems.

METHODS

Rainfall movement

For the calculation of the motion vector of rainfall events, the cross-correlation technique, first published by Zawadski (1973), was used. The motion vector is determined by the (k_x, k_y) displacement which gives the highest cross-correlation in a Cartesian coordinate system defined by the X_1- and Y_1-axes (east–west and north–south directions, respectively).

An important advantage in using this method to calculate the velocity and direction of movement is that very small spatial lags (k_x, k_y) can be used. The smallest lag used in this study equals 0.1 pixel length.

Out of the 20 series of radar images, 1603 pairs of consecutive images could be selected. On each of these pairs the translation method was used to calculate the velocity of the rainfall event U_{xy} (km h^{-1}) and the direction of the rainfall movement θ (rad), which is taken relative to the eastern direction. A histogram of the original and transformed data was made for each characteristic and tested for different bell-shaped theoretical probability density functions (pdfs). Therefore the normal distribution $N(\mu,\sigma)$, the Weibull distribution $W(\alpha,\beta)$ and the Gamma distribution $G(k)$ were considered.

The different transformations that were applied to the original data were the square root transformation, the log-transformation and the inverse of the data. To estimate the parameters for the distributions the least squares approach was used.

Determination of a rainstorm

A new concept, namely the *APratio* (km), is introduced:

$$APratio = \frac{\sum_{\text{rainy areas}} Area_i}{\sum_{\text{rainy areas}} Perimeter_i} \tag{1}$$

Based on the temporal evolution of the *APratio*, a technique was developed to reject images which contain rainy systems in development or decay.

When a rainstorm is growing (or dissipating), several small rainy areas can be detected in the image. This results in a relatively low *APratio*, since the total perimeter is much larger compared to the situation where the same rainy area is part of one connected system. So every image with a high *APratio* represents a rainy system which is fully developed. The area and the perimeter for the different rainy groups in every image of the time series are determined, the *APratio* is calculated and its evolution is plotted (e.g. Fig. 1). The next step includes the determination of arbitrary thresholds for the *APratio*, above which the rainy area can be considered to be fully developed. These thresholds are different for every time series of radar images and are mainly based on the visual interpretation of the plot. Only images with an *APratio* above these thresholds are further used in this study. Finally, in every radar image that is retained, the largest rainy area is selected as the rainstorm.

Rainstorm dimensions

To determine the dimension of a rainstorm, three different methods are proposed. Therefore two coordinate systems are defined. The first system uses the east–west (X_1) and north–south (Y_1) directions in the image, while the axes in the other system are chosen parallel (X_{23}) and perpendicular (Y_{23}) to the direction of movement of the rainstorm. Using different criteria, three rectangles can be built for every rainstorm (Fig. 2). The lengths of their sides gives an estimation for the size of the rainstorm. Two methods build the smallest possible rectangle around the rainstorm with sides parallel to coordinate axes: method 1 uses the X_1,Y_1 system, while method 3 uses the X_{23}, Y_{23} system. Method 2 builds its rectangle using the intersection points of the rainstorm with the X_{23},Y_{23} coordinate system.

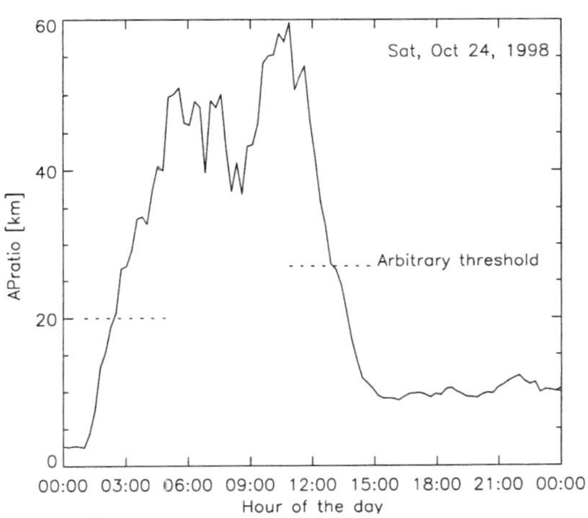

Fig. 1 The evolution of the *APratio* on 24 October 1998 determined using 96 radar images. Images with an *APratio* below the dotted lines are supposed to have a rainstorm in development or decay and are not subjected to any further analysis.

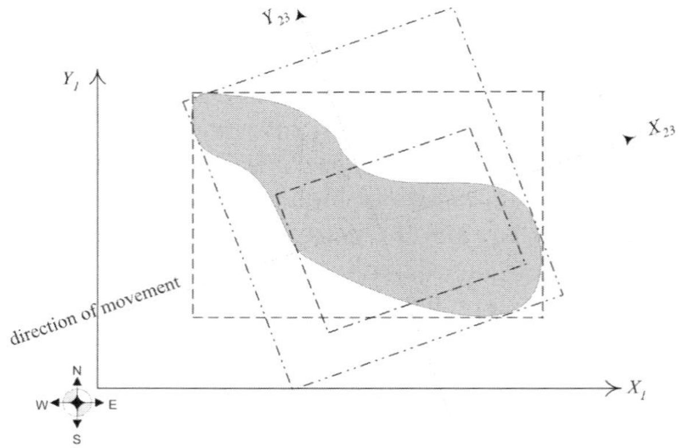

Fig. 2 Schematic overview of the three methods to determine the extension of a rainstorm. The different rectangles give the maximum size of the rainstorm for method 1 (– –), method 2 (– – ·) and method 3 (– · ·).

Rain cell clusters

Within a rainstorm, small areas with a high rainfall intensity can appear. These areas are rain cells for which earlier research could already conclude that they are clustered together (Bacchi *et al.*, 1991). Due to the spatial resolution of the radar images used in this research, the individual rain cells cannot be detected accurately, so only rain cell clusters will be considered. To simplify the analysis, we assume that the location of the cluster centres can be used as the location of the clusters. In order to locate the cluster mass centres in the radar image, the level slicing technique is used to define rain cell clusters. The connected areas which have a rainfall intensity above a preset threshold and have an minimum area of 46 km^2 (i.e. 8 pixels) are considered to be rain cell clusters. For each cluster, the threshold used depends on the rainfall intensity within the cluster. After the delineation of the clusters, the centre of each cluster is determined.

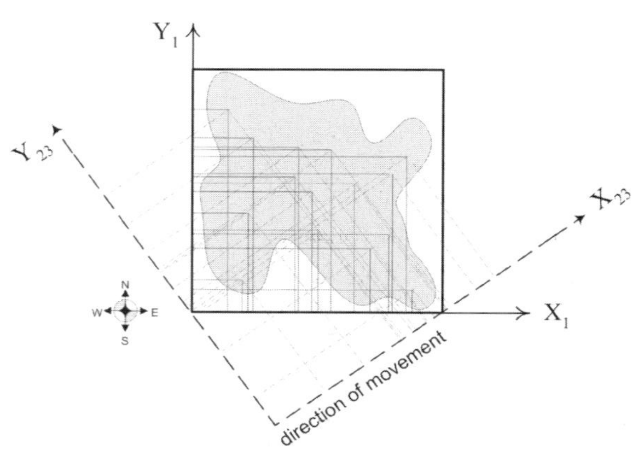

Fig. 3 Schematic overview of the 1-D method to calculate the Poisson parameter λ: after projection of the cluster centres, the intercluster distances are calculated.

We assumed that the spatial distribution of clusters is a homogeneous Poisson distribution with only parameter λ, which means that the intercluster distance follows an exponential distribution with a probability density function that is given by $f_x = \lambda\ e^{-\lambda x}$. The parameter λ is calculated using a one-dimensional (1-D) and a two-dimensional (2-D) method.

For the 1-D method, the cluster centres are projected on a straight line and the intercluster distances are calculated. The histogram of the intercluster distance is plotted and an exponential distribution is fitted whereby the Poisson parameter λ is directly found. We used four different lines to project the centres on: the X_1, Y_1, X_{23} and Y_{23} coordinate axes (Fig. 3). In this way it could be tested if there is a difference in cluster density for the different directions in the image.

In the 2-D method a pixel in the image is selected, a square is created around that pixel and the cluster centres within the square are counted. This procedure is repeated for increasing areas of the squares and for all pixels within the rainstorm. Finally, a plot of the number of counted cluster centres against the area of the used square is created (Fig. 4). The first part of the plot is linear and the slope of the trend line equals the λ-parameter, which represents the number of clusters per unit area.

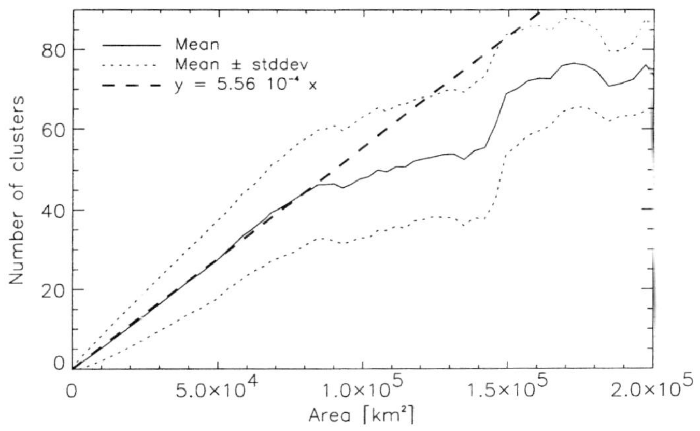

Fig. 4 Determination of the Poisson parameter λ using the 2-D method, where λ is the slope of the trend line in the linear part of the graph.

RESULTS

The velocity of the studied rainfall events varied between 1.8 and 185.3 km h^{-1}, with a mean of 51.78 km h^{-1}. Statistical analysis (Kolmogorov–Smirnov analysis and QQ-plots) of the histogram of rainfall velocity led to the selection of a Weibull distribution characterized with a scale parameter α equal to 0.98 and a shape parameter β equal to 2.10, as the most optimal probability distribution function. Via transformation of the data a better fit to the theoretical pdfs could not be achieved..

The direction in which the rainfall event is moving varied between –2.88 and 2.90 radians, with a mean of 0.327 radians. The best fit for the distribution of this rainfall movement characteristic is a normal distribution with a mean μ equal to 0.33 radians and a standard deviation σ of 0.75 radians. This means that the rainfall systems move mainly in the E–NE direction.

From the 1632 available radar images, only 934 were considered to represent a rainstorm as it is defined in this work. Once the rainstorms are defined, the total area and the perimeter of each rainstorm can be calculated (Table 1). Statistical analyses revealed that the distribution of the area and the perimeter, calculated using all the 934 processed rainstorms can be best approached by a normal distribution after a square root transformation of the data.

For the three proposed methods the statistical characteristics of the rainstorm dimensions are summarized in Table 1. Again theoretical distributions are fitted to the observed distributions of the different x and y variables, except for the distributions obtained with method 1. Because the histograms of x_1 and y_1 are truncated at 480 km (the maximum width of the radar image) we did not fit pdfs for these variables. For the four other variables (x_2, x_3, y_2 and y_3) a Weibull distribution gave the best fit, with shape parameters in the range 2.99 to 5.12 and scale parameters varying between 352 and 450.

To test whether the dimensions of the rainstorm are significantly different in x- and y directions, a non-parametric Mann-Whitney test was performed on the x_2,y_2 and the x_3,y_3 data, respectively. A significant difference was found between the dimensions of the rainstorm in the direction of the rainfall movement and the direction perpendicular to the direction of movement ($P < 0.05$). This result was obtained for both methods 2 and 3.

The results of the analysis of the cluster centres are summarized in Table 2. Using the 2-D method we obtained an idea about the cluster density within the rainstorm but no conclusions could be made about how these clusters are spatially organized. There the 1-D method offers a solution: it can be noticed that the clusters seem to be more distributed along the direction of the rainstorm movement compared with the distribution of the clusters in the direction perpendicular to the rainfall movement.

Table 1 Statistical information of some rainfall characteristics.

Characteristic	Mean	Standard deviation	Fitted distribution
1603 pairs of radar images:			
Velocity (km min^{-1})	0.863	0.412	W(α = 0.98 ,β = 2,10)
Direction angle (rad)	0.327	0.755	N(μ = 0.33 ,σ = 0.75)
934 rainstorms:			
Area (km^2)	87178	35707	N($\mu_{\surd}$ = 288.90, $\sigma_{\surd}$ = 60.96)
Perimeter (km)	3654	1174	N($\mu_{\surd}$ = 59.63, $\sigma_{\surd}$ = 9.90
x_1 (km)	382	90	-
y_1 (km)	405	75	-
x_2 (km)	328	117	W(α = 367.47, β = 3.06)
y_2 (km)	314	111	W(α = 351.66, β = 2.99)
x_3 (km)	410	104	W(α = 450.01, β = 4.43)
y_3 (km)	392	89	W(α = 425.71, β = 5.12)

W(α =, β =): Weibull distribution with scale parameter α and shape parameter β
N(μ =, σ =): Normal distribution with mean μ and standard deviation σ. A square root transformation of the data is indicated with a √–subscript.

Table 2 Calculated Poisson parameters λ for the spatial distribution of the cell cluster centres.

	Coordinate axis:			
	X_1	Y_1	X_{23}	Y_{23}
1-D method (cluster centres km^{-2})	0.199	0.211	0.135	0.145
2-D method (clusters km^{-2})	5.56×10^{-4}			

CONCLUSIONS

The cross-correlation technique has been used successfully to determine the velocity and the direction of movement of rainfall events. Statistical analysis revealed that the velocity of rainfall events follows a two-parameter Weibull distribution and that the distribution of the direction of rainfall movement can be approached by a normal distribution.

A technique to discriminate single rainstorms within radar images has been optimized, making use of the total area and the total perimeter of all rainy areas in the radar image. Three methods to determine rainstorm dimensions were proposed. Statistical analysis revealed that the dimensions of the rainstorms in *x*- and *y*-directions both follow a two-parameter Weibull distribution and are significantly different.

Using a level slicing technique rain cell clusters could be detected in the radar images. After the application of a 1-D and a 2-D method it could be proved that the spatial distribution of rain cell cluster centres follows an homogeneous Poisson distribution. A good interpretation of the spatial distribution of clusters can be made when using the results of the two proposed methods.

REFERENCES

Bacchi, B, Ranzi, R. & Borga, M (1996) Statistical characterization of spatial patterns of rainfall cells in extratropical cyclones. *J. Geophys. Res.* **101**, 277–286.

Krajewski, W., Lakshmi, V., Georgakakos, K. & Jain, S. (1991) A Monte Carlo study of rainfall sampling effect on a distributed catchment model. *Water Resour. Res.* **27**, 119–128.

Obled, C., Wendling, J. & Beven, K. (1994) The sensitivity of hydrological models to spatial rainfall patterns: an evaluation using observed data. *J. Hydrol.* **159**, 325–333.

Willems, P. (2001) A spatial rainfall generator for small spatial scales. *J. Hydrol.* **252**, 126–144.

Willems, P. & Berlamont, J. (1999) Probabilistic modeling of sewer system overflow emissions. *Water Sci. Tech.* **39**, 47–54.

Zawadski, I. (1973) Statistical properties of precipitation patterns. *J. Appl. Met.* **12**, 459–471.

Weather Radar Information and Distributed Hydrological Modelling (Proceedings of symposium HS03 held during IUGG2003 at Sapporo, July 2003). IAHS Publ. no. 282, 2003.

Stochastic modelling of the error structure of real-time predicted rainfall and rainfall field generation

YASUTO TACHIKAWA, YOSHIMITU KOMATSU, KAORU TAKARA
Disaster Prevention Research Institute, Kyoto University, Uji, Kyoto 611-0011, Japan
tatikawa@mbox.kudpc.kyoto-u.ac.jp

MICHIHARU SHIIBA
Department of Urban and Environmental Engineering, Kyoto University, Sakyo-ku, Kyoto 606-8501, Japan

Abstract To evaluate the uncertainty of a real-time river discharge prediction with a distributed rainfall–runoff model, an error structure of real-time rainfall prediction by a translation model is modelled as a spatial random field, and then predicted rainfall fields are simulated according to the characteristics of the prediction error structure. At first, prediction error fields are statistically analysed and an isotropic lognormal spatial random field is selected to model relative prediction error fields. Next, a method to generate a lognormal spatial random field is developed, which uses a matrix factorization technique that a covariance matrix is decomposed into its square root matrix. Using this method, rainfall field realizations that take into account the prediction error structure are well simulated.

Key words precipitation forecasting; random field; translation model; uncertainty

INTRODUCTION

For real-time rainfall and discharge prediction it is important to not only provide a forecasting value but also to give the prediction accuracy. So far, many methods of short-term rainfall prediction using weather radar data have been proposed, and some researchers have extended these methods to provide the uncertainty of real-time rainfall forecasting (Takasao *et al.,* 1994; Georgakakos & Krajewski, 1995; Georgakakos, 2000), in which a stochastic state-space form of a rainfall prediction model using operationally available radar data are presented. In this study, to evaluate the uncertainty of a river discharge prediction by using a physically based distributed hydrological model, the idea to solve a stochastic distributed hydrological system numerically on a Monte Calro simulation framework using rainfall prediction field simulations is adopted. If simulated alternative realizations of a predicted rainfall field having common statistical properties are provided as the input to a physically-based distributed hydrological model, the output discharges will produce the probability distribution of predicted discharges at any point in space. This is the objective of the research presented here.

In the next section, a radar data used here and a short-term rainfall prediction method using a translation model (Shiiba *et al.*, 1984; Takasao & Shiiba, 1985) are briefly described. After that, prediction error fields calculated with radar observed rainfall fields and predicted rainfall fields by the translation model are statistically analysed and it is found that the distribution of a relative prediction error fits to a lognormal distribution having a spatial correlation. Therefore an isotropic lognormal spatial random field model is selected to represent the spatial prediction error structure. Then a method to generate the lognormal spatial random field is developed, which uses a matrix factorization technique that a covariance matrix is decomposed into its square root matrix approximately by using the Chebyshev polynomials. Finally, rainfall field generations by using the method are demonstrated.

RADAR RAINFALL DATA AND TRANSLATION MODEL

Radar rainfall data

The radar data used are observed at the Miyama radar raingauge located at the centre of Kinki district, Japan, which is managed by the Ministry of Land, Infrastructure and Transport. The radar measures rainfall fields at 5-min intervals with the horizontal resolution of 3 km, covering the area of 120 km radius. In this study, the radar data of 80 × 80 grid cells (240 × 240 km with 3 km spatial resolution) observed between 5 September at 10:00 and 6 September at 6:00 in 1989 are used. By using the radar data, 12 kinds of averaged data with the combinations of time resolutions of 5, 15, 30, and 60 min and spatial resolutions of 3, 6 and 12 km are generated for statistical analyses.

Translation model

The horizontal rainfall distribution $z(x, y, t)$ with the spatial coordinate (x, y) at time t is modelled as:

$$\frac{\partial z}{\partial t} + u\frac{\partial z}{\partial x} + v\frac{\partial z}{\partial y} = w \tag{1}$$

where u and v are advection velocity along x and y, and w is growth–decay rate. Here u, v and w are assumed to have the forms:

$$u = c_1 x + c_2 y + c_3, \quad v = c_4 x + c_5 y + c_6, \quad w = c_7 x + c_8 y + c_9 \tag{2}$$

The values of parameters c_1, ..., c_9 are identified by the least square method using observed radar rainfall data and they are updated on a real-time basis (Shiiba *et al.*, 1984; Takasao & Shiiba, 1985). Three consecutive spatial rainfall distributions in 15 min are used to determine u and v, and rainfall fields with 5 min and 3 km resolution are predicted for each 5 min interval. After that, 12 kinds of averaged predicted data with different time and spatial resolution stated above are generated to analyse prediction error structures. When forecasting rainfall fields, the growth decay rate w is always assumed to be zero.

MODELLING OF REAL-TIME RAINFALL PREDICTION ERROR

Model of prediction error field

The spatial distribution of an absolute prediction error $\boldsymbol{E}_a$ and the one of a relative prediction error $\boldsymbol{E}_r$ with the forms:

$$\boldsymbol{E}_a = \boldsymbol{R}_o - \boldsymbol{R}_p, \quad \boldsymbol{E}_r = (\boldsymbol{R}_o - \boldsymbol{R}_p) / \boldsymbol{R}_p \tag{3}$$

are calculated and statistically analysed, where $\boldsymbol{R}_o$ is an observed radar rainfall field and $\boldsymbol{R}_p$ is a predicted rainfall field by the translation model. The calculation in equation (3) is conducted for each corresponding grid cell of $\boldsymbol{R}_o$ and $\boldsymbol{R}_p$. As shown in equation (3), $\boldsymbol{E}_r$ cannot be obtained if $\boldsymbol{R}_p$ is zero, so $\boldsymbol{E}_r$ is assumed to be zero if $\boldsymbol{R}_p$ is zero.

If a random field model represents the spatial distributions of $\boldsymbol{E}_a$ and/or $\boldsymbol{E}_r$ and it is well simulated according to the statistical characteristics, a realization of rainfall field $\boldsymbol{R}$ is simulated with the equation:

$$\boldsymbol{R} = \boldsymbol{R}_p + \boldsymbol{E}_a \quad \text{or} \quad \boldsymbol{R} = \boldsymbol{R}_p \times \boldsymbol{E}_r + \boldsymbol{R}_p \tag{4}$$

The simulated alternative rainfall fields $\boldsymbol{R}$ are applied to a distributed rainfall–runoff model and the probability distribution of a discharge prediction will be obtained.

Statistical characteristics of prediction error

Figure 1 shows the frequency distributions of an absolute prediction error $\boldsymbol{E}_a$ for 5-min ahead prediction, which are those calculated using grid-cells in which the predicted

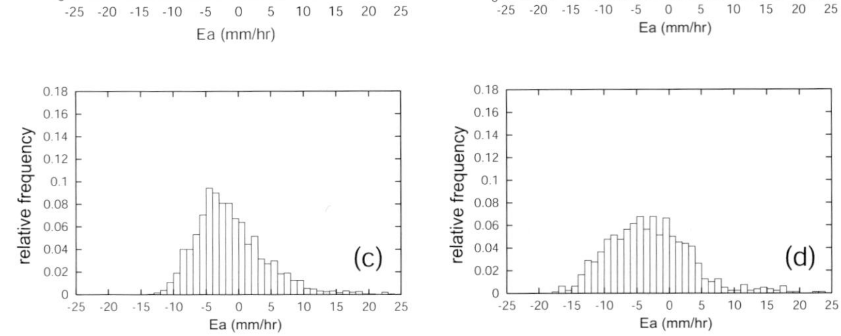

Fig. 1 Frequency distribution of absolute prediction error $\boldsymbol{E}_a$ for 5-min ahead prediction: frequency distributions of $\boldsymbol{E}_a$ for grid cells in which predicted rainfall intensity is (a) larger than zero; (b) between 5 and 10 mm h^{-1}; (c) between 10 and 15 mm h^{-1}; and (d) between 15 and 20 mm h^{-1}.

rainfall intensity is larger than zero, between 5 and 10 mm h^{-1}, between 10 and 15 mm h^{-1}, and between 15 mm h^{-1} and 20 mm h^{-1}, respectively. The shapes of frequency distributions are quite dependent on the predicted rainfall intensity and also $\boldsymbol{E_a}$ showed a positive spatial correlation. It is difficult to model a spatial random field having different probability distributions with a spatial correlation structure. Then, a relative prediction error $\boldsymbol{E_r}$ is tested. As shown in Fig. 2, every frequency distribution seems to fit to the same lognormal distribution with the lower bounded value, which takes –1 by equation (3) when the value of an observed rainfall is zero. By using the data, the spatial correlation coefficient of $\boldsymbol{E_r}$ is also calculated. In Fig. 3, it is found that $\boldsymbol{E_r}$ has a positive correlation within 10 km. From these results, $\boldsymbol{E_r}$ could be modelled as a lognormal random field with a spatial correlation structure.

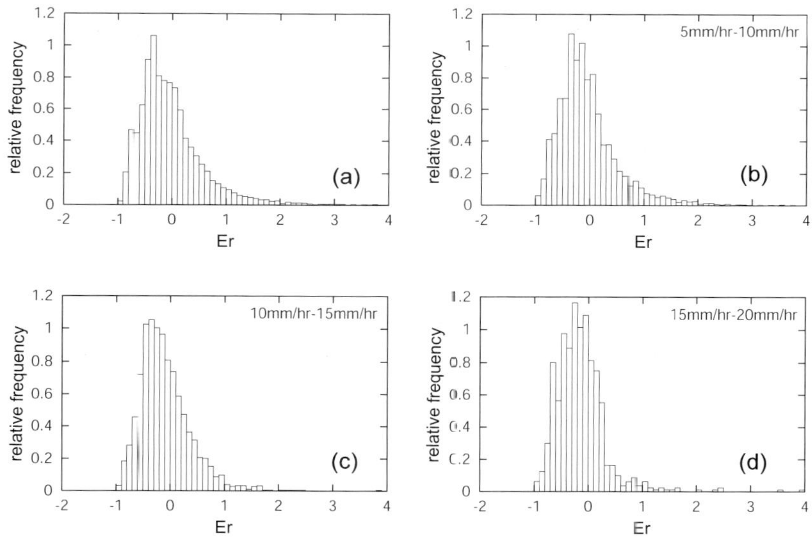

Fig. 2 Frequency distribution of relative prediction error $\boldsymbol{E_r}$ of 5-min ahead prediction. Frequency distribution of $\boldsymbol{E_r}$ for grid cells in which predicted rainfall intensity is (a) larger than zero; (b) between 5 and 10 mm h^{-1}; (c) between 10 and 15 mm h^{-1}; and (d) between 15 and 20 mm h^{-1}.

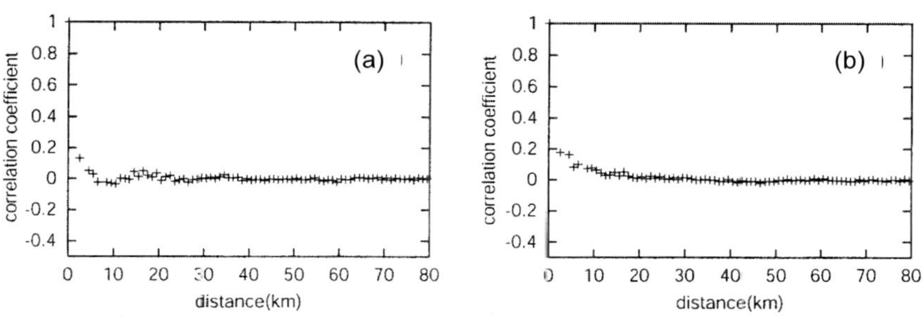

Fig. 3 Spatial correlation coefficient of relative prediction error $\boldsymbol{E_r}$ (a) case for 5-min ahead prediction, and (b) case for 60-min ahead prediction.

Table 1 Results of Kolmogorov-Smirnov-test of fit for relative prediction error to lognormal distribution. The significance level is 5%.

Time resolution of radar rainfall	Spatial resolution of radar rainfall	5-min ahead prediction	15-min ahead prediction	30-min ahead prediction	45-min ahead prediction	60-min ahead prediction
5 min	3 × 3 km	G	G	F	F	F
	6 × 6 km	G	G	G	F	F
	12 × 12 km	G	G	G	G	F
15 min	3 × 3 km	--	G	F	F	F
	6 × 6 km	--	G	G	F	F
	12 × 12 km	--	G	G	G	G
30 min	3 × 3 km	--	--	F	--	F
	6 × 6 km	--	--	F	--	F
	12 × 12 km	--	--	G	--	G
60 mim	3 × 3 km	--	--	--	--	G
	6 × 6 km	--	--	--	--	G
	12 × 12 km	--	--	--	--	G

G: good fit; F: failed fit; --: not applicable

By using the Kolmogorov-Smirnov test, 12 kinds of relative prediction errors with different time and spatial resolutions calculated by averaged data stated above are tested to fit a lognormal distribution. Each relative prediction error result is classified into three categories according to predicted rainfall intensity having less than 4 mm h^{-1}, between 4 and 10 mm h^{-1}, and more than 10 mm h^{-1}, respectively. Then the frequency distribution for each classified data is tested to fit to the frequency distribution for all grid cells with rainfall intensity larger than zero with the criterion of 5% significance level of the Kolmogorov-Smirnov test. If all the frequency distributions of the three classified data fit to the frequency distribution of all grid-cells with rainfall, the distribution is supposed to represent the distribution of the relative prediction error. The results of the Kolmogorov-Smirnov test are summarized in Table 1. The analysis proposes to use spatially-averaged data with 12-km resolution or time averaged data with 1 h if a prediction lead time is taken to be 1 h.

Characteristics of time persistency for relative precipitation error

To simulate a spatial field of $\boldsymbol{E_r}$ for future time, it is desirable that the statistical characteristics of $\boldsymbol{E_r}$ last for several hours. Figure 3 shows the mean and standard

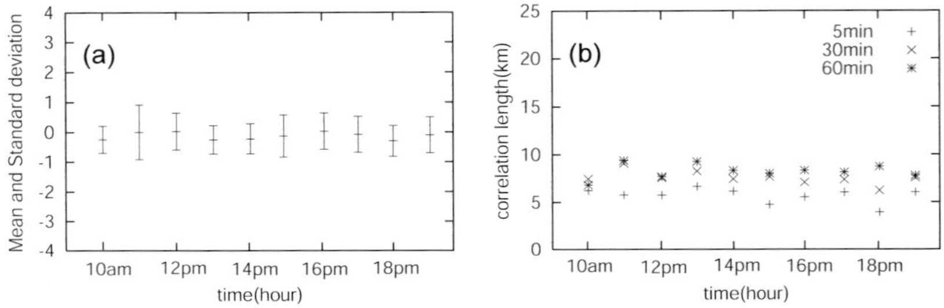

Fig. 4 (a) Time variation of mean and standard deviation of relative prediction $\boldsymbol{E_r}$ for 60-min ahead prediction. (b) Time variation of correlation length a.

deviation of $\boldsymbol{E}_r$ calculated at every hour from 10:00 to 19:00 on 5 September 1989 by using the data with 3-km spatial and 1-h time resolution. It shows that the changes in mean and standard deviation are small during one precipitation duration. Figure 4 shows the change of spatial correlation length of $\boldsymbol{E}_r$, α, the parameter of Gaussian spatial correlation function $\rho(h) = \exp(-h^2/\alpha^2)$, obtained by the least square method. The values of α are calculated every hour by using $\boldsymbol{E}_r$ of 5-, 30-, 60-min ahead prediction. The correlation length shows very slight changes in time for all cases. From the results, it is supposed that the statistical characteristics of $\boldsymbol{E}_r$ obtained at current time step will last several hours.

METHOD FOR LOGNORMAL RANDOM FIELD SIMULATION

The method to use a matrix factorization technique is adopted in which a covariance matrix is decomposed into its square root matrix approximately (Davis, 1987; Tachikawa & Shiiba, 2000).

Let $\boldsymbol{R}$ be a given covariance matrix, and consider $\boldsymbol{R}$ to decompose into $\boldsymbol{R} = \boldsymbol{SS}$. Here, $\boldsymbol{S}$ is a symmetric positive-definite matrix defined as $\boldsymbol{S} = \boldsymbol{Q}\boldsymbol{\Lambda}^{1/2}\boldsymbol{Q}^T$, where $\boldsymbol{Q}$ and $\boldsymbol{\Lambda}^{1/2}$ are an orthogonal and a diagonal matrix defined by means of eigenvalue decomposition of $\boldsymbol{R}$ into $\boldsymbol{R} = \boldsymbol{Q\Lambda Q}^T$. Then a simulation generated by $\boldsymbol{X} = \boldsymbol{Sw}$ is a Gaussian random field where w is a vector of independent standard Gaussian random variables, because:

$$E[\boldsymbol{XX}^T] = E[\boldsymbol{Sww}^T\boldsymbol{S}^T] = \boldsymbol{Q}\,\boldsymbol{\Lambda}^{1/2}\boldsymbol{Q}^T E[\boldsymbol{ww}^T]\,\boldsymbol{Q}\boldsymbol{\Lambda}^{1/2}\boldsymbol{Q}^T = \boldsymbol{Q\Lambda Q}^T = \boldsymbol{R}$$

To obtain $\boldsymbol{S}$, $\boldsymbol{\Lambda}^{1/2}$ is numerically expanded by using the Chebyshev polynomials as:

$$\boldsymbol{\Lambda}^{1/2} = \alpha_0\boldsymbol{I} + \alpha_1\boldsymbol{\Lambda} + \alpha_2\boldsymbol{\Lambda}^2 + ... + \alpha_p\boldsymbol{\Lambda}^p + \boldsymbol{\Delta}$$

and $\boldsymbol{S}$ is approximately obtained by:

$$\boldsymbol{S} = \boldsymbol{Q}(\alpha_0\boldsymbol{I} + \alpha_1\boldsymbol{\Lambda} + \alpha_2\boldsymbol{\Lambda}^2 + ... + \alpha_p\boldsymbol{\Lambda}^p + \boldsymbol{\Delta})\boldsymbol{Q}^T =$$

$$\alpha_0\boldsymbol{I} + \alpha_1\boldsymbol{R} + \alpha_2\boldsymbol{R}^2 + ... + \alpha_p\boldsymbol{R}^p + \boldsymbol{Q\Delta Q}^T \qquad (5)$$

Here $\boldsymbol{R}^n = (\boldsymbol{Q\Lambda Q}^T)^n = \boldsymbol{Q\Lambda}^n\boldsymbol{Q}^T$ is used to derive equation (5).

After getting a homogeneous Gaussian random field $\boldsymbol{X}$ with mean m_x, variance σ_x^2, and covariance function $C_x(h)$ using the above method, a lognormal random field $\boldsymbol{Y} = \exp[\boldsymbol{X}]$ with mean m_Y, variance σ_Y^2, and covariance function $C_Y(h)$ is finally obtained. The procedure for simulating a lognormal random field giving m_Y, σ_Y^2, and $C_Y(h)$ is as follows: (a) calculate σ_x^2 using the relationship $\sigma_Y^2 = m_Y^2[\exp\{\sigma_x^2\} - 1]$; (b) calculate m_x using the relationship $m_Y = \exp[m_x + \sigma_Y^2/2]$; (c) derive $C_x(h)$ using the relationship $C_Y(h) = m_Y^2[\exp\{C_x(h)\} - 1]$; (d) generate a homogeneous Gaussian random field $\boldsymbol{X}$ with mean m_x, variance σ_x^2, and covariance function $C_x(h)$; and (e) calculate $\boldsymbol{Y}$ using $\boldsymbol{Y} = \exp[\boldsymbol{X}]$.

EXAMPLE OF RAINFALL FIELD GENERATION

By using the simulation method, predicted rainfall fields are simulated and used according to the following procedure: (a) obtain observed radar rainfall field $\boldsymbol{R}_o(t_1)$ at

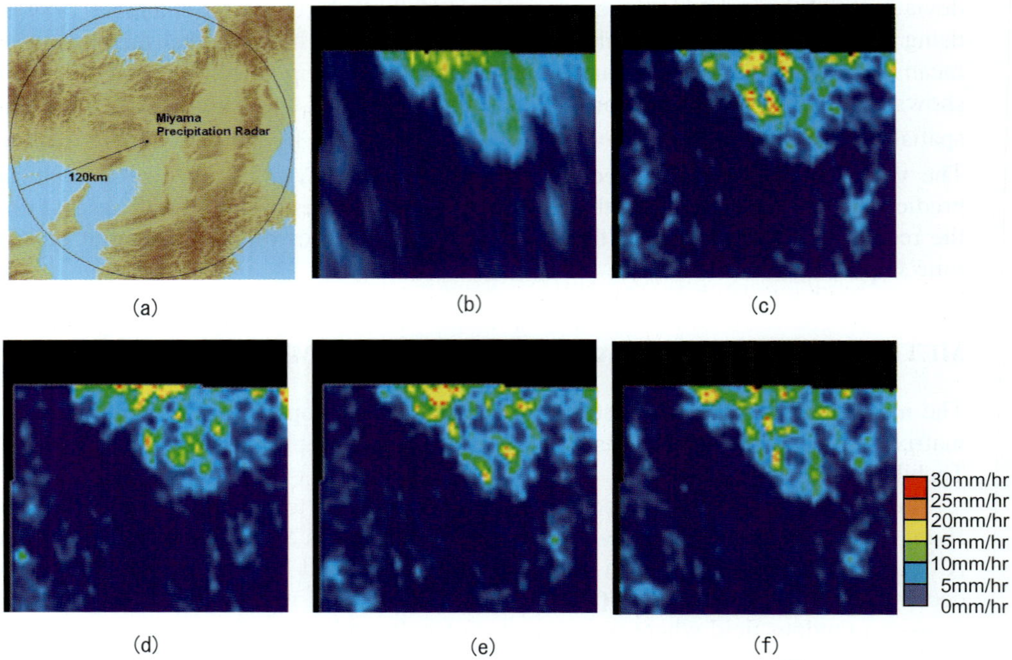

Fig. 5 (a) Miyama radar observation area. (b) Observed 60-minute mean precipitation. (c)–(f) Generated spatial fields of 60-minute mean predicted precipitation.

current time t_1; (b) calculate the mean, variance and correlation length of the relative error field $\boldsymbol{E_r}(t_1)$ by using $\boldsymbol{R_o}(t_1)$ and a rainfall field $\boldsymbol{R_p}(t_1)$ predicted for t_1 at previous time step; (c) forecast a rainfall field $\boldsymbol{R_p}(t_2)$ for time t_2 ($>t_1$) by using the translation model; (d) simulate an error field $\boldsymbol{E_r}(t_2)$ with the statistical characteristics obtained at step (b); (e) calculate $\boldsymbol{R}$ by using $\boldsymbol{R_p}(t_2)$ and $\boldsymbol{E_r}(t_2)$ using the equation (4); and (f) return to step (d) and generate alternative $\boldsymbol{R}$. After generating required rainfall fields and obtaining a probability distribution of river discharge prediction though a distributed runoff model, go to next time step and return to step (a).

Figure 5 shows examples of simulated rainfall fields at 12:00 on 5 September 1989. In this case, a precipitation field with 1-h and 3-km resolution shown in Fig. 5(b) is predicted by using the translation model and realizations of rainfall fields are generated based on it. The black coloured area in the figure shows the unpredictable area because the predicted rainfall field in the area moves from the outside of the radar measured region.

RESULTS

By using operational radar precipitation data and predicted rainfall field by the translation model, the error structure of predicted precipitation field was analysed. As a result, a relative prediction error was modelled as an isotropic lognormal spatial random field, and a method to generate a lognormal spatial random field was developed. In the method, a matrix factorization technique is adopted, where a given covariance matrix is decomposed into its square root matrix approximately by using the Chebyshev polynomials. Finally, using the method, generated realizations of rainfall fields and the simulated results were demonstrated. According to the Kolmogorov–Smirnov test of fit of a relative prediction error to a lognormal distribution, to use time or spatially-averaged radar data is suggested to apply the method presented here.

For further works, we need to devise an algorithm considering R_p is zero. In the method proposed here, if a predicted rainfall in a grid cell is zero, the relative prediction error for the grid cell is supposed to be zero. Also probability distributions of predicted river discharge should be evaluated through a distributed hydrological model with the use of simulated rainfall fields.

Acknowledgements The authors thank Yodo River Dams Control Office, Ministry of Land, Infrastructure and Transport, Japan for providing radar rainfall data and Associate Professor Nakakita at the Department of Urban and Environmental Engineering, Kyoto University, for providing the FORTRAN source code of the translation model.

REFERENCES

Davis, M. W. (1987) Generating large stochastic simulations—the matrix polynomial approximation method, *Math. Geol.* **19**(2), 99–107.

Georgakakos, K. P. & Krajewski, W. F. (1995) Worth of radar data in the real-time prediction of mean areal rainfall by nonadvective physically-based models. In: *New Uncertainty Concepts in Hydrology and Water Resources* (ed. by Z. W. Kundzewicz), 168–180. Cambridge University Press, Cambridge, UK.

Georgakakos, K. P. (2000) Covariance propagation and updating in the context of real-time radar data assimilation by quantitative precipitation forecast models, *J. Hydrol.* **239**, 115–129.

Shiiba, M., Takasao, T. & Nakakita, E. (1984) Investigation of short-term rainfall prediction method by a translation model. (Proc. 28th Annual Conf. on Hydraulics, JSCE) 423–428 (in Japanese). Japanese Society of Civil Engineers, Tokyo, Japan.

Takasao, T. & Shiiba, M. (1985) Development of techniques for on-line forecasting of rainfall and runoff. *Natural Dis. Sci.* **6**(2), 83–112.

Takasao, T., Shiiba, M. & Nakakita, E. (1994) A real-time estimation of the accuracy of short-term rainfall prediction using radar. In: *Stochastic and Statistical Methods in Hydrology and Environmental Engineering* (ed. by K. W. Hipel) **2**, 339–351. Kluwer Academic Publishers, The Netherlands.

Tachikwa, Y. & Shiiba, M. (2000) Gaussian and log normal random field generation based on the square root decomposition of a covariance matrix. *J. Hydraul. Coastal Environ. Engng JSCE* **656**(II-52), 39–46 (in Japanese).

Weather Radar Information and Distributed Hydrological Modelling (Proceedings of symposium HS03 held during IUGG2003 at Sapporo, July 2003). IAHS Publ. no. 282, 2003.

On the modelling of temporal correlations in spatial-cascade rainfall downscaling

ASSELA PATHIRANA
Department of Civil Engineering, Faculty of Science and Engineering, Chuo University, 1-13-27 Kasuga, Bunkyo-ku, Tokyo 112-8551, Japan
asselapathirana@yahoo.co.jp

SRIKANTHA HERATH
Environment and Sustainable Development Programme, United Nations University, 53-70 Jingumae 5-chome, Shibuya-ku, Tokyo 150-8925, Japan

TADASHI YAMADA
Department of Civil Engineering, Faculty of Science and Engineering, Chuo University, 1-13-27 Kasuga, Bunkyo-ku, Tokyo 112-8551, Japan

Abstract Multiplicative cascade schemes are extensively used to distribute the large-scale forcing of precipitation imposed by global atmospheric models for a limited spatial domain into its constituent grids. A model is proposed to introduce temporal persistence to one of the cascade based rainfall disaggregation models used for this purpose. A time series scheme based on the Markov process is used to relate the cascade properties to those of the previous time step. The resulting distributions could mimic the pixel level temporal persistence of spatial rainfall effectively, making cascade distributed spatiotemporal rainfall more suitable for applications in surface hydrology.

Key words multiplicative cascade scheme; rainfall disaggregation; rainfall downscaling

INTRODUCTION

Modern-day hydrologists involved in solving diversified water resources and environmental issues are increasingly recognizing the importance of incorporating high resolution spatial and temporal variability of rainfall as one of the most important parameters in applied hydrology applications. Spatial measuring techniques like weather radar reveal a much more complicated picture of rainfall variability than those revealed by traditional gauge networks. In spite of being complicated, the understanding of the relationships between different spatial scales has become crucial for the successful utilization of available spatial rainfall data sources for hydrological problem solving. One such area is the resolving of the aerial distribution of rainfall under specified forcing conditions. This particular scaling problem finds its uses in many areas including: (a) utilization of global and regional weather forecasts as data sources for locations that do not have adequate local monitoring facilities; and (b) understanding of the consequences of global climatic changes as predicted by global climatic models on the local hydrology. Perhaps the most widely researched solution for the problem is the use of limited-area atmospheric model outputs with dynamical equations (Pielke, 2002). Downscaling model outputs based on stochastic properties provide much simpler and computationally efficient solutions without the physical basis. There are also hybrid models that represent a compromise between the physical basis and simplifications based on stochastic models (e.g. Georgakakos & Krajewski, 1996). The present research using the fractal theory falls into the category of stochastic models.

Using cascade theories to spatially distribute global or mesoscale rainfall has a number of attractive features, where the adequacy of the ability to mimic the spatial variability and discontinuity as in observed rainfall is a major one. In the last two decades, a number of multiplicative cascade-based rainfall downscaling models and techniques have been developed. However, in spite of these numerous advances, the applications of these products in hydrological applications like runoff studies have been rare. Most of the successful efforts of multifractal analysis and cascade modelling of rainfall have been limited to either temporal or spatial dimensions. On the other hand, even a crude maintenance of both spatial and temporal correlations is a prerequisite for distributed run off modelling. Hydrological processes such as infiltration, evapotranspiration and overland flow critically depend on the temporal distribution of a rainfall event, in addition to the quantity of precipitation. Due to this sensitivity of the governing equations of watershed hydrology to the memory of the past states, a rainfall product needs to be accurate in the temporal persistence, in addition to the spatial features, for it to qualify as a candidate for surface response studies. The effects of temporal correlations can be relatively small at temporal scales like a day or a week, due to the smaller temporal limits of the prevailing rainstorm evolution. However, at hourly or smaller scales the effects of rainfall evolution in time are predominantly exhibited. Thus, the unavailability of means to maintain the temporal persistence in spatial models is perhaps a main reason for the lack of appeal to apply cascade downscaling of rainfall to runoff and other hydrological studies.

One theoretically attractive way of overcoming this inadequacy is to model rainfall as a full spatiotemporal process (Pegram & Clothier 1999; Deidda, 2000) utilizing the famous Taylor hypothesis. This involves transformation of the temporal dimension into a third spatial dimension by multiplying with a hypothetical velocity factor, derived empirically from observed rainfall. On the other hand, Lammering & Dwyer (2000) proposed an empirical scheme where the early steps of cascade simulations are exactly duplicated in the successive time steps. Further, they showed that the introduction of even a crude representation of temporal persistence could significantly improve the performance of rainfall products in hydrological analysis.

Alternatively, we propose a simpler time series approach combined with the cascade schemes, which can be used with relative ease to introduce the temporal correlations observed in rainfall into spatial downscaling schemes. We interpret the scaling of rainfall in space as the combined effect of a spatial multiplicative cascade process based on large-scale forcing and a number of time series processes that relate rainfall intensity at a given cascade level to that of the past time steps or the history. Thus, the model is capable of mimicking the temporal persistence at all spatial scales involved in the problem at hand. We use a number of Markov processes (Haan, 1977) to model the autocorrelations.

The ability of a cascade scheme to treat the zero rainfalls explicitly (e.g. Over & Gupta, 1996) may be of particular interest to the run off analyst. In addition to the intensity, data analyses shows that zero rainfalls at various spatial scales, also have a strong temporal correlations. In the present model, zero rainfall persistence is modelled based on a simple conditional probability scheme.

After developing the theoretical framework for the proposal based on the features of spatial rainfall, the model's ability to mimic temporal persistence is demonstrated using hourly radar-based spatial rainfall observations for the central part of Japan. The radar-based rainfall estimates calibrated by raingauges known as radar-AMeDAS data (Makihara, 1996) was used for the present study. A rain storm from 14 May 1997 12:00 h to 17 May 1997 12:00 h, covering an area of 128 × 128 pixels bounded by 39.65°N, 134.5°W, 33.3°S and 142.4375°E was selected to demonstrate the operation of the model.

CASCADE DOWNSCALING

Multiplicative cascade schemes were first used to describe turbulent energy relations between eddies of different sizes in fluid flows. During the last two decades, the theory has increasingly been used to describe and model rainfall process in space and time dimensions. While the quantity of interest of the former case is energy, that of the latter is the quantitative rainfall. Figure 1 gives a schematic representation of a multiplicative cascade scheme in two dimensions. The process starts with a uniform intensity in the process domain. At each step, a uniform field is subdivided into four equal parts and each is multiplied by a cascade weight W that is derived from a specific probability distribution. The 128 × 128 data space used in the present study involves seven cascade steps.

In the present research, the β-lognormal model proposed by Over & Gupta (1996) was used to model the scaling in spatial dimension. A key feature of this model is the explicit modelling of non-rainy areas in addition to the rainfall intensity. In this model it is assumed that the zero–nonzero partitioning and the intensity process are independent of each other, which is not completely realistic. Typical uses of this model

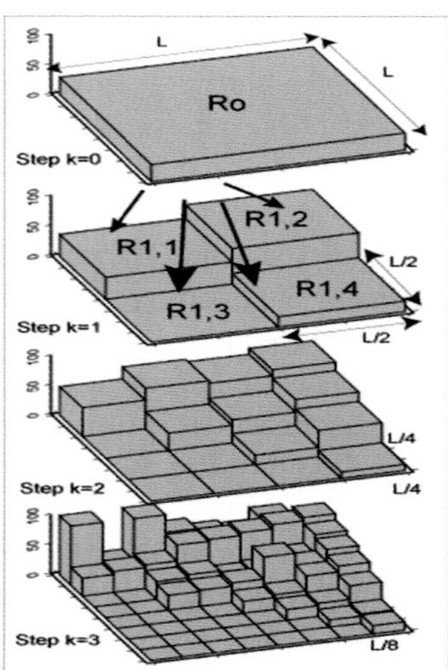

Fig. 1 A multiplicative cascade process in two-dimension. Uniform fields are divided into four and multiplied with the cascade weight W to obtain the next step.

for distributing a large-scale forcing amount into finer grids can be found in Jothityangkoon *et al.* (2000) and Pathirana & Herath (2002).

TEMPORAL CORRELATIONS

Physical understanding of the rainfall process, as well as numerous observations, confirms that it is one that evolves in space over time. In most cases storm movement in a specific direction is involved. However, these patterns are very much dependent on the atmospheric conditions, landforms, and the wind patterns involved in the specific location and time (for example, see Fig. 2). There were numerous attempts to improve the realistic representation of rainfall-related atmospheric phenomena using physically-based atmospheric models. However, in the context of a stochastic model these are hard to represent with an accuracy that justifies the increased complexity of model. The approach proposed in this paper is far simpler, with the assumption that the correlations of cascade weights can adequately capture the essential effects of storm evolution.

In order to understand the nature and magnitude of the correlations at different cascade levels, the following analysis was done: each spatial field of hourly series of rainfall snapshots was degenerated (the exact reverse of the process shown in Fig. 1) to obtain cascade weights W at each branch at each cascade level. The autocorrelations were calculated using the time series of weights corresponding to each pixel (i, j) at each cascade level k. Figure 3 illustrates these time series for a cascade process in one-dimensional space. A single value of autocorrelation for each cascade level was

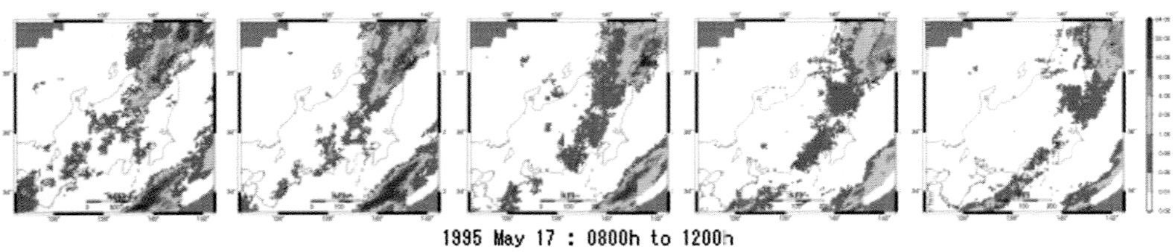

Fig. 2 A selected portion of a hourly spatial rainfall series of a storm over the central part of Japanese archipelago. Rainfall patterns show a clear evolution over time and a non-random directional movement.

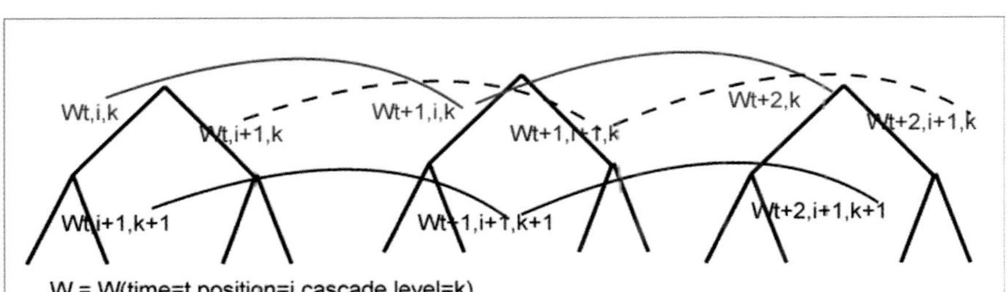

Fig. 3 The cascade weights at each cascade level are modelled as correlated processes. A one-dimensional cascade is shown for simplicity.

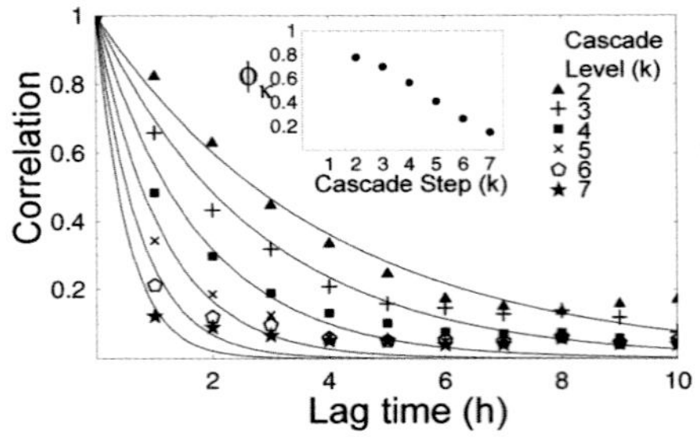

Fig. 4 Autocorrelations of cascade weights between adjacent rainfall fields. Fittings of a Markov process are also shown.

obtained by considering the time series for each pixel (i, j) for the same level k are realizations of the same time series process. In the case of the nth cascade step of a two-dimensional cascade, there would be 2^{2n} number of weights involved and thus, 2^{2n} number of correlation values will be averaged. The results of the analysis are shown in Fig. 4. The analysis shows that the cascade weights have significant autocorrelations and that they diminish in magnitude with the cascade step. Further, it is clear that a Markov process given by $W_t = \phi\ W_{t-1} + W$ can adequately represent the correlation process, where $0 < \phi < 1$, W is a random variable and t is time.

MODEL FORMULATION

The generator (the function which generates the weights W) of a typical β log-normal cascade model, adopting a notation slightly different from the original of Over & Gupta (1996), can be expressed in the following form:

$$P(W = 0) = 1 - e^{-\gamma};\ P(W = e^{\gamma-\sigma^2/2+\sigma X} = e^{G(r-\sigma^2/2\sigma)}) = e^{-\gamma} \tag{1}$$

where X is a standard normal variable and $G(\mu,\sigma)$ indicates a normal distribution with mean μ and standard deviation σ. The model parameter γ is used to explicitly partition rainy and non-rainy areas and σ handles the intensity distributions by controlling the generator for non-zero W. Note that the cascade process is statistically conserved due to the fact that expected value, $E(W) = 1$. In order to introduce the temporal persistence to cascade modelling, this model should be modified to incorporate correlations among cascades adjacent in time. Since the cascade model requires the weights to be positive-definite, the logarithms of the weights, instead of cascade weights, were modelled. This does not cause a problem since the logarithmically transformed weights also show a good autocorrelation structure (Fig. 5). Then, with a Markov process as the basis for autocorrelations, the following generator can be proposed for the intensity formulation:

$$W_i = e^{\phi_k \log(W_{i-1}) + G(\mu_i',\sigma_i)} \tag{2}$$

where i is the time step, ϕ_k the Markov parameter for cascade level k and:

$$\mu'_i = \gamma_i - \sigma^2_i/2 - \phi_k(\gamma_{i-1}/2);\ \sigma'^2_i = \sigma^2_i - \phi^2_k\sigma^2_{i-1}\ \text{ for }\ \sigma^2_i > \phi^2_k\sigma^2_{i-1} = 0\ \ o.w. \tag{3}$$

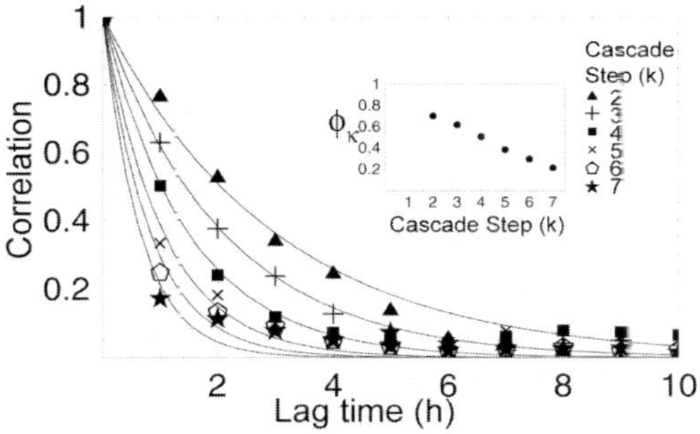

Fig. 5 Autocorrelations of logarithmically transformed cascade weights of observed rainfall.

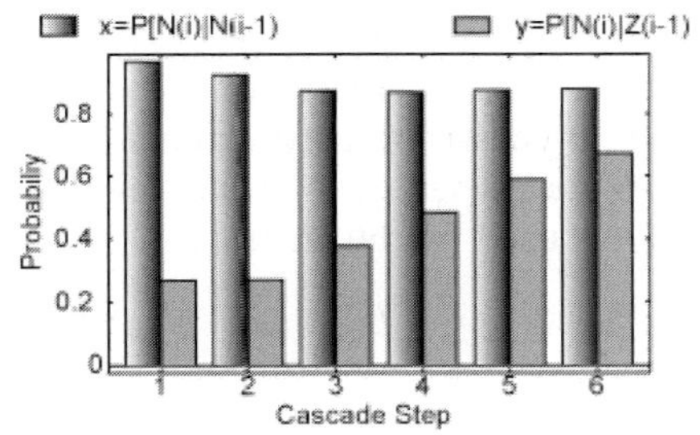

Fig. 6 Probability of rain in a previously rainy pixel $x = P_{N(i)} \mid P_{N(i-1)}$ and probability of rain in a previously dry pixel $y = P_{N(i)} \mid P_{Z(i-1}$ at different cascade levels.

The last statements assume that $(G(\mu_i',\alpha_i')$ is independent of W_{i-1}. Note that the new generator is also conserved. In addition to the intensities, the rainy/non-rainy areas also have temporal correlations. Analysis of the conditional probabilities of rain occurrence at pixels that had/did not have rainfall in previous time steps showed that the probability of rainfall is significantly high if there was rainfall in previous step (Fig. 6). Thus, the conditional probability:

$$x = P[N(i) \mid N(i-1)] \tag{4}$$

where i is the time step, N denotes non-zero rainfall, was used as a single value for the whole model to represent the persistence. Since there was no strong dependency of the quantity x with the cascade level (in contrast of ϕ) a constant value of x was used for all levels. For mathematical consistency, the model should make sure that the probability $y = P[N(i) \mid Z(i-1)]$ (where z indicates zero rainfall) should satisfy $0 \leq y \leq 1$. These details are not presented.

MODEL CALIBRATION

The calibration of the model should have two stages: the first stage is obtaining the parameters for the model given in equation (1). Since the typical use of the cascade model is to distribute a given uniform rainfall amount by a large-scale model among

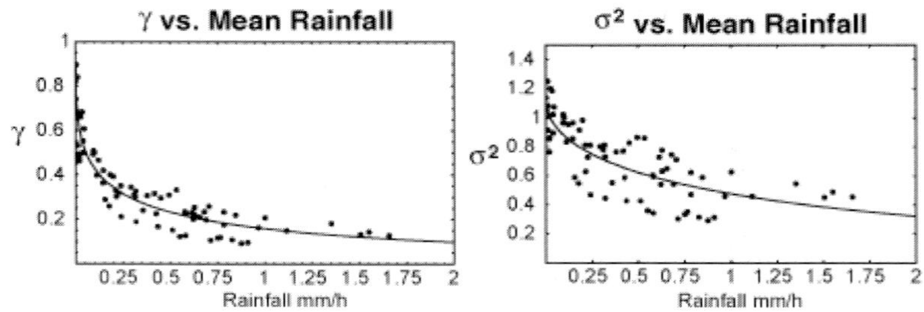

Fig. 7 Regression relationships of σ^2 and γ with large-scale forcing, obtained by data analysis.

the constituent pixels, these model parameters are expressed as regression functions of the large-scale forcing (Fig. 7). The second stage of calibration involves the determination of the temporal persistence parameters σ_k and x in equations (2) and (4). The former is found by fitting Markov processes to autocorrelation functions by least square fitting as shown in Fig. 5. The mean estimate of conditional probability given by data analysis (for all cascade levels) is used for the latter, x.

MODEL OPERATION AND RESULTS

Once a cascade model is calibrated, it can be used to spatially distribute a rainfall amount specified by large-scale forcing in a way that is statistically similar to observed data. Cascade models can accurately represent the spatial variability, intermittency and, sometimes, extra features such as zero value fractions and spatial heterogeneity (Jothityangkoon *et al.*, 2000; Pathirana & Herath, 2002). The additional feature of the proposed model is its ability to represent the temporal persistence that is present in observed data.

The procedure of downscaling a time series of large-scale forcing values using the model is as follows: First, values of model parameters σ^2 and γ are obtained from the regression relationships (Fig. 7). If the present time step is the first in the series, or if there was a zero rainfall amount in the previous step, the typical cascade downscaling based on weights obtained from equation (1) is used. If this is not the case, then equation (4) is used to divide the rainy fraction given by γ among previously rainy and previously non-rainy areas. Then equation (3) is used to determine weights for the non-dry portion of the cascade level. It should be noted that, according to the present model formulation, persistence of dry–wet partitioning x does not depend on the cascade level, though that of the intensity formulation ϕ_k does.

Figure 8 shows the autocorrelations found in simulated cascades. The Markov model could preserve the cascade level persistence comparable to those found by data analysis. Figure 9 shows two time series of spatial distributions, one downscaled using the typical cascade simulation without considering the autocorrelations and the other using the model proposed in this paper. The present model could introduce the temporal persistence similar to that found in observed data.

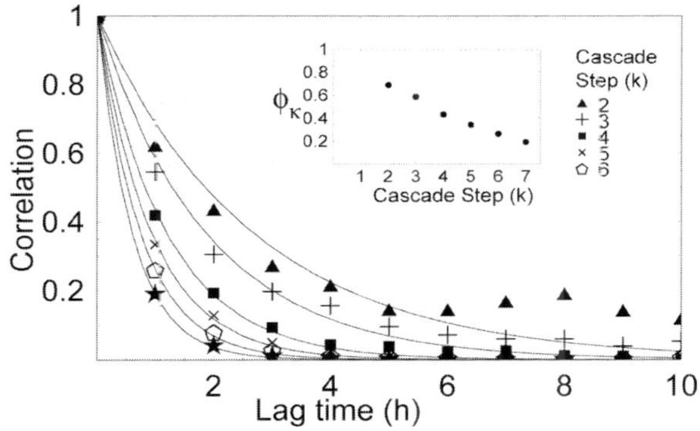

Fig. 8 Autocorrelations of logarithmically transformed cascade weights of simulated rainfall.

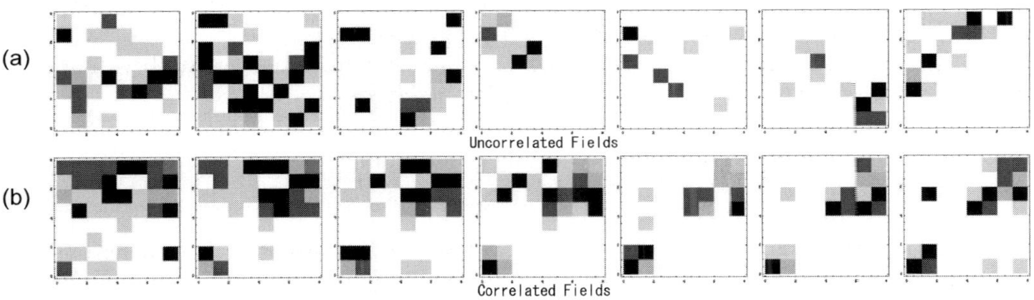

Fig. 9 Sample rainfall distributions using cascade theory: (a) typical cascade simulations, without considering temporal correlations, and (b) simulations using present model. Results produced at three cascade steps for clarity.

DISCUSSION

Spatial multiplicative cascade theory can be effectively used to distribute a rainfall amount prescribed by large-scale forcing on a spatial domain, into its constituent grids. However, when the cascade simulations of the large-scale forcing values on the temporal succession are done independently of each other, the resulting spatial distributions lack the temporal persistence that is shown by observed rainfall (Fig. 2). While these distributions faithfully reproduce the characteristics of variability and intermittency of spatial rainfall, they are not suitable as input for applications in surface hydrology for the above reason. The correlation technique proposed in the present paper modifies the cascade simulation scheme so that the pixel-level temporal persistence is introduced into cascade simulations using a set of simple time series models to determine cascade weights. The modified model could maintain in the simulated spatial rainfall fields and the temporal correlations that are present in observed spatial rainfall.

The temporal persistence in real spatial rainfall is maintained by physical mechanisms like wind such as caused cloud movement and rainstorm evolution over time. Since those mechanisms are not explicitly represented in the present model, it cannot realistically mimic the movement of rainstorms over time. The present method is significantly less complicated and easier to calibrate than the full three-dimensional treatment of rainfall. On the other hand, it is fully based on the empirical properties of the observed spatial rainfall and configurable to suit the rainfall patterns of a particular geographical location or season.

Acknowledgements The research leading to this paper was funded by a grant from the Japan Society for Promotion of Science.

REFERENCES

Deidda, R. (2000) Rainfall downscaling in a space-time multifractal framework. *Water Resour. Res.* **36**, 1779–1794.

Georgakakos, P. & Krajewski, W. F. (1996) Statistical-microphysical causes of rainfall variability in the tropics. *J. Geophys. Res.* **101**(D31), 26, 165–26, 180.

Haan, C. T. (1997) *Statistical Methods in Hydrology,* Iowa State University Press, Iowa, USA.

Jothityangkoon, C., Sivapalan, M. & Viney, N. R. (2000), Tests of a space-time model of rainfall in southwestern Australia based on nonhomogeneous random cascades. *Water Resour. Res.* **36**, 267–284.

Lammering, B. & Dwyer, I. (2000) Improvement of water balance in land surface schemes by random cascade disaggregation. *Int. J. Climatol.* **20**, 681–695.

Makihara, Y. (1996) A method for improving radar estimates of precipitation by comparing data from radar and raingauges. *J. Met. Soc. Japan* **74**, 459–480.

Over, T. M. & Gupta, V. K. (1996) A space-time theory of mesoscale rainfall using random cascades. *J. Geophys. Res.* **101**(D21), 26319–26331.

Pathirana, A. & Herath, S. (2002) Multifractal modelling and simulation of rain fields exhibiting spatial heterogeneity. *Hydrol. Earth System Sci.* **6**(4), 695–708.

Pegram, G. G. S. & Clothier, A. N. (1999) *Space-Time Modelling of Rainfall in Fine Intervals: The "String of Beads" Model*, Water Res. Comm. Report no. 752/1/99, Pretoria, South Africa.

Pielke, R. A. (2002) *Mesoscale Meteorological Modelling*, 2nd edn. International Geophysics Series, Academic Press, New York, USA.

2 Assessment of Uncertainty in Distributed Hydrological Modelling

A methodology for assessing the utility of distributed model forecast applications in an operational environment

KONSTANTINE P. GEORGAKAKOS & THERESA M. CARPENTER

Hydrological Research Center, 12780 High Bluff Drive, Suite 250, San Diego, California 92130, USA

KGeorgakakos@hrc-lab.org

Abstract Distributed models have long been viewed as beneficial for hydrological forecasting. However, errors in the precipitation data and model parameters may diminish any gains in prediction accuracy realized by accounting for the spatial variations in precipitation and parameters. Central to these considerations is the resolution of the issue of balancing model complexity with parameter and input estimation uncertainty. The present paper exemplifies a probabilistic methodology to address this issue. A distributed model with components used in the operational forecasting of streamflow in the USA is utilized to produce flow simulations for a catchment in the south-central USA. Models of uncertainty for radar-rainfall data and model parameters are developed. Ensemble flow simulations are produced for a number of input and parameter uncertainty scenarios and for a high and low spatial resolution model configuration. Kolmogorov-Smirnov testing is then performed to assess whether the peak-flow simulations of the low- and high-resolution models can be distinguished with high confidence. Assessments of the implications regarding performance under operational forecast conditions are given.

Key words distributed hydrological modelling; ensemble flow prediction; parameter estimation; radar rainfall uncertainty

INTRODUCTION

The research question this paper addresses is: are ensemble flood flow simulations from models with (a) spatially aggregate parameters and (b) spatially distributed parameters distinguishable on scales of the order of 10^3 km^2 under present-day operational parametric and radar-rainfall uncertainty? Consideration of flow simulations from models is a necessary first step prior to the consideration of lumped and distributed hydrological model operational forecasts that incorporate substantial precipitation forecast uncertainty. Premises of the analysis are:

(a) parametric and radar-rainfall uncertainties are significant under operational conditions;
(b) ensemble flow simulation provides a more complete and useful representation of model response than a single nominal simulation;
(c) the runoff generation component is the same for lumped and distributed models;
(d) the scale of order 10^3 km^2 is significant for operational flow forecasting.

The focus of the paper is to exemplify the approach used to answer the research question. This approach is applicable to any type of distributed hydrological model. In the examples to follow, we use the Sacramento soil moisture accounting model for runoff generation in soil columns for the distributed model and the lumped model, and a kinematic channel routing component for streamflow simulation and time distribution of generated runoff. The distributed model used is a sub-catchment based (rather than grid based) model with spatially distributed parameters and input. The distributed model formulation, validation and sensitivity analyses with a variety of data sets from various basins in the USA are described in Carpenter *et al.* (2001) and Carpenter & Georgakakos (2002).

METHODOLOGY

The target of the methodology is to distinguish the lumped from the distributed model simulations when these simulations are in the form of ensembles that incorporate the uncertainty due to erroneous model parameter values and noisy radar rainfall input.

The methodology consists of the following consecutive steps:

(a) Using long historical records, calibrate lumped and distributed models for the basin of interest.

(b) Develop generic models of uncertainty for parameters based on attributes of spatial data used and on calibration error statistics.

(c) Develop generic models of uncertainty for radar rainfall estimates at the radar-bin scale and upscale uncertainty models to the scale resolved by the distributed and lumped models.

(d) For each model type (lumped *vs* distributed) sample from the respective parameter uncertainty distribution and develop ensemble simulations for several significant events in the test catchment for the catchment outlet (area of order 10^3 km^2).

(e) For each event, assess whether the ensemble of simulated flows by the lumped model at the time of the observed flow peak belongs to the same distribution as the ensemble generated for the same time and spatial scale by the distributed model. A Kolmogorov-Smirnov statistical test is used to compare the distributions.

(f) Repeat (d) and (e), sampling from both the parametric and radar-rainfall uncertainty developed for each model type in (b) and (c).

TEST BASIN AND MODEL PERFORMANCE

The 1232-km^2 Blue River basin with outlet near Blue, Oklahoma, is used as the test basin for this analysis. The basin is one of four basins used at the US Distributed Modeling Intercomparison Project (DMIP) organized by the US National Weather Service. It is in a semiarid environment, receiving much of its warm season rainfall from convective storms. The spatial data used in distributed modelling for the basin are: USGS Digital Elevation (90-m resolution); USGS LULC (200-m resolution); STATGO soils data; EPA Reach File 3 (RF3). The distributed model subdivides the basin into 21 sub-catchments with an average size of 59 km^2. Estimates of parameters

Table 1 Performance of lumped and distributed model for historical period.

	Distributed model	Lumped model
Calibration period	8 May 1993–31 May 1999	8 May 1993–31 May 1999
Hourly cross-correlation	0.86	0.79
Hourly flow bias	1.40%	5.80%

and input were obtained for each of these sub-catchments. The lumped model considers the basin as an aggregate unit and applies parameters and input to the entire area.

The hourly historical hydrometeorological data used consisted of: discharge (USGS site 07332500); NEXRAD (WSR-88D) Stage III Precipitation; and energy forcing provided by DMIP databases (air temperature, air pressure, solar radiation, relative humidity). The historical data period was of a 6-year duration. Table 1 indicates the performance of the lumped and distributed models for the historical calibration period. The performance of the two models is comparable and it generally suggests that both models reproduce the historical flows well (less than 10% bias with an explanation of variance of hourly observed flows that is in the range 63–74%). There is slightly better performance shown for the distributed model. Georgakakos & Carpenter (2002) show simulations of individual events.

MODELS FOR PARAMETER UNCERTAINTY

The generic formulation for parametric uncertainty for both the lumped and the distributed model is:

$$\alpha = \mu_\alpha + \varepsilon_\alpha$$

where μ_α is the mean value of parameter α, and ε_α is a uniformly distributed error in range $[-\alpha_L, \alpha_L]$.

For the distributed model, spatially uniform parameters were estimated with an adjustment made for the upper soil parameters to account for the different spatial scale of the various sub-catchments. The median of the range of the STATSGO database available water content and permeability was then used to distribute the parameter values in space, keeping the basin spatial mean equal to the uniform parameter estimate obtained earlier. In all cases, the ranges of the STATSGO database estimates were used to define the limit α_L of the uniform distribution of the parametric error.

The constant and the exponent parameters of the soil model percolation function were adjusted in space from their uniform estimates using the distribution of texture and a table of association of texture and percolation parameters developed by the Staff of the Hydrology Laboratory, US National Weather Service (reproduced here for convenience in Table 2). The ranges indicated in the table were used to estimate α_L in each case. In the soil model the percolation rate normalized by the maximum possible baseflow, p_n, is given by:

$$p_n = 1 + C\, x_n^{\,M}$$

Table 2 Association of soil texture and percolation parameters.

	Percolation exponent, M	Percolation constant, C
Sand	1.4–1.8	5–20
Silt	1.8–2.5	20–75
Clay	2.5–3.5	75–200

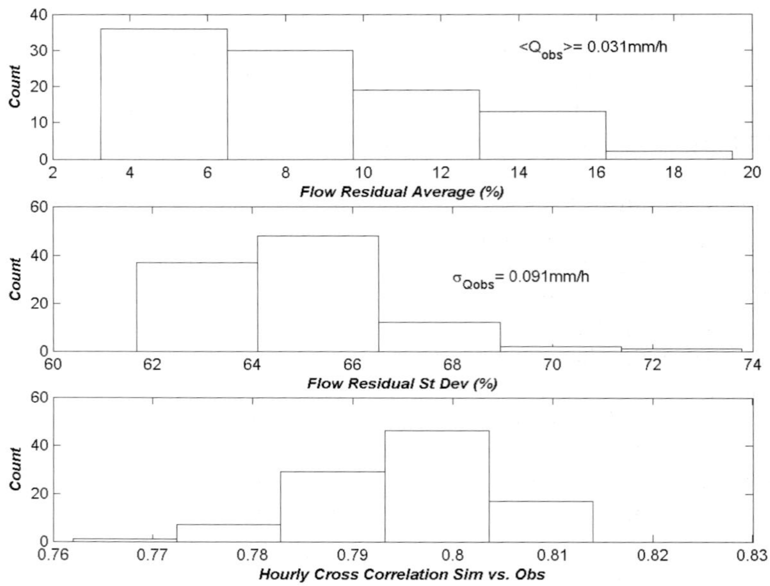

Fig. 1 Performance indices used in calibration of lumped model with Blue River data.

where C and M are the percolation parameters and x_n is the lower soil zone storage available for percolation normalized by lower soil zone capacity (sometimes called the lower zone deficiency ratio).

For the lumped model, the range in parameter error $[-a_L, a_L]$ is determined by degree-of-belief estimates of the hydrologist who calibrated the lumped model for the Blue River basin. A ± 25% error range was used for the uncertain parameters of the upper soil zone. The diagrams of Fig. 1 show the effect on various model performance indices of using this uncertainty range during the calibration period. It is noted that the model performance indices have ranges that are within the range of model performance considered adequate when calibrating with multiple objectives, so the 25% parametric uncertainty estimate appears reasonable for this basin and this model.

MODELS FOR RADAR RAINFALL UNCERTAINTY

The basis of the hourly mean areal rainfall estimates used for sub-catchments in this study is radar rainfall data, Stage III, from the NEXRAD (WSR-88D) radar. The radar rainfall data are available for radar bins of approximate size 3.5 × 3.5 km^2. For error estimation of the mean areal rainfall for each sub-catchment of the distributed model and for the entire basin for the lumped model, an error model was postulated for radar-bin rainfall estimates. The parameter of the radar-bin error model represents an error half-range that is a function of the radar rainfall estimate magnitude for the radar bin. The postulated model for the relative error half-range (E/R_m: error half-range E divided by observed rainfall R_m) is as follows: largest relative errors correspond to lighter rainfall amounts with a linear decrease of the relative error half-range with increasing observed rainfall amount up to 25 mm h^{-1}. For rainfall amounts greater than 25 mm h^{-1}, the relative error half-range remains constant and equal to 0.5. The distribution of the radar-bin error is assumed uniform within the range limits $[Rm - E, Rm + E]$, and no space correlation of the radar-bin error is assumed in this initial phase of the study. The latter postulate may result in underestimation of the rainfall error variance for sub-catchments that contain several radar bins, if in fact significant spatial correlation in radar-bin rainfall error exists.

Using Monte Carlo sampling from the radar-bin error model for each of the events, parameters of error functions were developed for each sub-catchment of the distributed model as well as for the entire basin area as used by the lumped model. The error functions assumed uniform mean areal rainfall distribution for each sub-catchment, with a distribution half-range that is given relative to the mean areal rainfall of the sub-catchment in Fig. 2. Dependence of the relative half-range of the error for sub-catchments is still linear for mean areal rainfall rates less than 25 mm h^{-1}, but the slope and intercept of the line decrease with increasing sub-catchment area. The lowest fitted line in Fig. 2 corresponds to the entire Blue River basin area. The errors for each sub-catchment of the distributed model and for the entire basin for the lumped model were

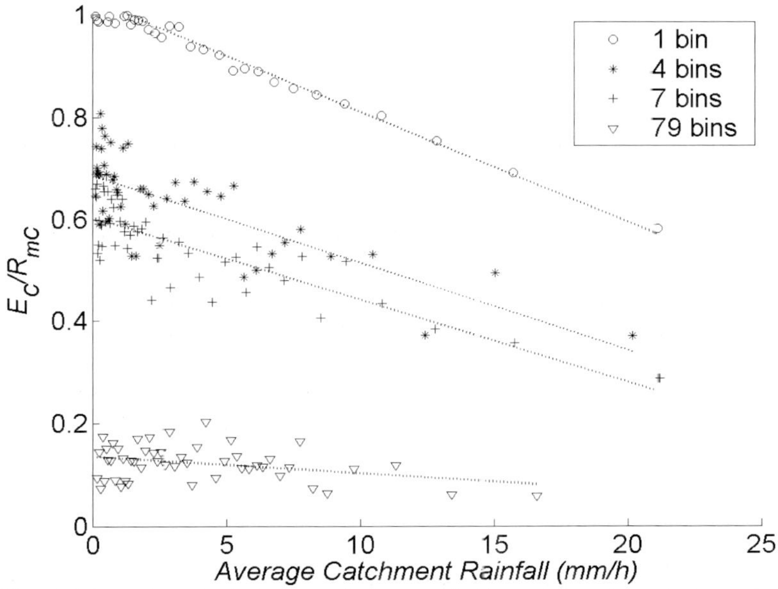

Fig. 2 Sub-catchment relative error half-range E_c/R_{mc} as a function of sub-catchment rainfall R_{mc}. The legend shows the number of bins for the sub-catchments of the Blue River.

assumed uniformly distributed in the range $[R_{mc} - E_c, R_{mc} + E_c]$ with R_{mc} representing the sub-catchment rainfall mean (obtained by averaging the enclosed radar-bin observations), and E_c representing the error half-range.

RESULTS OF APPLICATION

A total of 25 events were selected from the historical database that spanned the period from December 1993 through March 1999. The methodology was applied as described above for each of these events. For each event the models to run in a Monte Carlo fashion (sampling from parametric and input uncertainty distributions) for 2 months prior to the event to establish a stable distribution of the initial soil moisture field at the beginning of the event and then the Kolmogorov-Smirnov test was applied during the event period. The Kolmogorov-Smirnov test statistic D_o is determined by:

$$D_o = \max \left| P_{\text{Distr}}(Q_p) - P_{\text{Lump}}(Q_p) \right|$$

$$\text{Prob}\,(D > D_o) = F(N_e, D_o)$$

$$N_e = \frac{N_D N_L}{N_D + N_L}$$

with $N_D = N_L = 100$ and with the following nomenclature:
N_D, number of distributed model ensemble members; N_L, number of lumped model ensemble members; Q_p, simulated discharge at the time of the observed peak flow for each event; P_{Distr}, cummulative frequency distribution of distributed model ensemble flows at peak time; P_{Lump}, cummulative frequency distribution of lumped model ensemble flows at peak time.

The Kolmogorov-Smirnov test does not depend on the type of distribution but it does require that the distributions involved be continuous. It is most effective for measuring shifts in distributions (most sensitive around the median value of distributions). The null hypothesis in this application is that the two sets of flow samples, resulting from the distributed and the lumped models at the time of the observed flow peak are drawn from the same distribution (and therefore are indistinguishable). The significance level probability for accepting the null hypothesis is 1%. For parametric uncertainty only, the test indicates that the distributions at the time of the observed peak are different at the 1% significance level. That is, $Prob(D > D_o)$ is less than 1% for all 25 events. When radar rainfall input uncertainty is added to the parametric uncertainty, then in all but one event $Prob(D > D_o)$ is less than 1%. These results imply that for this catchment and intercomparison scale, the distributed and lumped model ensemble flow simulations of high flows are statistically different with high confidence.

Given that the distributed and lumped model simulations of event peaks are statistically different for virtually all events under parametric and input uncertainty, the question arises as to which model has better performance. This question requires the definition of a suitable assessment metric that explicitly takes into consideration that the flow “simulation” is in the form of an ensemble due to parametric and input uncertainty. The sample frequency that the simulated flow is within $\pm 20\%$ of the

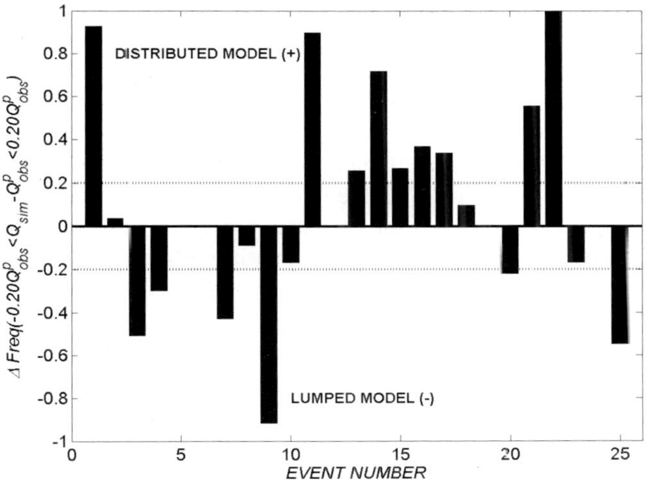

Fig. 3 Differences in sample frequencies between distributed and lumped model of the event that the simulated ensemble flows at the time of the observed peak flow is within ±20% of the observed peak flow.

observed peak flow is used as our assessment metric for this analysis. Figure 3 shows the differences in these sample frequencies between the distributed and the lumped model. Positive numbers indicate higher sample frequency for the distributed model, which is interpreted to mean better performance. Negative numbers indicate better performance for the lumped model. There are 100 samples in each ensemble of simulated flows for each of 25 events. Because the assessment metric depends on the number of samples, only values outside the interval $[-0.2, +0.2]$ are considered indicative of better performance. Values of the assessment metric within the interval were considered uncertain. The results in Fig. 3 show that in 44% of the events the distributed model had a better performance, in 36% of the events the lumped model had a better performance and in 20% of the events best performance could not be determined with the given assessment metric. Study of the dependence of the model performance with respect to magnitude of the observed peak flow indicated that the distributed model is clearly better in the medium peak-flow range. For the highest peak flows there is indication that the lumped model performs better, but more data is necessary for a more definitive statement in this peak-flow range.

CONCLUSIONS AND RECOMMENDATIONS

A new methodology is described and applied in this paper for determining whether distributed model performance is different from lumped model performance under parametric and input uncertainty. A probabilistic metric is also proposed to assess which model has better performance, once it is established that the model simulations are distinguishable with high confidence. The methodology was applied to the 1232 km^2 Blue River basin with outlet near Blue, Oklahoma. The results of the application show that the ensemble flow simulations from distributed and lumped models are statistically different with a high degree of confidence, and that for medium flow events, the distributed model has a better performance than the lumped model.

It would be useful for future research to improve the input error models used herein. Allowing spatio-temporal dependence on the radar-bin error model is an especially important research area. If such dependence is high, then it is likely that in terms of being able to distinguish lumped from distributed model simulations in the present operational environment the results presented in this paper are rather optimistic. The use of various levels of spatial aggregation for the distributed hydrological model and application of the proposed methodology is another area of future research, which would help determine the spatial scale of model discretization that is most suitable for use with operational estimates of parameters and input. Lastly, application of the methodology to other basins and different types of distributed models should also be made.

Acknowledgements This research was supported by the Office of Hydrological Development (OHD) of the US National Weather Service, National Atmospheric and Oceanic Administration under Award no. NA07WH0368. The authors wish to thank Drs Michael Smith and Victor Koren of OHD for several useful comments and suggestions. The ideas and opinions expressed herein are those of the authors and do not necessarily reflect those of NOAA and of its sub-agencies.

REFERENCES

Carpenter, T. M. & Georgakakos, K. P. (2002) Distributed model flow sensitivity to uncertainty in radar-rainfall input. In: *Information from Weather Radars and Distributed Hydrological Modelling* (ed. by Y. Tachikawa *et al.*) (Proc. Symp., Sapporo HS03, July 2003), 114–120. IAHS Publ. no. 282 (this volume).

Carpenter, T. M., Georgakakos, K. P. & Sperfslage, J. A. (2001) On the parametric and NEXRAD-radar sensitivities of a distributed hydrological model suitable for operational use. *J. Hydrol.* **253**, 169–193.

Georgakakos, K. P. & Carpenter, T. M. (2002) Probabilistic analysis of model complexity versus parametric and input uncertainty in operational streamflow forecasting. HRC Tech. Note no. 20. Hydrological Research Center, San Diego, California, USA.

Weather Radar Information and Distributed Hydrological Modelling (Proceedings of symposium HS03 held during IUGG2003 at Sapporo, July 2003). IAHS Publ. no. 282, 2003.

Propagation of precipitation uncertainty through a flood forecasting model

SHREEDHAR MASKEY
IHE-Delft, PO Box 3015, 2601 DA Delft, The Netherlands
maske@ihe.nl

VINCENT GUINOT
Maison des Sciences de l'Eau, Université Montpellier 2, F-34095 Montpellier Cedex 5, France

ROLAND K. PRICE
IHE-Delft, PO Box 3015, 2601 DA, Delft, The Netherlands

Abstract Uncertainty in the precipitation forecast is one of the major sources of uncertainty in real-time flood forecasting. Precipitation uncertainty generally consists of uncertainty in: (a) the quantity, (b) the temporal distribution, and (c) the spatial distribution of precipitation. This paper presents a methodology for propagating the three forms of precipitation uncertainty through a rainfall–runoff routing type deterministic flood forecasting model. It uses fuzzy set theory combined with a genetic algorithm. The methodology is particularly useful where the probabilistic forecast of precipitation is not available. The results show that the output uncertainty due to uncertain temporal and spatial distributions of precipitation can be significantly dominant over the uncertainty due to the uncertain quantity of precipitation.

Key words flood forecasting; fuzzy set; genetic algorithm; precipitation; uncertainty

INTRODUCTION

Forecasting a flood using rainfall–runoff routing type models requires the forecasting of precipitation for the forecast period. Over the last decade significant progress has been made in the quantitative forecast of precipitation using sophisticated radar technology. The uncertainty in forecast precipitation remains, however, one of the major sources of uncertainty in real-time flood forecasting. It consists of uncertainty in: (a) the quantity, (b) the temporal distribution over the forecast period, and (c) the spatial distribution over the watershed. This paper presents a methodology for propagating the three forms of precipitation uncertainty through a rainfall–runoff routing type deterministic flood forecasting model. The extension principle of the fuzzy set theory is used to propagate the uncertainty. This principle is implemented applying the so-called α-cut method. A genetic algorithm (GA) is used to determine the maximum and minimum of the model output, which is an essential part of the extension principle by α-cuts. The uncertainty in temporal and spatial distributions of precipitation is treated using the concept of the principal precipitation pattern (PPP). This methodology is independent of the structure of the forecasting model. That is, the methodology can be used with any rainfall–runoff routing type deterministic models. The method is applied to a watershed model of Klodzko (Poland), built with HEC-1 (USACE, 1998) and HEC-HMS (USACE, 2000).

METHODOLOGY

The methodology for the propagation of precipitation uncertainty introduces the notions of a three-valued quantitative precipitation (3-VQP) and a principal precipitation pattern (PPP).

Three-valued quantitative precipitation (3-VQP)

Let P_i represent the total precipitation for a sub-basin i ($i = 1, \ldots, m$; m = number of sub-basins) accumulated during the forecast period (T). P_i is assumed to be obtained from a quantitative precipitation forecast (QPF). Since this value is not absolutely certain, assume it has an error of $\pm e$ percent. In the absence of a probabilistic forecast of precipitation, the notion of the 3-VQP is introduced. This is defined as the most credible, minimum and maximum forecast precipitations, where:

The most credible forecast precipitation, $P_{i,mc} = P_i$ (1)

The minimum forecast precipitation, $P_{i,\min} = P_i - eP_i / 100$ (2)

The maximum forecast precipitation, $P_{i,\max} = P_i + eP_i / 100$ (3)

A triangular membership function (Fig. 1) is constructed from the 3-VQP. The qualitative meaning of the triangular membership is the following. The "true" value of precipitation, P_i, is certainly included between $P_{i,\min}$ and $P_{i,\max}$ and is likely to be close to $P_{i,mc}$. The key words are "included" and "close". These words constitute the only information that one has about the problem (Rivelli & Ridolfi, 2002).

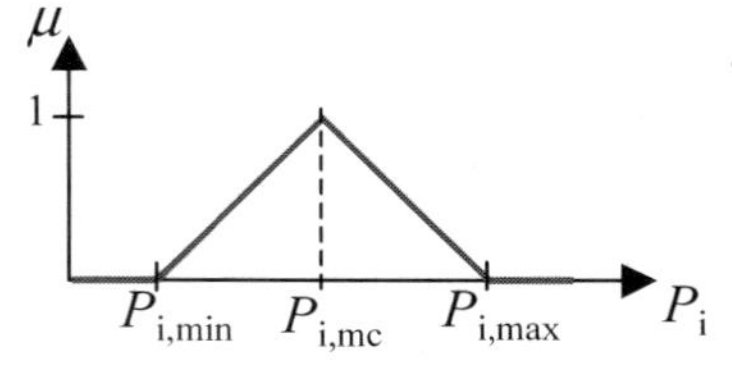

Fig. 1 Fuzzy triangular membership function based on the 3-VQP.

Principal precipitation pattern (PPP)

To take into account the uncertainty in temporal and spatial distribution of precipitation, the notion of the PPP is introduced. A forecast period is divided in to n subperiods where the length of a subperiod, $t = T/n$. Then the PPP is the pattern such that the total precipitation for a sub-basin occurs during one of the subperiods and no precipitation occurs during the remaining subperiods. For example, assuming three subperiods the 3 PPPs (Fig. 2) will be (P_i, 0, 0), (0, P_i, 0) and (0, 0, P_i). It should be noticed however that the PPP is not used directly as inputs to the model. The actual input precipitation pattern will be based on the PPP and randomly generated temporal

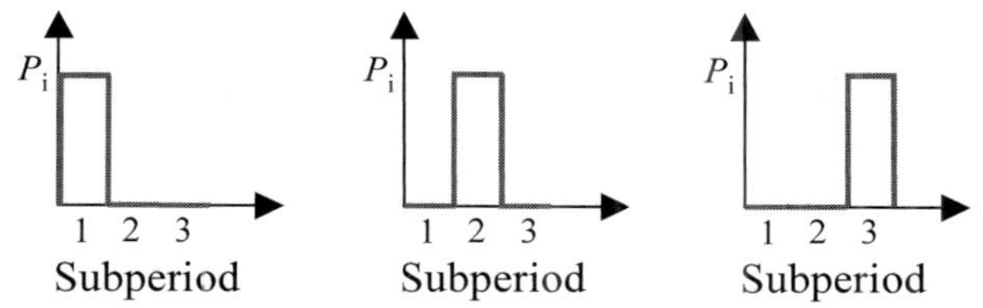

Fig. 2 Principle precipitation patterns (PPPs) considering three subperiods of forecast.

and spatial distribution factors (pattern coefficients). A matrix of temporal and spatial distribution factors, $\boldsymbol{B}$, is introduced such that:

$$\boldsymbol{B} = \begin{bmatrix} b_{1,1} & \cdots & b_{1,n} \\ \vdots & b_{i,j} & \vdots \\ b_{m,1} & \cdots & b_{m,n} \end{bmatrix} \tag{4}$$

where $0 \leq b_{i,j} \leq 1$ (for all i, j) and $b_{i,1} + ... + b_{i,n} = 1$ (for all i), and the matrix of the precipitation pattern, PT, is given by $PT = P_i\boldsymbol{B}$. Thus by generating different values of $b_{i,j}$, as many numbers of precipitation patterns as necessary can be generated. Normally, the temporal distribution is incorporated by considering a fixed number of predefined precipitation patterns. The advantage of the present approach is that there are equal chances that all possible precipitation patterns (the subperiod level) are checked. In the precipitation uncertainty processor (PUP) of Kelly & Krzysztofowicz (2000), the temporal disaggregation of precipitation was treated using the notions of wet subperiods and timing patterns.

Propagation of precipitation uncertainty

Uncertainty represented by a fuzzy membership function can be propagated through a model using the extension principle of fuzzy set theory. First introduced by Zadeh (1975) the extension principle enables us to extend the domain of a function on fuzzy sets. The extension principle is generally carried out at different level α-cuts. An α-cut of a fuzzy set A, denoted as A^α is the set of elements which have memberships greater than or equal to α. For each α-cut two values of the output are required called the lower bound (LB) and the upper bound (UB). The LB is the minimum and the UB is the maximum of the model output. In the present example, representing the model function by f and the model output discharge by Q, the extension principle gives:

$$Q_{LB}\alpha = \min[f([P_{1,LB}, P_{1,UB}]^\alpha, \quad [f([P_{1,LB}, P_{1,UB}]^\alpha, \boldsymbol{B} \quad \alpha \in [0,1] \tag{5}$$

$$Q_{UB}\alpha = \max[f([P_{1,LB}, P_{1,UB}]^\alpha, \quad [f([P_{1,LB}, P_{1,UB}]^\alpha, \boldsymbol{B} \quad \alpha \in [0,1] \tag{6}$$

where Q^α_{LB} and Q^α_{UB} are the lower and upper bounds of the model output discharge, respectively; $P^\alpha_{i,LB}$ and $P^\alpha_{i,UB}$ are the lower and upper bounds of the forecast precipitation (accumulated for the forecast period), respectively, for the given α-cut.

Thus, finding the membership function of an output using the extension principle is a problem of finding the minimum/maximum of functions (equations (5) and (6)) in which the constraints are the bounds in the input variables defined for the given α-cut. The optimization problem is therefore subject to the constraints:

$$\left.\begin{aligned} &P^{\alpha}_{i,LB} \le P^{\alpha}_{i} \le P^{\alpha}_{i,UB} && \forall i \\ &0 \le b_{i,j} \le 1 && \forall (i,j) \\ &\sum_{j=1}^{n} b_{i,j} = 1 && \forall i \end{aligned}\right\} \tag{7}$$

A genetic algorithm is used to generate a range of precipitation patterns in order to deduce the lower and upper bounds of the model output discharge.

Algorithms for minimum and maximum determination

Algorithms for solving optimization problems range from linear, nonlinear to global. The global optimization algorithms (GOAs) possess a particular advantage in solving optimization problems using off-the-shelf software, for which the details of the underlying algorithms may not be known (Maskey *et al.*, 2002). One of the most famous GOAs is the genetic algorithm (GA). A comprehensive evaluation of optimization algorithms is not the intent of this study. However, the successful and extensive use of GAs in various fields of engineering, including water-related problems, inspired the authors to choose a GA scheme as an optimizer.

Genetic algorithms are search algorithms based on the mechanics of natural selection and natural genetics (Goldberg, 1989). The version of the GA used in this study consists of the tournament selection, uniform crossover, two children per pair of parents, jump and creep mutations, elitism and niching.

APPLICATION EXAMPLE

Model overview

A flood forecasting model of the Klodzko watershed has been considered. HEC-1 (USACE, 1998) and HEC-HMS (USACE, 2000) by the Hydrologic Engineering Centre of the US Army Corps of Engineers were used to build the model. HEC-HMS (its Calibration Module) was used for model calibration and HEC-1 was used for the simulation. HEC-1 is one of the most frequently used rainfall–runoff models in the United States (Melching *et al.*, 1991). The methods used to represent different processes in the present model are: (a) The Soil Conservation Service (SCS) Curve Number (CN) method for the direct runoff volume (precipitation excess) computation; (b) Clark's unit hydrograph (UH) method for the transformation of excess precipitation to runoff; (c) the exponential recession method for baseflow computation; and (d) the Muskingum method for the flow routing.

A schematic diagram of the model is shown in Fig. 3. The basin consists of nine sub-basins (SBs) and covers the total area of 1744 km^2 with sub-basin areas ranging

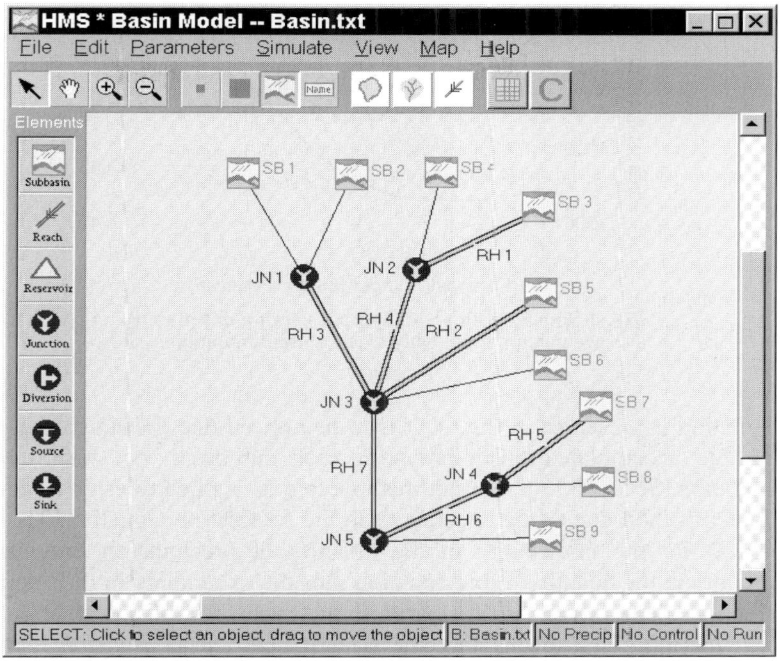

Fig. 3 Schematic diagram of the watershed (SB = sub-basin, RH = reach, and JN = junction).

from 64 km^2 to 280 km^2. It has seven reaches (RHs) and five junctions (JNs). Observed precipitation data (cumulative for every 3 h) are available for the flood of July 1997. Due to the lack of forecast precipitation data, the observed precipitation is assumed as the forecast precipitation. Since the frequency of the available precipitation is 3 h, the size of the forecast period is also taken as 3 h. The period is divided into three subperiods of 1 h each. For a practical application in real time, the 3 h forecast period may be too short. Given the abrupt rise of the flood in a short period, this is fairly appropriate for the application of the methodology. For each forecast, only the uncertainty in the forecast precipitation during the same forecast period is considered. For example, to forecast the flood for the time 6 h, no uncertainty is assumed in the precipitation up to the time 3 h, and so on. An error of 30% is assumed in the given precipitation to estimate the 3-VQP. Although the error is taken arbitrarily, discussion with experts suggested that the assumption is not too far from reality. The starting date for the simulation is 4 July 1997 at 06:00 h, while the forecast of precipitation started on July 7 1997 at 06:00 h.

RESULTS

All results presented and discussed here are based on the forecast downstream of JN 5 (Fig. 3). Firstly, the simulation is carried out to compute the flood discharge without considering the uncertainty. The simulated flood together with the basin-average

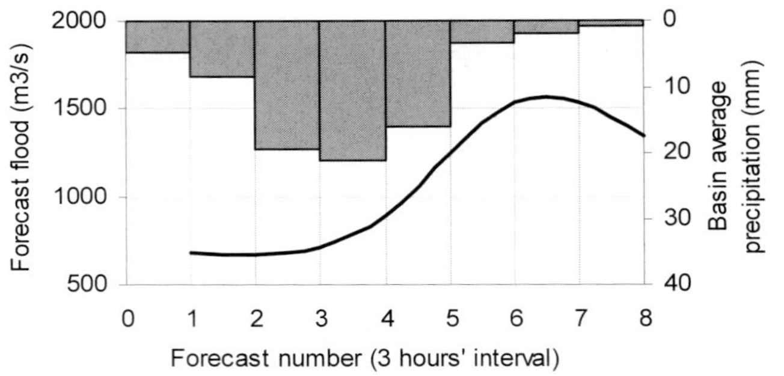

Fig. 4 Simulated flow without considering uncertainty in precipitation with the basin average precipitation. Only a part of the simulation is shown.

precipitation is presented in Fig. 4. It is to be noticed that the simulation was carried out taking the accumulated precipitation in each sub-basin, not with the basin-average precipitation. Secondly, the present methodology is applied to estimate the uncertainty in the forecast flood due to the uncertainty in the forecast precipitation. This is done in two steps: (a) for the uncertainty in the quantity of precipitation only, and (b) for the uncertainty in the quantity of precipitation plus the uncertainty in the temporal and spatial distributions. The former assumes a uniform distribution of precipitation throughout the forecast period. This means that the pattern coefficients $b_{i,1} = b_{i,2} = b_{i,3} = 1/3$. For simplicity, these two cases are referred to as precipitation quantity uncertainty (PQU) and precipitation quantity and pattern uncertainty (PQPU), respectively.

For each forecast, $Q_{UB} - Q_{LB}$ is calculated from the results at a zero level α-cut for both cases: PQU and PQPU. This quantity expresses the maximum amount of estimated uncertainty in the forecasted flood (Fig. 5). It is clearly seen that in all forecasts the output uncertainty is dominated by the uncertainty in the precipitation patterns. As can be seen in Fig. 5, the maximum uncertainty occurred for the fourth and fifth forecasts. The precipitation patterns that give the maximum flow for the fifth forecast is shown in Fig. 6. The results shown in Figs 5 and 6 are for a zero level α-cut only. To construct a complete membership function (MF) of the output, the computation needs to be carried out at different level α-cuts. Four more α-cuts at α = 0.25, 0.5, 0.75 and 1 were selected. The MFs obtained for the fifth forecast due to PQU and PQPU are shown in Fig. 7.

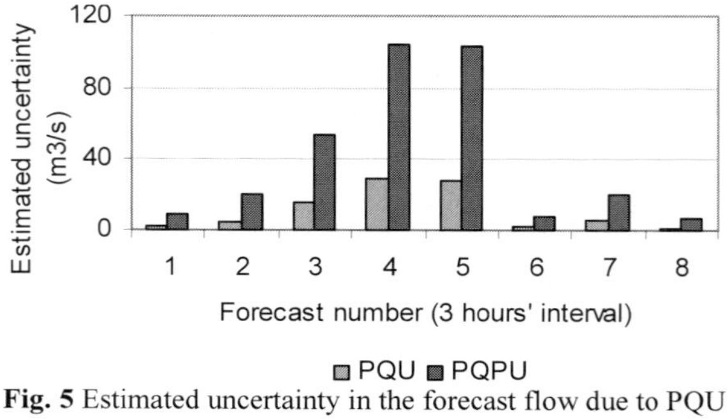

Fig. 5 Estimated uncertainty in the forecast flow due to PQU and PQPU.

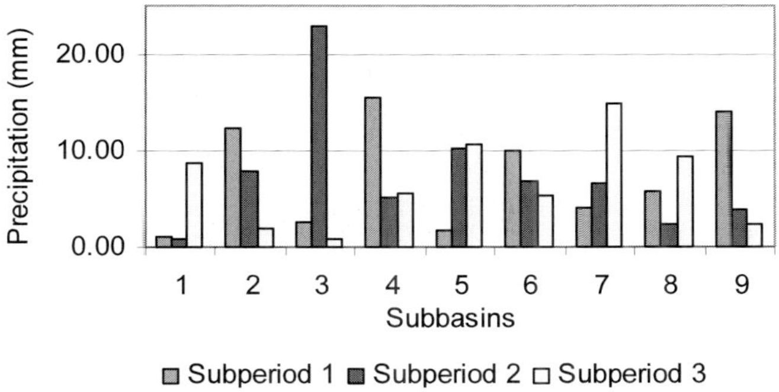

Fig. 6 Precipitation pattern resulting in the maximum flow for the fifth forecast.

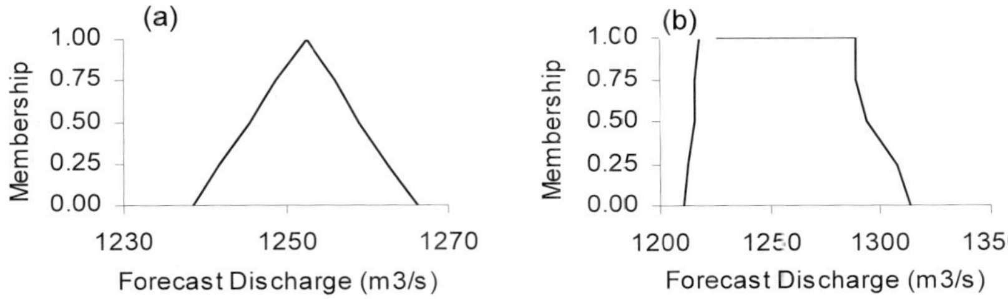

Fig. 7 Membership function of the forecast flow (fifth forecast): (a) due to PQU and (b) due to PQPU.

CONCLUSION

The presented methodology for the propagation of precipitation uncertainty can be very useful in the absence of a probabilistic precipitation forecast. This methodology is independent of the structure of the forecasting model. In other words, it can be used with any rainfall–runoff routing type deterministic model. The methodology uses two important concepts: 3-VQP and PPP. The concept of 3-VQP is used to represent uncertainty in the quantity of forecast precipitation by a fuzzy membership function from limited information about uncertainty. The concept of PPP is used to represent uncertainty in the temporal and spatial distributions of the precipitation. The results show the good potential of the fuzzy set theory combined with a GA for the propagation of uncertainty. The results also show that the output uncertainty due to the uncertainty in the temporal and spatial distributions can be significantly dominant over the uncertainty due to the uncertain in the quantity of precipitation. Therefore, it suggests that we might be making a big mistake by taking basin average precipitation uniformly distributed throughout the forecast period for forecasting, which is generally the case in many forecasting systems. The estimated uncertainty in the output (Fig. 5) may seem small compared to the magnitude of the flood (Fig. 4). This is due to the relatively small forecast period (3 h) considered. Obviously, increasing the forecast period significantly increases the uncertainty in the forecast precipitation and thereby increases the output uncertainty.

Acknowledgements This research was initiated by the EC project Operational Solutions for the Management of Inundation Risks in the Information Society (OSIRIS), contract IST-1999-11598. The authors extend their gratitude to the IMWM, Poland for providing the data used in this study.

REFERENCES

Goldberg, D. E. (1989) *Genetic Algorithms in Search, Optimization, and Machine Learning*. Addison-Wesley, Reading, Massachusetts, USA.

Kelly, K. S. & Krzysztofowicz, R. (2000) Precipitation uncertainty processor for probability river stage forecasting. *Water Resour. Res.* **36**(9), 2643–2653.

Maskey, S., Jonoski, A. & Solomatine, D. P. (2002) Groundwater remediation strategy using global optimization algorithms. *J. Water Resour. Plan. Manage. Div. ASCE* **128**(6), 431–440.

Melching, C. S., Yen, B. C. & Wenzel, H.G., Jr (1991) Output reliability as guide for selection of rainfall-runoff models. *J. Water Resour. Plan. Manage. Div. ASCE* **117**(3), 383–398.

Revelli, R. & Ridolfi, L. (2002) Fuzzy approach for analysis of pipe network. *J. Hydraul. Engng* **128**(1), 93–101.

USACE (1998) HEC-1 *Flood Hydrograph Package User's Manual*. Hydrologic Engineering Centre, USACE, Davis, California, USA.

USACE (2000) *Hydrologic Modelling System HEC-HMS Technical Reference Manual*. Hydrologic Engineering Centre, USACE, Davis, California, USA.

Zadeh, L. A. (1975) The concept of a linguistic variable and its application to approximate reasoning. Part I. *Inform. Sci.* **8**, 199–249.

Combining physically-based and conceptual approaches in the development and parameterization of a distributed system

VICTOR KOREN, SEANN REED, MICHAEL SMITH & ZIYA ZHANG

Office of Hydrologic Development, National Weather Service, 1325 East-West Highway, Silver Spring, Maryland 20910, USA

victor.koren@noaa.gov

Abstract A grid-based Hydrology Laboratory Research Modelling System (HL-RMS) that combines lumped conceptual and distributed model features has been developed by the US National Weather Service (NWS). HL-RMS consists of a well-tested conceptual water balance model applied on a regular spatial grid and physically-based kinematic routing models. A parameter estimation procedure that combines spatially-distributed and "integrated" at basin outlets properties is discussed. Initial tests using high-resolution radar precipitation estimates show that HL-RMS yields results comparable to well-calibrated lumped model simulations and outperforms a lumped model over basins where rainfall variability effects are significant. Combining outlet information and spatially-variable basin properties provides a reliable procedure to estimate distributed model parameters. While uniform adjustment of water balance model parameters provides reasonable results at basin outlets, the problem of removing scale effects in nested basins is still a challenge.

Key words distributed model; kinematic routing; radar rainfall estimates; scaling effects; spatially-variable parameters; water balance

INTRODUCTION

The growing availability of radar precipitation estimates and other sets of spatial data have intensified research in distributed modelling. Numerous distributed models and modelling approaches have been developed to address a wide variety of issues. Models ranging in complexity from the so-called "physically-based fully distributed" to "semi-distributed" or conceptual lumped models applied at smaller scales have been constructed. The range of issues that must be addressed in distributed modelling can be found in the most recent special issue of *Hydrological Processes* (Beven & Feyen, 2002), a collection of more than 20 papers. As Beven stated in the preface to the issue, "if the reader perceives that this issue reveal the problems to be addressed in the application of distributed models more than providing solutions, then that properly reflects the current state of the art". The main problem is how to apply point process laws to the basin scale given the tremendous vertical and horizontal heterogeneities of basin properties. Beven (1995) argues that the aggregation approach towards large-scale hydrological modelling is an inadequate approach to the scale problem. As a result, parameter estimation procedures for physically-based models are not well defined, and the physically-based model may be essentially reverting to a type of elegant black box (Loague, 1990). Vieux & Moreda (2002) suggested a manual "ordered physics-based parameter adjustment" procedure assuming that there are reasonable *a priori* estimates of a limited number of spatially-variable model parameters.

In light of these modelling trends, concerns, and the specific needs of the US National Weather Service (NWS), the use of simpler approaches was investigated. This study echoes the sentiment of Robinson & Sivapalan (1995) who stated that there are not enough analyses in finding connections between physically-based and conceptual models, "although this is precisely what is required for the advancement of hydrological modelling for predictive purposes". We believe that more comprehensive analysis of conceptual lumped and physically-based distributed models on large basins is needed to fully benefit from existing lumped modelling experience. In this paper we present a preliminary attempt to combine lumped conceptual and distributed model features in the development and parameterization of the Hydrology Laboratory Research Modelling System (HL-RMS).

HL-RMS STRUCTURE

HL-RMS is a flexible modelling system, able to use grid cells or sub-basins as the computational elements for rainfall–runoff modelling. Currently, HL-RMS is defined on a regular rectangular grid. Each grid cell consists of a water balance component and a hillslope and channel routing component. A number of conceptual hillslopes at each grid cell are defined to make overland flow distances physically realistic for the relatively large cell size (16 km^2). A drainage density parameter is used to subdivide a cell into equally sized overland flow planes. Conceptual hillslopes drain water to a conceptual channel within the same grid cell. A conceptual channel usually represents the highest order stream of a selected grid cell. It is assumed that all hillslopes have the same properties inside each grid cell but they may be different from cell to cell. Cell-to-cell channel routing is done using a flow direction grid. An automatic procedure was developed to generate a coarser resolution flow direction grid from higher resolution DEM data. To facilitate efficient routing calculations, the drainage network is translated into a computational sequence of grid cells in an upstream to downstream order.

Fast response runoff from the water balance model is routed over conceptual hillslopes within each cell to a conceptual channel. Slow response runoff is assumed to enter the channel system directly from the soil and therefore bypass the hillslope routing. There is no physical connection between soil moisture states in adjacent grid cells. The conceptual channel is the only source of water exchange between neighboring pixels.

The water balance component of the current version of HL-RMS uses the Sacramento Soil Moisture Accounting Model (SAC-SMA) (Burnash, 1995), and hillslope-channel routing employs the kinematic wave model. Several factors played a role in this selection. Use of the SAC-SMA is a practical choice because NWS hydrologists have great experience with lumped applications of the model. Also, the work of Koren *et al.* (2000) established relationships between SAC-SMA parameters and soil properties, making it possible to run simulations using parameter estimates that vary within a basin. The kinematic wave model is well tested, and it provides reasonable accuracy under a wide range of flow conditions.

The SAC-SMA represents conceptually heterogeneous runoff processes over a range of spatial scales from tens to a few thousand square kilometres. There are strong physical arguments and application results to support the model structure. The basic design of SAC-SMA centres on a two layer structure: a relatively thin upper layer, and usually a much thicker lower layer which supplies moisture to meet the evapotranspiration demands. Each layer consists of tension and free water storages that interact to generate soil moisture states and six runoff components. Partitioning of rainfall into surface runoff and infiltration is governed by the upper layer soil moisture deficit and a percolation rate to the lower layer.

A fairly general numerical scheme (Smith, 1980) is used to integrate kinematic wave equations. It employs weighting factors α and β in the spatial and temporal dimensions, respectively. The scheme stability ranges from unconditionally stable ($\alpha = 0$ and $\beta = 1$) to unconditionally unstable ($\alpha > \beta$). The best approximation is obtained at $\alpha = \beta = 0.5$, when the scheme has second order accuracy. However, it is practically impossible to satisfy the scheme stability criteria (the Courant number $C_r = 1$) at this value of weighting factors under variable channel properties. Because HL-RMS is applied over a wide range of hillslope-channel hydraulic conditions, values of 0 and 1 are used for temporal and spatial weight factors, respectively, to provide the unconditional stability. While this will lead to some reduction of simulation accuracy, our tests suggest that accuracy degradation is less significant than input data and model/parameter uncertainties for basins studied. It is also important that truncation errors of the scheme increase independently of the space–time increments ratio as α and β depart from the value of 0.5 that allows a flexible selection of space–time increments to compensate some accuracy reduction.

HL-RMS PARAMETERIZATION

Water balance and routing model parameters are assumed to be constant within each grid cell; however, they can vary from cell to cell. The approach adopted here is a two-step procedure: derivation of *a priori* parameter grids, and adjustment of these grids using observed outlet hydrographs. The basic idea used to derive parameter grids is to combine distributed grid cell data (e.g. slope, soil properties) with integrated basin properties observed at the outlet (e.g. flow measurement information, discharge time series). While this approach was applied in the first step of the routing parameter estimation procedure, it was used only in the second step of water balance model parameterization.

To account for the spatial variability within a basin, *a priori* SAC-SMA parameter grids developed by Koren *et al.* (2000) were used. High-resolution soil data (texture, hydrological group, and depth) for the conterminous US were applied. Results from lumped simulations using basin averaged *a priori* parameters (Koren *et al.*, 2000, 2002) suggest that while *a priori* estimates cannot outperform results from well calibrated parameters on gauged basins, these values provide reasonable initial estimates for ungauged basins. To improve the initial estimates, HL-RMS scales *a priori* grids over selected basin by the ratio of the SAC-SMA parameter from lumped calibration to the average parameter value from the *a priori* grid (Koren *et al.*, 2002). Because lumped-calibrated SAC-SMA parameters are scale-dependent (Koren *et al.*,

1999), some further fine tuning of selected scaled parameters might be required in distributed simulations. Fortunately, the scale effect is reduced significantly if a ratio of calibration and distributed simulation scales decreases. Koren *et al.* (1999) observed about a 3% runoff difference when calibrated SAC-SMA parameters over 1000 km^2 basin were applied at a grid scale of 16 km^2.

The kinematic wave relationship between discharge and channel cross-section was approximated by a power function defined by two parameters, specific discharge, Q_s, and exponent value, m. Two options have been developed to define these parameters: (a) use of the Chezy–Manning approximation assuming a prismatic channel, referred as a channel parameterization method, and (b) direct estimation of these parameters from an empirical discharge and cross-section relationship, referred to as a rating curve method. The basic idea for both options is to disaggregate information from outlet measurements representing basin-integrated properties into interior grid cell parameters using local geomorphologic properties. Channel shape parameters (channel parameterization option) or specific discharge and exponent (rating curve option) are first estimated at the basin outlet by fitting a curve to a plot of cross-section *vs* top width data, or to a plot of cross-section *vs* discharge data, respectively.

Two geomorphic assumptions that follow from channel geometry laws are used to estimate channel shape parameters at upstream cells: (a) the ratio of channel-forming flows at different cells, $r_{Q,i}$, equals the ratio of drainage areas, F, and (b) the ratio of cross-sectional areas of different channels, $r_{A,i}$, is a known function of stream orders. These assumptions lead to simple estimation procedures. In the channel parameterization method, when a two-parametric power approximation of channel cross-section was used, distributed channel top width parameter, a_i, can be estimated from:

$$a_i = (b+1)^{-b}\left[\frac{\sqrt{S_i}}{n_i}\frac{(r_{A,i}A_o)^m}{r_{Q,i}Q_o}\right]^{3(b+1)/2} \tag{1}$$

where S_i and n_i are channel slope and roughness, Q_o and A_o are discharge and cross-section at the outlet, and b is a shape parameter assumed to be constant within a basin, equaling an estimated value at the outlet. With known channel geometry at each cell, values of channel specific discharge, Q_s, and exponent parameter m are estimated from:

$$Q_{s,i} = \frac{\sqrt{S_i}}{n_i}\left[a_i(b+1)^b\right]^{-2/3(b+1)}, \quad \text{and} \quad m = \frac{b+5/3}{b+1} \tag{2}$$

In the rating curve method, specific discharge at each grid cell is estimated from:

$$Q_{s,i} = Q_{s,o}\frac{F_i}{F_o}r_{A,i}{}^m \tag{3}$$

As in the channel parameterization method, the exponent parameter m is a constant that is equal to an estimated outlet value. An advantage of this method is that it does not require estimation of channel slope and roughness at upstream points. However, because of this, the method may poorly represent channel geometry if there are distinct portions of basin, which have significant differences in average channel slopes (e.g. transition from mountain region to a plain). Note that values of the specific discharge and exponent parameters from these two options may differ because of empirical curve fitting to different experimental data. It can be shown that the specific discharge ratios are equal if the exponent parameters of these options are equal. Comparison of specific discharge estimates from these two methods for a few tested headwater basins shows a linear relationship indicating a good agreement between the exponent parameter values.

RESULTS AND DISCUSSION

The tests described here have been performed on watersheds within the Arkansas–Red River basin in Oklahoma. The main reason is that this region has the longest archive of 4-km NEXRAD Stage III precipitation grids (NWS archive) derived from radar-gauge analyses (Seo & Breidenbach, 2002), and these rainfall estimates have been evaluated more thoroughly than those produced in other parts of the country. In addition, these basins have few complications such as snow, reservoirs, and complex topography such as mountainous areas.

A priori SAC-SMA parameter grids at the 4 km scale were generated over the entire region. These grids display a large variability of basin properties that affect runoff generation processes over this region. Initial routing parameter grids were derived using measured discharge data from only a few stations on the main stem of the Arkansas River. A parameter estimation procedure to derive values for upstream grid cells is applied sequentially from downstream to upstream gauges independently. At each estimation loop, grid cell parameters above a selected gauge, estimated previously from the next downstream gauge, are overwritten by new estimates from the selected upstream gauge. For analysis, these grids were then refined over a few selected headwater basins.

Two critical questions to consider in evaluating the potential benefits of distributed modelling are: (a) whether a distributed model can produce simulations that are comparable to or better than simulations from existing lumped models, and (b) whether it is possible to define a distributed parameter model calibration strategy that is robust across spatial scales. We consider the ability to produce simulations that are comparable to lumped results a positive result because there are other potential benefits from running a distributed model: the ability to simulate flows at small, ungauged outlets within a watershed, and the ability to incorporate future sources of spatial data are good examples.

Continuous discharge simulations at an hourly time step were generated for the 8-year period. HL-RMS results are evaluated against both hourly observed hydrographs and simulated hydrographs from a lumped version of the SAC-SMA model paired with a unit hydrograph routing procedure. First, "reference" results were generated using *a priori* estimates of water balance and routing parameters for both HL-RMS and a lumped application of SAC-SMA. Second, SAC-SMA parameters were calibrated in a lumped mode for all parent basins. *A priori* SAC-SMA parameter grids of HL-RMS were then adjusted as discussed above, using lumped calibration results for defined watersheds. Major adjustments to gridded parameter values were

Table 1 Accuracy statistics for selected headwater basins in the Arkansas–Red basins.

Watershed	Area (km^2)	HL-RMS simulations $*_{vol.}$	HL-RMS simulations $*_{peak.}$	HL-RMS simulations $*_{RMS}$	HL-RMS simulations R	Lumped simulations $*_{vol.}$	Lumped simulations $*_{peak.}$	Lumped simulations $*_{RMS}$	Lumped simulations R
Calibrated parameters:									
Blue	1232.8	25.0	***25.0***	***139.0***	***0.87***	***23.0***	35.0	141.0	0.86
Eldon	795.1	***16.4***	***25.7***	131.0	0.90	18.5	26.0	***114.5***	***0.92***
Tahlequah	2483.8	***11.3***	***20.5***	70.8	0.92	12.6	25.8	***64.9***	***0.94***
Watts	1644.6	***11.9***	***26.4***	86.5	0.92	12.9	30.2	***82.8***	***0.93***
Savoy*	432.8	***19.9***	52.2	***185.6***	***0.86***	20.9	***52.0***	196.7	0.85
Kansas*	284.9	23.8	***53.0***	***161.7***	***0.81***	***23.7***	55.8	189.6	0.73
Uncalibrated parameters:									
Blue	1232.8	38.0	***40.3***	187.0	0.81	***31.0***	42.8	***163.0***	***0.83***
Eldon	795.1	***27.4***	***45.3***	***160.0***	***0.84***	30.2	53.4	181.9	0.79
Tahlequah	2483.8	***13.4***	***19.2***	***84.5***	***0.90***	23.7	25.6	97.2	0.86
Watts	1644.6	***13.8***	***26.0***	***105.8***	***0.89***	23.1	30.5	109.9	0.87
Savoy	432.8	***22.4***	***49.8***	***194.2***	***0.85***	29.1	54.5	228.6	0.79
Kansas	284.9	***24.2***	***52.2***	***185.1***	***0.75***	26.9	57.1	221.1	0.62

Notes:

(a) Watersheds marked by * are nested basins (Savoy and Kansas are sub-basins of Tahlequah, Illinois River). These sub-basins actually were not calibrated; instead parent basin calibrated parameters were used in simulations.

(b) Values in bold italic mean that this statistic is better for distributed (lumped) simulations compared to lumped (distributed).

achieved by direct scaling from lumped calibrated parameters. Further minor adjustments of a few upper zone parameter scale factors were made using a manual trial-and-error process similar to an approach suggested by Vieux & Moreda (2002).

Table 1 presents statistical analyses of these simulations. Shown are two statistics for selected flood events (20–25 flood events for each basin were included in the analyses): percent of mean absolute error of flood runoff, $*_{vol}$, and flood peak, $*_{peak}$. Also presented are two overall statistics describing the entire simulation period: percent of root mean square error, $*_{RMS}$, and a correlation coefficient of hourly discharges, R. A few observations can be made about the results in Table 1.

While distributed and lumped simulations can both produce reasonable flood event simulations for calibrated watersheds, distributed model results are slightly better (compare columns 2–3 *vs* 6–7; note that better statistics are in bold italic). The distributed model benefit can also be seen in Fig. 1, which is a plot of flood peak errors for all selected floods from distributed and lumped runs over three basins. However, only the Blue River simulations show improvements over lumped simulations for the overall statistics of continuous runs (compare columns 4–5 *vs* 8–9). Other watersheds yield overall results comparable to lumped simulations with a slight decrease in accuracy.

It is important to note that simulations for two nested basins, Savoy and Kansas (not calibrated directly, but parameters from the calibration of the parent basin are applied instead), outperformed lumped simulations even more consistently than we would expect from a distributed model.

Consistent improvements from the uncalibrated parameter version of HL-RMS over uncalibrated lumped simulations are achieved for most watersheds (compare

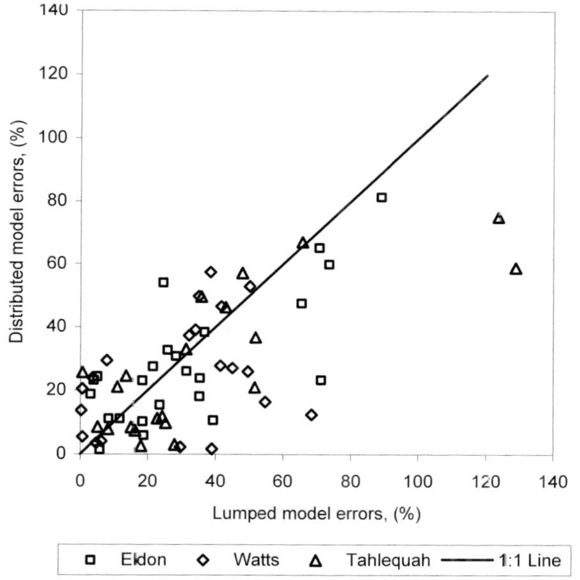

Fig. 1 Absolute flood peak errors (%) from distributed and lumped runs over three parent basins.

columns 2–5 *vs* 6–9 of the Uncalibrated parameters section of Table 1). It means that spatial patterns of *a priori* parameters and rainfall interact reasonably well, and as a result, the parameter scaling procedure used in this analysis may be a reasonable component in a multi-step calibration.

Tests suggest that HL-RMS provides a flexible framework for rainfall–runoff analysis and practical applications of distributed models, and that it is computationally feasible to run the system over large regions. The approach adds more practicality to the process of model parameterization, and facilitates an easier transition from current lumped model-based operational systems to more powerful distributed systems. Schemes developed to estimate distributed routing parameters based on local grid cell and basin outlet integrated properties produce reasonable results for the range of spatial scales without any calibration. Progress has been made in quantitatively estimating spatially-variable rainfall–runoff parameters by combining soil properties and lumped calibration results, but a robust method to calibrate the distributed parameter model remains to be developed.

REFERENCES

Beven, K. (1995) Linking parameters across scales: sub-grid parameterizations and scale dependent hydrological models. In: *Scale Issues in Hydrological Modelling*, (ed. by J. D. Kalma & M. Sivapalan), 263–282. Wiley, New York.

Beven, K. & Feyan, J. (eds) *The Future of Distributed Hydrological Modelling* (2002) Special Issue of *Hydrol. Processes* **16**(2) 169–574.

Burnash, R. J. C. (1995) The NWS river forecast system—catchment modelling. In: *Computer Models of Watershed Hydrology*, (ed. by V. P. Singh), 311–366. Water Resources Publications, Littleton, Colorado, USA.

Koren, V. I., Finnerty, B. D., Schaake, J. C., Smith, M. B., Seo, D. J. & Duan, Q. Y. (1999) Scale dependencies of hydrologic models to spatial variability of precipitation. *J. Hydrol.* **217**, 285–302.

Koren, V. I., Smith, M., Wang, D., Zhang, Z. (2000) Use of soil property data in the derivation of conceptual rainfall–runoff model parameters, 15th *Conference on Hydrology*, AMS, 103–106. Long Beach, California, USA

Koren, V., Smith & M., Duan, Q. (2002) Use of *a priori* parameter estimates in the derivation of spatially consistent parameter sets of rainfall–runoff models. In: *Calibration of Watershed Models, Water Science and Application 6* (ed. by Q. Duan, S. Sorooshian, H. Gupta, A. Rosseau & R. Turcotte), 239–254. Am. Geophys. U., Washington DC, USA.

Loague, K. (1990) R-5 Revisited 2. Reevaluation of a quasi-physically-based rainfall–runoff model with supplemental information. *Water Resour. Res.* **26**(5) 973–987.

Robinson, J. S. & Sivapalan M. (1995) Catchment-scale runoff generation model by aggregation and similarity analysis. In: *Scale Issues in Hydrological Modelling*, (ed. by J. D. Kalma & M. Sivapalan), 311–330. Wiley, New York, USA.

Seo, D.-J. & Breidenbach, J. P. (2002) Real-time correction of spatially nonuniform bias in radar rainfall data using rain gauge measurements. *J. Hydromet.* **1**, 222–240.

Smith, A. A. (1980) A generalized approach to kinematic flood routing. *J. Hydrol.* **45**, 71–89.

Vieux, B. E. & Moreda, F. G. (2002) Ordered physics-based parameter adjustment of a distributed model. In: *Calibration of Watershed Models, Water Science and Application 6* (ed. by Q. Duan, S. Sorooshian, H. Gupta, A. Rosseau & R. Turcotte), 267–281. Am. Geophys. U., Washington DC, USA.

Improving *a priori* estimates of hydraulic parameters in a distributed routing model via variational assimilation of long-term streamflow data

DONG-JUN SEO, VICTOR KOREN & SEANN REED
Hydrology Laboratory, Office of Hydrologic Development, National Weather Service, 1325 East-West Highway, Silver Spring, Maryland 20910, USA
dongjunseo@noaa.gov

Abstract Due to very large dimensionality of the problem, exhaustive estimation of grid cell-specific parameters in distributed models is highly impractical (if not infeasible). As such, much attention has been given in recent years to accurate *a priori* estimation of distributed parameters from all available data sources, to the degree possible. Though they may capture the spatial variability of the parameters reasonably well, such *a priori* estimates of distributed parameters are subject to various sources of systematic and scale-dependent errors that can potentially be of rather large magnitude. The purpose of this work is to explore improving, or refining, *a priori* estimates of distributed hydraulic parameters of a kinematic wave routing model via variational assimilation of long-term streamflow observations.

Key words distributed parameters; hydraulic routing; streamflow; variational assimilation

INTRODUCTION

With the wide availability of digital elevation models (DEM), various spatial data sets and GIS tools, one can now routinely produce *a priori* estimates (whatever the goodness and physically-basedness may be) of grid-specific hydraulic parameters in distributed routing models. Though they may capture the spatial variability of the parameters reasonably well, such *a priori* estimates are necessarily subject to various sources of systematic and scale-dependent errors that can potentially be of rather large magnitude. Accordingly, if viable techniques can be found, it may be possible to significantly improve the goodness of the *a priori* estimates with only a modest amount of computational expense. Here, we explore variational assimilation, i.e. gradient-based least squares minimization based on adjoint of the (forward) dynamic model under consideration, as a potential tool for improving, or refining, the *a priori* estimates of distributed routing parameters. The reasoning behind the use of variational assimilation here is that, if the model dynamics is realistic and the linearized dynamics does capture the sensitivity of the distributed parameters to the observed streamflow at the basin outlet, the adjoint technique offers a practical way of identifying the parameters (and their locations) that exert the greatest influence on the streamflow simulation, and of quantifying, under the least squares criterion, the magnitude of adjustment necessary.

ROUTING MODELS

The routing models considered are the kinematic wave formulations for hillslope and channel flows as implemented in the US National Weather Service Hydrology Laboratory's (NWS/HL) Research Modeling System (RMS; Koren *et al.*, 2002). Soil moisture accounting is carried out, as in RMS, by a gridded application of the Sacramento model (Burnash *et al.*, 1973) with no lateral exchange. Below, we reproduce the description of the routing models from Koren *et al.* (2002).

The hillslope flow is modelled as:

$$\partial h / \partial t + L_h \partial q / \partial x = R_s, \qquad 0 \le x \le L_h \tag{1}$$

$$q = q_s h^{5/3}, q_s = 2DS_h^{1/2} n_h^{-1} \tag{2}$$

where h is the depth of the overland flow, q is the discharge per unit area of hillslope, R_s is the fast-responding runoff from the soil moisture accounting model, S_h is the hillslope, n_h is the roughness coefficient, D is the drainage density (1/[L]) and L_h is the hillslope length (= 1/(2D)). The three parameters, D, S_h and n_h, in q_s are specified for each grid cell. A no-flow boundary condition is assumed at $x = 0$ and L_h.

The channel flow is modelled as:

$$\partial A / \partial t + \partial Q / \partial x = (q_{Lh} + R_g) f_c / L_c, \qquad 0 \le x \le L_c \tag{3}$$

$$Q = Q_s A^{Mc} \tag{4}$$

Where A is the (wetted) channel cross section, Q is the discharge, q_{Lh} is the lateral inflow per unit length of the channel, R_g is the slow-responding runoff from the soil moisture accounting model, f_c is the area of the grid cell, L_c is the channel length within the grid cell, Q_s is the channel specific discharge, and M_c is the exponent parameter. The upstream boundary condition at each grid cell is the total discharge draining into it. Assuming the power-law relationship of $W = a\ H^b$ between the top width W and depth of the channel H, where a and b denote the width and shape parameters, respectively, the *a priori* estimates of Q_s and M_c are given by:

$$Q_s = S^{1/2} n^{-1} (1+b)^{-2b/\{3(b+1)\}} a^{-2/\{3(b+1)\}} \tag{5}$$

$$M_c = (b + 5/3)/(b+1) \tag{6}$$

The parameters to be adjusted are then the specific hillslope discharge, q_s, the exponent, M_c, and the channel specific discharge, Q_s.

PARAMETER ADJUSTMENT

To refine the *a priori* estimates, we solve the following least squares minimization problem subject to the routing model dynamics (equations (1)–(4)):

$$\text{Minimize}\, J = 1/2[Z_Q - H(\boldsymbol{X})]^T \boldsymbol{R}^{-1} [Z_Q - H(\boldsymbol{X})] \tag{7}$$

In the above, Z_Q denotes the vector of long-term streamflow observations, H denotes the structure function that relates the distributed routing parameters to the streamflow at the basin outlet, $\boldsymbol{X}$ denotes the vector of all three routing parameters at all grid cells, and $\boldsymbol{R}$ denotes the covariance matrix associated with the error, $Z_Q - H(\boldsymbol{X})$. Assuming temporal independence among the errors, we may rewrite equation (7) as:

$$\text{Minimize}\, J = 1/2\sum\{z_{Q,i} - h_i(\boldsymbol{X})\}^2 / \sigma_Q^2 \qquad (8)$$

In the above, $z_{Q,i}$ denotes the streamflow observation for the ith hour, $h_i(\boldsymbol{X})$ denotes the structure function that relates the routing parameters to the simulated streamflow at the basin outlet at hour i, σ_Q^2 denotes the error variance (assumed to be flow magnitude-independent), and the summation is over all hours in the duration of the streamflow record of choice. In equations (7) and (8), the absence of the background error term (which levies penalty for adjusting the *a priori* estimates "too much") is a reflection that we have little *a priori* knowledge as to how "good" the *a priori* estimates are.

The minimization is carried out using the conjugate gradient method of Fletcher-Reeves-Polak-Ribiere (Press *et al.* 1986). The gradients necessary for the minimization were evaluated via the adjoint of the routing models, equations (1)–(4), generated by the Tangent linear and Adjoint Model Compiler (TAMC; Giering, 1999). Significant modifications were necessary to the automatically generated adjoint code to improve computational efficiency.

RESULTS AND DISCUSSIONS

To test the procedure, the *a priori* estimates of routing parameters (see Koren *et al.*, 2002 for details) for the basin near Watts, Oklahoma, USA, were refined based on a 3-year period of hourly streamflow observations. Extensively modelled in the Distributed Model Intercomparison Project (DMIP, http://hsp.nws.noaa.gov/oh/hrl/dmip), the basin has 100 grid cells, the size of the cell being approximately 4×4 km^2. The total number of control variables in the minimization was hence 300. Figure 1 shows the scatter plots of the parameters Q_s and M_c between the *a priori* and the refined estimates. Note that a uniform *a priori* estimate was used for m. Sensitivity to the hillslope specific discharge, q_s, was found to be rather small in that the adjusted estimates differ little from the *a priori* estimates (not shown). The channel specific discharge, Q_s, tends to be larger in the upstream areas due to higher slopes (not shown) hence the scatter plot of Q_s in Fig. 1 is an indication that the streamflow at the outlet is more sensitive, as one might expect, to the channel-specific discharge along the main stem closer to the outlet than at the upstream branches. The scatter plot of Q_s also indicates that, in the upstream areas where the parameter exerts little influence to the outlet streamflow, there is little point, as one might also expect, in adjusting the *a priori* estimates in the absence of any other observations. Figure 2 shows the scatter plots of streamflow between the observed and the simulated flows, before and after the adjustment over the 3-year parameter estimation period. A significant improvement in streamflow simulation at the outlet is apparent. In addition to independent validation of the simulation at the outlet, which is under way, a true test of the procedure is to perform independent validation of streamflow at interior locations. For the parameter adjustment procedure to be operationally viable, flow simulations at interior locations based on adjusted parameters should consistently be at least as good as those based on the *a priori* estimates. Such a "blind" independent validation is also under way, and the results will be reported in the near future.

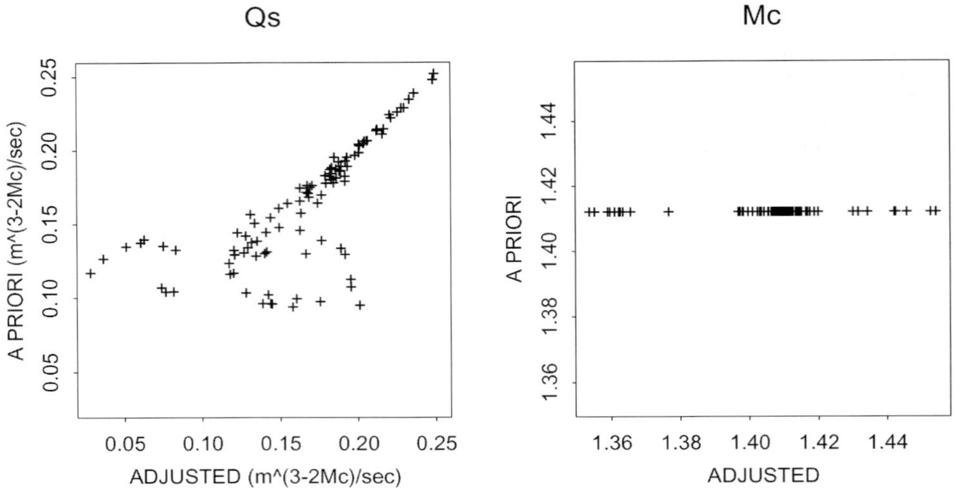

Fig. 1 Scatter plots of distributed routing parameters; the channel-specific discharge, Q_s and the exponent, M_c, at all grid cells between the *a priori* and the adjusted estimates.

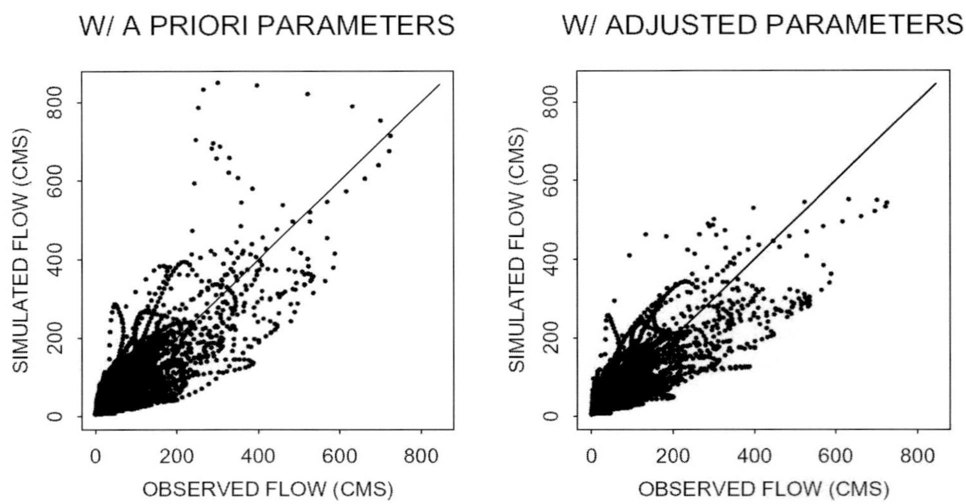

Fig. 2 Scatter plots of hourly streamflow at the outlet between the observed and the simulated flows before and after the adjustment.

Acknowledgement This work is supported by the Advanced Hydrologic Prediction Service (AHPS) Program of the National Weather Service.

REFERENCES

Burnash, R. J. C., Ferral, R. L. & McGuire, R. A. (1973) A generalized streamflow simulation system—Conceptual modeling for digital computers. National Weather Service, NOAA, and the State of California Department of Water Resources Technical Report. Joint Federal-State River Forecast Center, Sacramento, California, USA.

Chao, W. C. & Chang, L.-P. (1992) Development of a four-dimensional variational assimilation analysis system using the adjoint method at GLA Part 1: Dynamics. *Mon. Weath. Rev.* **120**, 1661–1673.

Giering, R. (1999) *Tangent Linear and Adjoint Model Compiler User's Manual, TAMC Version 5.2* (available from http://www.autodiff.com/tamc).

Koren, V., Smith, M., Reed, S., Zhang, Z. & Seo, D. J. (2002) Hydrology laboratory research modeling system of the National Weather Service. *J. Hydrol.* (submitted).

Press, W. H., Flannerty, B. P., Teukolsky, S. A. & Vetterling, W. T. (1986) *Numerical Recipes.* Cambridge University Press, Cambridge, UK.

Weather Radar Information and Distributed Hydrological Modelling (Proceedings of symposium HS03 held during IUGG2003 at Sapporo, July 2003). IAHS Publ. no. 282, 2003.

Distributed model flow sensitivity to uncertainty in radar-rainfall input

THERESA M. CARPENTER & KONSTANTINE P. GEORGAKAKOS
Hydrological Research Center, 12780 High Bluff Drive, Suite 250, San Diego, California 92130, USA
tcarpenter@hrc-lab.org

Abstract The aim of the current research is to assess the sensitivity of the range of flow simulations produced by a distributed hydrological model, suitable for use in an operational environment, under various definitions of uncertainty in the radar rainfall estimates. Monte Carlo sensitivity analysis is used, sampling from various assumed distributions of mean areal precipitation error, and the variation in resulting ensembles of flow simulations is measured using a normalized range of 10% high minus 10% low flows. The results show that the average flow ensemble range over all events in a given sub-catchment possesses significant spatial-scale dependence for all assumed radar rainfall uncertainty characterizations.

Key words distributed hydrological model; ensemble flow forecasts; radar rainfall data

INTRODUCTION

The widespread implementation of weather radar has been accompanied by increased interest and research into the development and application of hydrological models that can use to advantage the high spatial and temporal resolution of the precipitation estimates from the radars. Such distributed hydrological models, several of which are summarized in Ogden *et al.* (2001), allow for spatially-distributed precipitation input and to varying degrees, spatially distributed parametric input. Often, digital databases of terrain elevation, soils, and land use characteristics are used in defining the spatial distribution of parametric input. Although more complex than traditional spatially-lumped hydrological models, distributed models have the potential to provide additional insight on hydrological conditions, such as soil moisture and streamflow, at locations without existing flow observations. However, significant uncertainty exists both in radar rainfall estimates obtained from radar reflectivity, and in hydrological parameters derived from available spatial databases, which exhibit great spatial heterogeneity. How do these uncertainties affect the output of distributed models? Does such an effect vary for different locations within a given watershed?

This paper examines the influence of uncertainty in radar rainfall estimates on distributed hydrological model simulations under operational environment data conditions. The model HRCDHM (Hydrologic Research Center Distributed Hydrologic Model) was applied to four rivers in the central United States. The study watersheds ranged in area from approximately 1000 km^2 to 2500 km^2. For distributed modelling, sub-catchments were defined within each study watershed of area in the range 45–85 km^2. HRCDHM was developed as a transitional step between existing operational hydrological models, which are spatially-lumped in precipitation and parametric input, and future distributed hydrological models. It is built upon model components that are adaptations of existing operational models, and features a GIS link to allow for ease of data ingest. The model is shown to reproduce observed flows well for all the case study watersheds under the assumption of no uncertainty in the input radar rainfall.

This paper presents a brief description of the model and study watersheds, followed by a discussion of the sensitivity analysis framework with explicit uncertainty characterization. The sensitivity analysis output is summarized in terms of flow variability for several locations within each study watershed. Conclusions and potential future studies are discussed in the final section of the paper.

MODEL OVERVIEW

HRCDHM is a catchment-based model. The hydrological model components are applied on a sub-catchment basis within a given watershed of interest. These components include: (a) ingest of radar precipitation estimates from the US NWS operational weather radar (WSR-88D, also called NEXRAD) and computation of sub-catchment mean areal precipitation based on the gridded radar precipitation estimates; (b) soil moisture accounting and runoff generation using the Sacramento Soil Moisture Accounting model for each sub-catchment; (c) upland channel routing for streams within each sub-catchment based on geomorphic consideration; and (d) channel routing for the main stem rivers using kinematic routing.

HRCDHM is linked with a GIS to ingest digital terrain databases, to delineate watershed and sub-catchments boundaries, and to compute geometric properties of the sub-catchments that are used in the hydrological computations, e.g. sub-catchment drainage area, length, and slope. Channel cross-sectional characteristics, which are necessary for the routing computations, are determined through regional regression relationships with the GIS-computed sub-catchment properties. Additional databases of soil characteristics and land cover are utilized to define parameters of the hydrological model components. The use of a specific soils database in estimating the distribution of model parameters for the study watersheds will be described in the following section. For more details on model formulation the interested reader is referred to Carpenter *et al.* (2001).

CASE STUDY BASINS

This study focused on several watersheds within the south-central US, which fall under the umbrella of coverage of the NEXRAD radar at Tulsa, Oklahoma. The river basins include: the Blue River in south-central Oklahoma; the Illinois River in eastern Oklahoma and western Arkansas; the Baron Fork in Oklahoma; and the Elk River in Missouri. The Elk River watershed lies northeast of the Illinois River basin, which falls directly north of the Baron Fork. The Blue River lies to the south and away from the other basins and differs from the others in its fairly narrow, elongated shape. The land surface of all of the watersheds is a mixture of agricultural/pasture lands and forest, with small variations of soil properties across each watershed. Generally, the

Table 1 Study watersheds and calibration statistics.

Basin	Area (km^2)*	Avg sub. area (km^2)	of TS+	CCOR	Bias
Blue R, Blue OK	1232	59	49882	0.86	1.4%
Baron Fork, Eldon OK	795	45	52530	0.87	3.2%
Elk R, Tiff City MO	2258	86	51372	0.82	0.7%
Illinois R, Watts OK	1644	84	52615	0.87	–0.4%
Illinois R, Tahlequah OK	2483	84	51049	0.88	1.8%

*Drainage area at gauge location.
+Number of time steps varies by location due to missing observations.

terrain is mildly sloping, with silty-loam soils. The drainage area ranges from 1000 to 2500 km^2. The catchment delineation yielded sub-catchments with average sizes of the order of 45–85 km^2.

For the application of HRCDHM, outlet locations with existing flow observations were selected for each river basin. Using the GIS, the watershed boundaries for these locations, along with a sub-catchment delineation level, were determined. The locations and drainage areas are summarized in Table 1. An archive of the operational precipitation estimates from the Tulsa radar, along with historical observed flow records with hourly resolution, were available for the period May 1993–May 1999. Using these records, the parameters of the hydrological soil-channel model component of HRCDHM were calibrated over the historical period with hourly resolution for each of the five watersheds. The model parameters were calibrated assuming uniform parameters within each watershed. Once satisfied with the calibration, several soil model parameters were spatially distributed within the watershed based on properties of the soils from available databases.

Specific soil properties were extracted from the STATSGO database (NRCS, 1994) and related to parameters of the Sacramento model. The soil properties were the available water content, permeability, and soil texture classification. Average soil properties were computed for each sub-catchment and for various depth layers consistent with the Sacramento model definition of an upper and lower soil zone. These soil properties were then related to the following model parameters: storage capacities (upper soil zone), interflow rate, and percolation parameters. The distribution of sub-catchment model parameters ($PARAM_i$) was based on a simple scaling of the calibrated parameter (based on uniform parameters, $PARAM_{cal}$), by the sub-catchment average soil property ($SOIL\_PROP_i$) normalized by the average soil property within the watershed ($\Sigma SOIL\_PROP_i / N$):

$$PARAM_i = \frac{SOIL\_PROP_i}{\sum SOIL\_PROP_i / N} * PARAM_{cal} \tag{1}$$

This provides a consistent and objective method for distributing model parameters based on observed soil characteristics. As the STATSGO database has national coverage, this method could be applied to any watershed within the United States.

Table 1 includes statistics between the hourly observed and simulated flows over the entire calibration period for the simulation with distributed parameters. Due to fairly homogeneous soil properties, the statistics for the uniform parameter case were similar to those presented in Table 1. As evident in the statistics, HRCDHM was able to reproduce the observed hourly flows well for each location.

SENSITIVITY ANALYSES

Precipitation input in the model is given in terms of mean areal precipitation (MAP) values for each sub-catchment. Characterizing the error in radar rainfall in terms of mean areal precipitation at various scales is difficult given the lack of numerous precipitation observation stations needed to establish a "ground truth". Therefore, uncertainty in precipitation input was defined in two ways. The first assumes no knowledge of the rainfall error structure, and the degraded sub-catchment MAP has a simple additive noise:

$$P_e = P_o \times (1 + \alpha) \tag{2}$$

For this case, P_e is the degraded, or with uncertainty, sub-catchment MAP, P_o is the sub-catchment MAP based on observed radar precipitation, and α is random error. The value of α is selected from a uniform distribution in the range [–0.5, +0.5]. Thus the sub-catchment MAP has 50% uncertainty bounds. This uncertainty definition is termed "uniform" in the following discussion.

The second definition of precipitation uncertainty follows an exponential relationship (Krajewski & Georgakakos, 1985):

$$P_e = P_o \times 10^{\varepsilon} \tag{3}$$

where, again, P_e is the degraded sub-catchment MAP, and P_o is the "observed" sub-catchment MAP. The error term, 10^{ε}, assumes knowledge of the structure of the errors. The value of ε is selected from a uniform distribution in the range [–0.2, +0.2], thus yielding a ratio of degraded MAP to observed MAP of 0.6–1.6. This uncertainty definition is term "exponential" in the following discussion.

To assess the impact of such precipitation uncertainty on flow simulations, a Monte Carlo simulation framework was employed. Random perturbations in the sub-catchment precipitation values were introduced given the uncertainty definitions above. These perturbations were applied at each sub-catchment and at each time step over selected events in the historical period. Uncertainty was introduced only for non-zero precipitation, and therefore, no precipitation error is added during dry periods. Event dates were selected within the May 1993–May 1999 time period based on the occurrence of a flow event at the basin outlet. Using the simulated flows at the basin outlet, events were selected, starting approximately two days prior to the rising of the hydrograph, and extend until the basin flow condition was reached following the peak. Thus, events ranges from seven to 15 days in length, and in some cases, included multiple peaks. Given the proximity of the basins and types of storms common in the region, there was a significant overlap of events in the basins, with a total of 25–30 events selected for each watershed. In the sensitivity runs, uncertainty in precipitation input was introduced approximately two months prior to the start of each event to allow adjustment of initial conditions for the model soil water, and continued through each event.

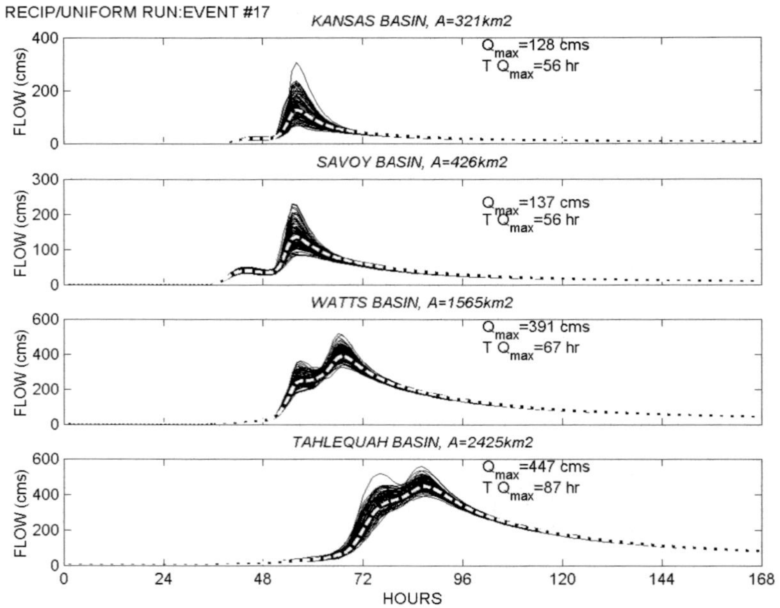

Fig. 1 Ensemble flow traces (100) generated by the Monte Carlo simulation methodology for each of four locations within the Illinois River watershed with outlet at Tahlequah and for the event of 19–25 February 1997. This is for uniform radar rainfall uncertainty. The nominal simulation is shown with the dashed white lines. $Q_{\max}$ and $T_{Q\max}$ denote the peak flow magnitude and timing for the nominal simulations.

For each watershed and each event, a total of 100 Monte Carlo simulations were performed. An example of the Monte Carlo output is illustrated in Fig. 1, showing 100 flow traces for the Illinois River at Tahlequah location, along with three interior watershed locations. For each of the five study watersheds listed in Table 1, similar graphics were produced for the outlet location and at least one other interior watershed location. The results were then summarized in terms of a measure of the variability in the ensemble of simulated flow traces. This measure, termed R_C, is defined as the difference between the 90th and 10th percentile cumulative flow, normalized by the median cumulative flow:

$$R_C = \frac{R_{C90} - R_{C10}}{R_{C50}} \tag{4}$$

The measure was computed for each time step over the selected events, and the maximum value was reported for each event and each uncertainty case for selected locations. The variability in R_C values among events is substantial, ranging from nearly 0.0 to 0.6 for individual locations. However, a tendency for larger R_C values to occur for smaller drainage areas emerged. The trend was observed often, but not for every event or for each basin. Deviations from this trend occurred more frequently for the Blue River basin.

An average value of R_C was computed over all events for each location. These average R_C values are plotted against drainage area for each uncertainty case (uniform

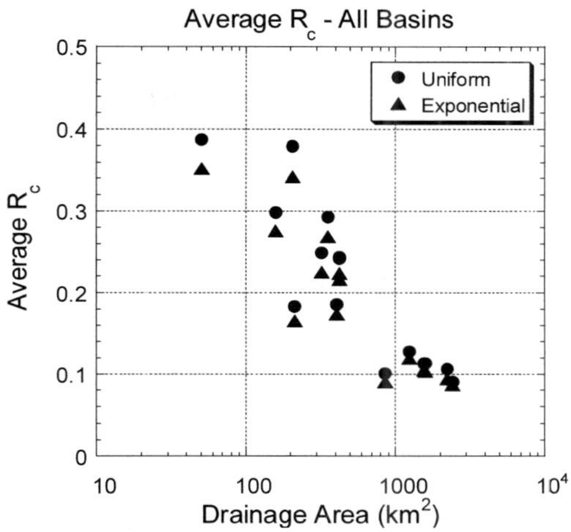

Fig. 2 Average flow ensemble range, R_C, as a function of drainage area for all watershed locations. ●: uniform mean areal rainfall uncertainty, ▲: exponential mean areal rainfall uncertainty.

and exponential) in Fig. 2. The plot shows that this trend of decreasing sensitivity with drainage area holds on average over all events and over all study watersheds. The plot also indicates that on average, the uniform uncertainty case produces higher sensitivity in ensemble range than the exponential uncertainty case.

CONCLUSIONS AND FUTURE WORK

This paper presents results of a sensitivity analysis of a catchment-based distributed hydrological model to uncertainty in rainfall input. The hydrological model allows for both distributed precipitation input and distributed hydrological model parameters, and for this work, the model is applied to five watersheds in the central United States. With calibration, and assuming no uncertainty in the input, the model is able to reproduce the historical observed flows quite well.

The sensitivity analysis was performed in a Monte Carlo framework, for selected events in each watershed. Two definitions of the rainfall uncertainty were used: one assuming no knowledge on the structure of precipitation errors, and the second, assuming the error structure has an exponential form. For each study watershed and each event, a total of 100 Monte Carlo simulations were performed for each uncertainty distribution. The sensitivity results were summarized in terms of an average measure of the flow ensemble range for several watershed locations. These average measures reveal a decreasing trend in normalized flow ensemble range as drainage area increases. This new finding is indicative of the ability of larger watersheds to act as low pass filters to precipitation input uncertainty, and it complements the already well known finding that large watersheds act as low pass filters on high-frequency temporal rainfall variability. This finding is consistent with that of earlier work by the authors (Carpenter *et al.*, 2001), which examined fewer locations and events.

An on-going research effort refines the uncertainty models by introducing rainfall uncertainty at the radar-pixel scale, as opposed to at the sub-basin MAP scale as presented in this work. Effort has begun to examine the relationship of rainfall uncertainty introduced at each radar pixel and the resulting MAP uncertainty for varying sized sub-basins and for various spatio-temporal dependence assumptions of the pixel-scale radar rainfall error for these case study watersheds. Application of the methodology presented in this work to other geographic locations and possibly other distributed hydrological model structures is warranted and should be attempted.

Acknowledgements The research reported herein has been supported by a cooperative agreement between HRC and the Hydrology Laboratory (HL) of the US National Weather Service, National Oceanic and Atmospheric Administration. The collaboration and comments of Drs Michael Smith, Victor Koren, and Seann Reed of HL have been valuable throughout this research effort. The watershed locations presented in this work were selected based on HRC's participation in the HL-organized Distributed Model Intercomparison Project (DMIP).

REFERENCES

Carpenter, T. M., Georgakakos, K. P. & Sperfslage, J. A. (2001) On the parametric and NEXRAD-radar sensitivities of a distributed hydrological model suitable for operational use. *J. Hydrol.* **254**, 169–193.

Krajewski, W. F. & Georgakakos, K. P. (1985) Synthesis of radar rainfall data. *Water Resour. Res.* **21**, 764–768.

NRCS (Natural Resources Conservation Service) (1994) State Soil Geographic (STATSGO) database: data use information. Publ. 1492, US Dept of Agriculture, Fort Worth, Texas, USA.

Ogden, F. L., Garbrecht, J., DeBarry, P. A. & Johnson, L. E. (2001) GIS and distributed watershed models, II. Modules, interfaces and models. *J. Hydrol. Engng* **6**(6), 515–523.

Distributed hydrological modelling in the Yata watershed using the WEP model and propagation of rainfall estimation error

YANGWEN JIA

Department of Water Resources, Institute of Water Resources & Hydropower Research, Che-Gong-Zhuang West Road 20, Beijing 100044, China

jiayw@iwhr.com

TSUYOSHI KINOUCHI & JUNICHI YOSHITANI

Hydrologic Engineering Team, Hydraulic Engineering Research Group, Public Works Research Institute, Minamihara 1-6, Tsukuba, Ibaraki 305-8516, Japan

Abstract The WEP model is validated in the Yata watershed (166 km^2) using the observation data in 2000 and 2001. The simulation is carried out in a time step of 1 h and a grid cell size of 100 m. The impact of rainfall estimation error on annual water budgets was studied by carrying out a sensitivity analysis and the impact of rainfall spatial distribution on discharge prediction was investigated by using gauged rainfall data and radar rainfall data. It was found that the spatial distribution of rainfall during typhoon seasons has a great impact on river discharge prediction. It is concluded that the WEP model is applicable to both short-term prediction of floods and long-term prediction of water budgets with satisfied accuracy.

Key words groundwater level; hydrological model; rainfall estimation error; radar; urbanization; water balance; WEP; Yata watershed

INTRODUCTION

Distributed hydrological modelling is required for hydrological predictions in watersheds with heterogeneous land surfaces, soils and aquifers under uneven distribution conditions of rainfall. A robust distributed hydrological model should be applicable to both long-term prediction of water budgets and short-term prediction of floods with satisfactory accuracy. However, long-term simulations using distributed hydrological models are very few in the literature.

The Yata watershed is located in the northwest of Ibaraki prefecture, Japan, and has an area of 166 km^2, of which agricultural land use and forest account for about 70%. Several institutions carried out field investigations of groundwater and river discharges in the watershed in the 1970s (Ichikawa *et al.*, 1976). The Public Works Research Institute (PWRI) and the local government in Japan began to monitor water and heat fluxes, river discharges and groundwater levels in 1999. However, there is no continuous long-term observation in the watershed; therefore, a hydrological simulation is required to understand the continuous variations in water budgets, river discharges and groundwater levels.

The WEP (Water and Energy transfer Processes) model is based on that of Jia & Tamai (1998a,b) and was improved at the Public Works Research Institute (Jia *et al.*, 2001). Although the WEP model simulates most of the hydrological processes using methods similar to the physically-based spatially distributed (PBSD) models like MIKE SHE (Refsgaard & Storm, 1995), its computation time is decreased through applying the Green-Ampt model to infiltration simulation during heavy rainfall and conducting multi-layered modelling of aquifers.

The purposes of this study are to validate the WEP model in the Yata watershed, and to evaluate the impact of rainfall estimation error and rainfall spatial distribution on river discharge prediction. A continuous long-term simulation of 24 years (1978–2001) is under study.

MODEL DESCRIPTION

Model structure

The vertical structure of the WEP model within a grid cell is shown in Fig. 1. The state variables include depression storage on land surfaces and canopies, soil moisture content, land surface temperature, groundwater level, and water stage in rivers, etc. A brief description of modelling approaches of hydrological processes follows with equations and modelling approaches of energy transfer processes referred to in Jia & Tamai (1998a) and Jia *et al.* (2001).

Modelling approaches of hydrological processes

Evaporation is calculated with the Penman equation and transpiration is calculated by using the Penman-Monteith equation. The average evapotranspiration in a grid cell is obtained by areally averaging those from each land use.

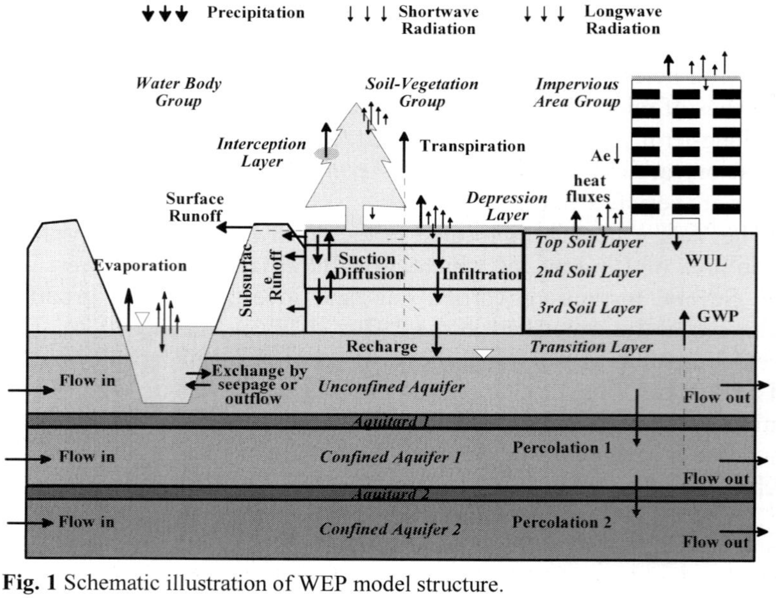

Fig. 1 Schematic illustration of WEP model structure.

Infiltration during heavy rains is calculated utilizing the generalized Green-Ampt model for infiltration into multi-layered soil profiles suggested by Jia & Tamai (1998b), whereas soil moisture movement in unsaturated soils during other periods is solved using the Richards model. A heavy rainfall period is defined as a period during which the rainfall intensity is greater than the saturated soil hydraulic conductivity.

Surface runoff from the soil-vegetation group consists of two parts, namely the infiltration excess during heavy rainfall periods and the saturation excess during the other periods. Infiltration excess occurs when the depression storage on the land surface surpasses its maximum value. The depression storage is balanced with rainfall as inflow and infiltration, evaporation and infiltration excess as outflows. Saturation excess during the remaining periods may occur if the groundwater level in the unconfined aquifer rises and the topsoil layer becomes nearly saturated. It is deduced by applying the Richards model. Subsurface runoff is calculated according to land slopes and unsaturated soil hydraulic conductivities in those grid cells adjacent to rivers.

Groundwater flow in multi-layered aquifers is simulated using the Boussinesq equation and the interactions between surface water and groundwater of the unconfined aquifer are considered through a source term. The source term includes the recharge from unsaturated soil layers, the groundwater outflow to rivers, the water use leakage, the pumped groundwater, the percolation to the lower aquifer, and the evapotranspiration from groundwater. Groundwater outflow is calculated according to the hydraulic conductivity of riverbed material and the difference between river water stage and groundwater head in the unconfined aquifer.

Treated as overland flow, runoff from a grid cell is routed along the steepest (down) of eight directions to its adjacent cells using the kinematic wave method in the 1-D scheme and the downhill Newton method. The computation sequence is determined from the lowest flow accumulation number (headwater area) to the largest flow accumulation number (river mouth). River flow is routed for every tributary and a main channel using the kinematic wave method or the dynamic wave method in the 1-D scheme and the Newton-Raphson method.

MODEL VALIDATION

The Yata watershed, input data and parameters

A diagram of the Yata watershed is shown in Fig. 2. There are three main rivers in the watershed, which flow into the Ushiku Lake. Inside or around the watershed, there are two meteorological gauge stations managed by the local meteorological bureau and one by PWRI. In October 2000, three continuous water stage gaugeing stations were installed and periodical discharge observations were carried out to obtain the relationship between water stage and discharge. Hourly meteorological data from January 1978 to December 2001 were collected.

The land-use data four periods are shown in Table 1. The data of 1994 show that 14% of the watershed is paddy field, other farmland and forest occupy about 50%, and the rest is covered by buildings, rivers, etc.

For aquifer data, borehole geology data at 119 sites within or near the watershed were picked up to clarify the soil stratigraphy of the watershed. The ground soil is

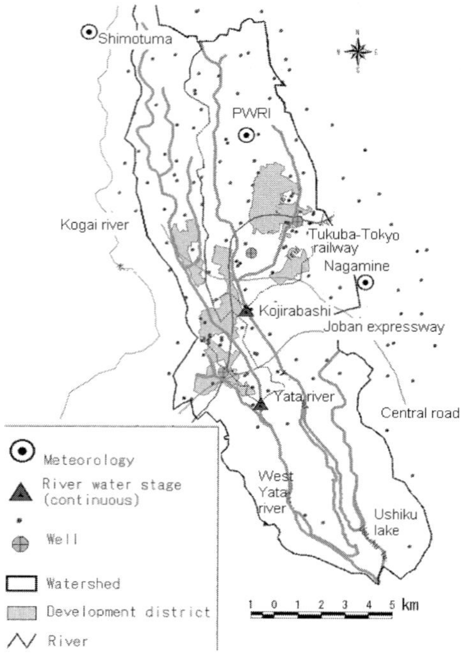

Fig. 2 The Yata watershed.

Table 1 Land use in the Yata watershed (km^2).

Types	1976	1984	1989	1994
Forest	36.13	29.00	27.06	26.63
Paddy	24.32	24.62	24.55	23.66
Field	65.56	56.96	57.23	55.39
Under preparation	0.00	0.38	0.30	0.30
Vacant land	19.55	4.23	2.88	3.45
Industrial use	0.00	2.48	2.87	3.01
Housing area	12.45	20.44	21.40	21.71
Road	0.78	5.52	5.87	6.86
Park	0.00	1.00	1.40	1.39
Public works	0.00	12.22	12.60	12.69
River/Lake	7.82	6.66	6.69	6.81
Sum Total	166.61	166.61	166.61	166.61
Impervious ratio	0.079	0.157	0.165	0.177

divided into five layers, i.e. three aquifers and two aquitards to separate each aquifer from another. Hydrogeological parameters of each aquifer and aquitard, i.e. hydraulic conductivity, specific yield, were set up referring to the results of *in situ* pumping tests (Geological Survey of Japan, 1988). The topsoil is mainly Kanto loam, except for alluvial silt in riparian zones. Experiments were carried out to obtain the characteristic curve of soil moisture and suction.

In addition, ground elevation and population were obtained from available databases; domestic water use, agricultural water use (mainly pumped from rivers outside the watershed) and industrial water use were deduced from the reports of the local government and organizations; and data of river sections were also collected.

Model validation

To validate the WEP model in the watershed, simulation was first performed for the period from January 1998 to December 2001. The initial and boundary conditions were determined according to the field observation results and "warming up" simulation of one year (1998) was first carried out to eliminate the possible effect of initial condition setting. The results after January 1999 were used for model validation.

Figure 3 shows a comparison of simulated discharge with observed discharge at the Kojirabashi station (see Fig. 2). The Kojirabashi station, with a command area of 47.6 km^2, is located at the middle reach of the Yata River. Figure 4 shows a comparison between simulated and observed groundwater levels for all the simultaneous observation wells within the watershed. Although there are some differences between the simulated and the observed results, all the comparisons show that the model gives quite satisfactory results.

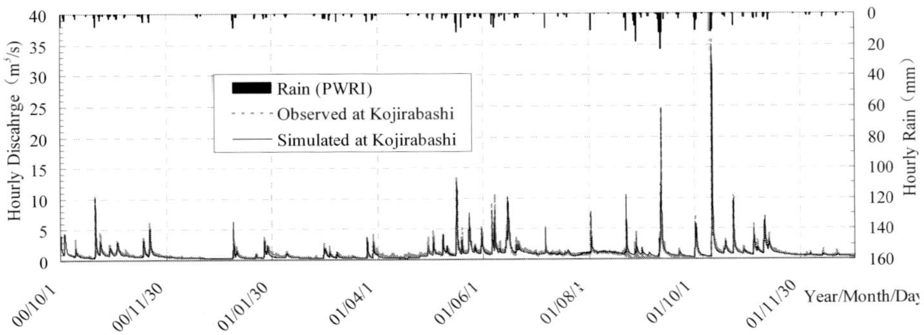

Fig. 3 Model validation using observed and simulated river discharges at the Kojirabashi.

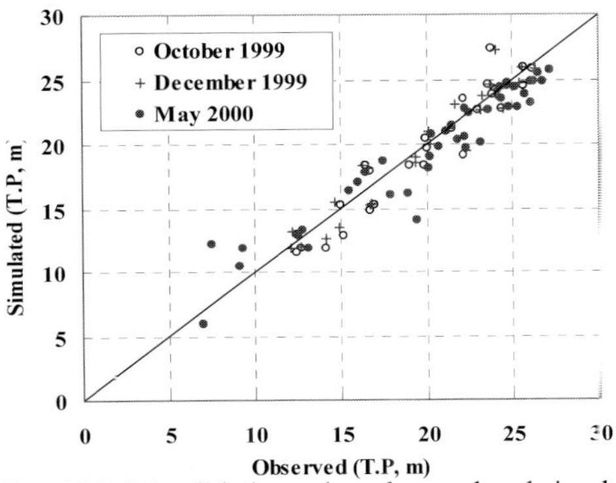

Fig. 4 Model validation using observed and simulated groundwater levels. T.P. means elevation above the averaged seawater level in the Tokyo Bay.

Influence of parameter estimation errors

The main parameters of the WEP model include the saturated hydraulic conductivity of soil layers, the permeability and specific yield of the unconfined aquifer, the hydraulic conductivity and thickness of riverbed material, the Manning's roughness of river flow and overland flow, vegetation parameters, and paddy field parameters. Some of the parameter sensitivity study of the WEP model has been performed in previous studies. For example, it was found that annual water budgets are quite sensitive to the saturated hydraulic conductivity, the annual groundwater outflow is very sensitive to the hydraulic conductivity of riverbed material, and the groundwater level is sensitively influenced by the aquifer permeability. In this study, we analysed the influence of the Manning's roughness on the river discharge hydrograph at the Kojirabashi station. It is found that the increase of the Manning's roughness of overland flow obviously decreases the peak discharge and delays the occurrence of the peak discharge, whereas the influence of the Manning's roughness of river flow is not so obvious in the Yata watershed. It is believed that the concentration time of overland flow is much longer than that of river flow in this watershed.

PROPAGATION OF RAINFALL ESTIMATION ERROR

Sensitivity of annual water budgets to rainfall input data

Annual water budget sensitivity is investigated by increasing/decreasing the hourly rainfall in 2000, which is equivalent to the systematic errors that occur in rainfall observation data. The result is shown in Fig. 5. The relative error means the estimation error of water budget divided by the budget corresponding to the precipitation in 2000. It can be seen that estimation errors of runoff and groundwater outflow are greater than the rainfall estimation error, though those of evapotranspiration and infiltration are smaller than the rainfall estimation error. This is because evapotranspiration is also influenced by energy budgets and infiltration is also affected by soil conductivity and groundwater level.

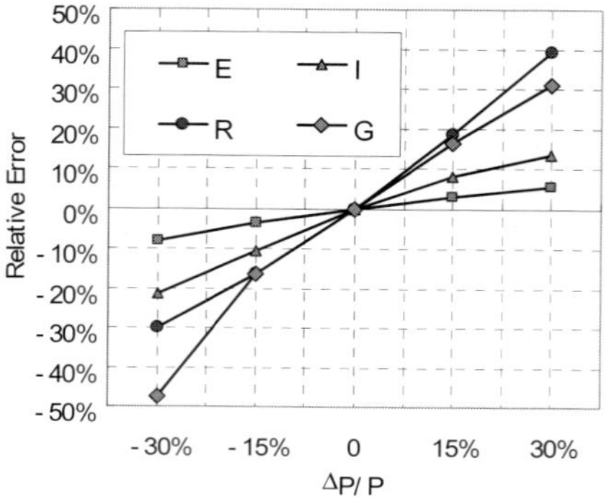

Fig. 5 Sensitivity of annual water budgets to rainfall input data. *E* is evapotranspiration, *I*, infiltration; *R*, direct runoff; *G*, groundwater outflow to rivers; and *P*, precipitation.

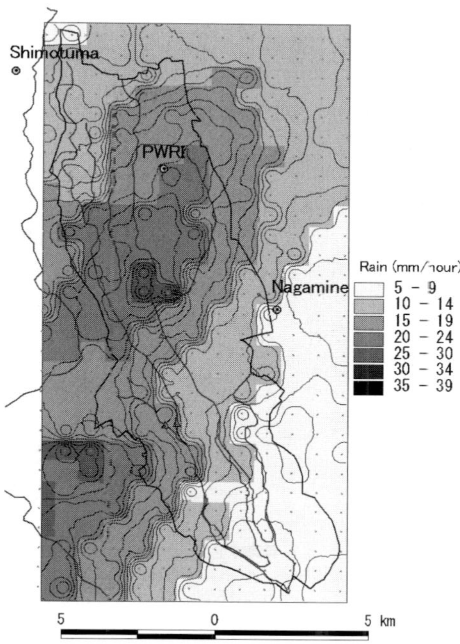

Fig. 6 Distribution of radar rainfall from 20:00 to 21:00 h on 10 October 2002.

Impact of rainfall distribution on river discharges

To study the impact of rainfall distribution on river discharges, hourly radar rainfall data from August to October 2001 were obtained from the Ministry of Land, Infrastructure and Transport (MLIT), Japan. The radar rainfall data have a resolution of about 1 km. During typhoon seasons, rainfall distribution is very uneven. Figure 6 shows an example of distribution of radar rainfall from 20:00 to 21:00 h on 10 October 2002 in the watershed.

Simulation results using the radar rainfall data are compared with those using the Thiessen method with two gauging stations (Shimotuma and Nagamine), and with those using three gauging stations (Shimotuma, Nagamine and PWRI). Figure 7 shows the comparisons of three rainfall events at the Kojirabashi station in 2001: (a) from 21 August 00:00 h to 23 August 24:00 h; (b) from 10 September 00:00 h to 12 September 24:00 h; and (c) from 10 October 00:00 h to 12 October 24:00 h. It can be seen that, in general, the results using radar rainfall data and those using three gauging stations are better matched with the observed than those using two gauging stations. The results using two gauging stations give a simulation error as big as 152% for the flood peak on 22 August, 4% for the flood peak on 11 September and 40% for the flood peak on 10 October. The reason the results using radar rainfall data do not show better matching with the observed data is that gauging stations downstream of and outside the watershed, rather than the three gauging stations mentioned above, were utilized by MLIT for the radar rainfall calibration against ground-gauged rainfall. The averaged discharges from 1 August to 31 October and relative simulation errors are shown in

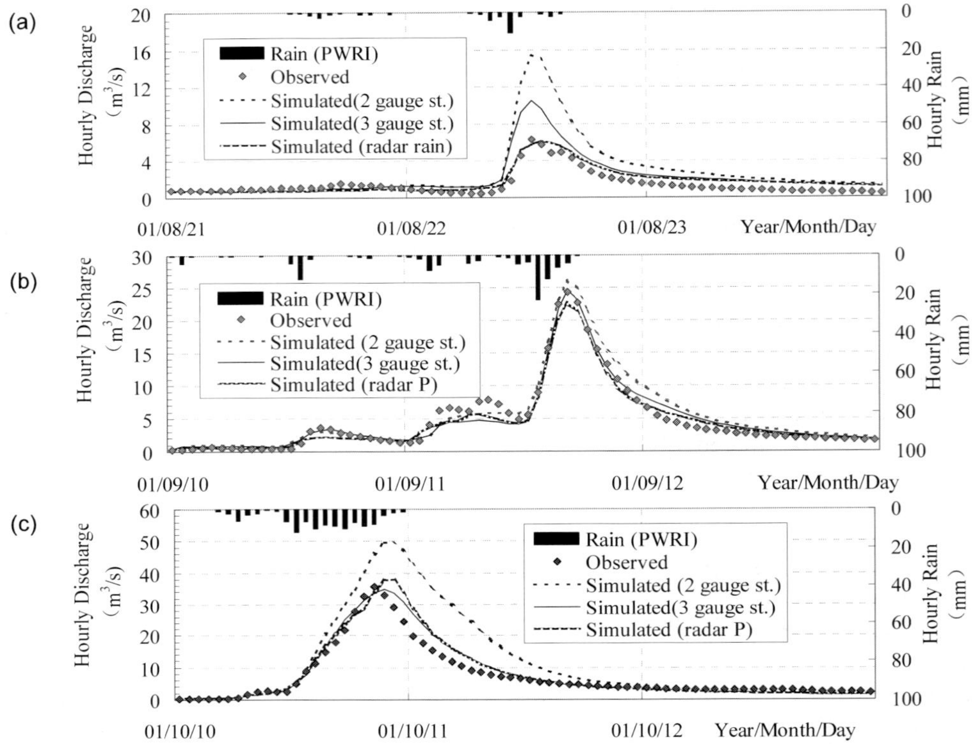

Fig. 7 Discharge comparisons during three rainfall events at the Kojirabashi in 2001: (a) from 21 August 00:00 h to 23 August 24:00 h; (b) from 10 September 00:00 h to 12 September 24:00 h; and (c) from 10 October 00:00 h to 12 October 24:00 h.

Table 2,from which the same remarks as above can be concluded but the results using two gauge stations do not show too big errors. It means that rainfall distributions may have less impact on the estimation of long-term water budgets.

It should be mentioned that the estimation errors of radar rainfall also influence river discharges. The radar rainfall errors can come from both radar observation and calibration based on gauge stations and they include the systematic error and the

Table 2 Averaged discharges from 1 August to 31 October 2001, at three sites.

Site	Observed ($m^3 s^{-1}$)	Simulated using two gauge st. rainfall ($m^3 s^{-1}$)	Simulated using three gauge st. rainfall ($m^3 s^{-1}$)	Simulated using radar rainfall ($m^3 s^{-1}$)
Kojirabashi	1.332	1.698	1.499	1.547
		(27.5%)	(12.5%)	(16.5%)
Sakaimatu	1.125	1.302	1.161	1.226
		(19.9%)	(3.2%)	(9.0%)
Takaoka	0.196	0.241	0.244	0.21
		(23.0%)	(24.5%)	(7.1%)

Note: Values in brackets are relative errors compared with the observed values.

random error. The influence of radar rainfall errors on river discharges is closely related to the location of storm centre and the direction in which the storm comes to the watershed, and it will be carried out in the study of the next step.

CONCLUSIONS

In this study, through analysing the sensitivity of annual water budgets to rainfall estimation error and investigating the impact of rainfall spatial distribution on river discharges, it was found that the spatial distribution of rainfall during typhoon seasons has a great impact on river discharge prediction. However, the impact of rainfall distribution on seasonally averaged discharges becomes smaller than that on flood discharges. It is concluded that the WEP model is applicable to both short-term prediction of flood and long-term prediction of water budgets with satisfactory accuracy and distributed hydrological modelling needs distributed rainfall data to give accurate predictions.

Acknowledgements Professor Yoshihisa Kawahara, Kagawa University, Japan, Dr Guangheng Ni Nippon, Koei Co. Ltd, Japan and Mr Koei Tojima, Docon Co. Ltd, Japan, are gratefully acknowledged for their work in the WEP Model Development and Application project. Sincere thanks are also given to Mr Katuyuki Koike of the Foundation of River and Basin Integrated Communications, Japan, and to Mr Tadashi Matuura in the Ministry of Land, Infrastructure and Transport, Japan, for providing information on radar rainfall.

REFERENCES

Geological Survey of Japan (1988) Environmental geological map of the Tsukuba Science City and its surroundings (in Japanese), 23-2. Tsukuba, Japan.

Ichikawa, M., Takayama, S., Tanaka, T., Taguchi, Y. & Sakura, Y. (1976) Study on the effects of urbanization on hydrological cycle (in Japanese). *Research on Tsukuba* **1**, 10–21.

Jia, Y. & Tamai, N. (1998a) Integrated analysis of water and heat balance in Tokyo metropolis with a distributed model. *J. Japan Soc. Hydrol. & Water Resour.* **11**(1), 150–163.

Jia, Y. & Tamai, N. (1998b) Modelling infiltration into a multi-layered soil during an unsteady rain. *J. Hydrosci. Hydraul. Engng JSCE* **16**(2), 1–10.

Jia, Y., Ni, G., Kawahara, Y. & Suetsugi, T. (2001) Development of WEP model and its application to an urban watershed. *Hydrol. Processes* **15**, 2175–2194.

Refsgaard, J. C. & Storm, B. (1995) Chapter 23 MIKE SHE. In: *Computer Models of Watershed Hydrology* (ed. by V. J. Singh), 809–846. Water Resources Publ., Littleton, Colorado, USA.

Weather Radar Information and Distributed Hydrological Modelling (Proceedings of symposium HS03 held during IUGG2003 at Sapporo, July 2003). IAHS Publ. no. 282, 2003.

The effects of radar-derived rainfall uncertainties on forecasts from a distributed hydrological model

JONATHAN J. GOURLEY & BAXTER E. VIEUX
School of Civil Engineering and Environmental Science, University of Oklahoma, Norman, Oklahoma 73019, USA
jonathan.gourley@nssl.noaa.gov

Abstract The advent of weather radar has provided the potential to estimate rainfall accurately at high spatial and temporal resolutions. When these estimates are input to a distributed hydrological model, forecasts of streamflow may be used to anticipate, and thus mitigate, the potential hazards associated with a flash flood. In hydrological modelling, forecast uncertainty has traditionally been a function of the uncertainty in the model parameters, and in some cases the model structure. The study presented here uses the physics-based, distributed *r.water.fea* hydrological model with modifications to address the impact of uncertainties in the input rainfall estimates on streamflow predictions using an extension of the Generalized Likelihood Uncertainty Estimation methodology. The ensemble modelling approach allows us to evaluate the accuracy of different rainfall algorithms independently at the scale of an integrating watershed. The study plan and some initial results are presented.

Key words distributed hydrological model; ensemble hydrological prediction; parameter estimation; radar quantitative precipitation estimation; rainfall uncertainty

INTRODUCTION

It is well known that the accuracy of streamflow predictions from a hydrological model is heavily dependent on the accuracy of the rainfall inputs. Several efforts are underway (e.g. Gourley *et al.*, 2002) to improve quantitative precipitation estimation (QPE) by understanding the situations in which radar estimates can be erroneous and utilizing data from multiple sensors (e.g. infrared satellite, raingauges, numerical weather model output, and lightning flashes). As QPE algorithms are being formulated from emerging radar technologies such as polarization diversity, it is vital to the developers to know the error characteristics associated with the estimates. Traditionally, this has been done by comparing the remotely-sensed QPEs to raingauges at collocated grid points. In addition to the measurement errors associated with raingauges, it has been noted that the sampling sizes between a typical radar pixel and a raingauge orifice differ by about eight orders of magnitude (Droegemeier *et al.*, 2000). A methodology is proposed herein that provides the framework to evaluate potential QPE improvements at the scale of application, a watershed. In addition, the probabilistic approach enables uncertainty estimates or confidence intervals to be assigned to the predicted hydrological variables.

BACKGROUND

A model calibration methodology has been devised that is based upon the premise that multiple combinations of parameter values may yield acceptable simulations of the system, and thus the likelihood or probability of a given set of parameters producing characteristic system behaviour is desired. The generalized likelihood uncertainty estimation (GLUE) method (Beven & Binley, 1992) has been shown to provide an assessment of the uncertainty in hydrological predictions resulting from uncertainty in the model parameters. This model-ensemble approach provides the framework to reduce the number of parameter sets by rejecting those that yield behaviour that are not characteristic of the physical system. In this sense, it is used for model calibration. It is recommended in Beven & Binley (1992) and in Freer *et al.* (1996) that the different parameter sets should be allowed to interact with different model structures and different precipitation inputs. Presently, there is a lack of information regarding the error characteristics associated with radar QPEs in relation to hydrological prediction accuracy. As pointed out in Droegemeier *et al.* (2000), hydrological forecast uncertainty cannot be reasonably assessed until the uncertainty in the rainfall observations has been determined *a priori*. Entekahbi *et al.* (2002) identify the precipitation inputs as one of the major limitations to improved hydrological predictability. The study reported herein focuses on the influence of uncertainty in radar-derived QPEs on hydrological prediction accuracy of streamflow. In doing so, a methodology is devised to provide a consistent framework to compare the relative accuracy of different precipitation inputs at the scale of an integrating watershed.

DATA AND STUDY AREA

The study undertaken utilizes radar-derived precipitation estimates from a rainfall event that resulted in significant flow in the Blue River near Blue, Oklahoma, USA, beginning on 12 November 1994. Hourly rainfall estimates were obtained from the National Weather Service—Arkansas Basin River Forecast Center for a 12-day period. The rainfall estimates are derived from radar reflectivity data (using $Z = 300R^{1.4}$) and applied as a mosaic onto a $4 \times 4\ \text{km}^2$ grid. A domain-wide mean field bias using raingauge data is applied to the estimates on an hourly basis. For the purpose of this paper, a single event is examined in order to highlight the proposed methodology of evaluating the effects of differing precipitation inputs on hydrological simulations. Future research will extend the methodology described herein to additional types of multisensor quantitative precipitation estimates, other basins, and a series of rainfall events. The goal of this is to be able to use a model of streamflow at the watershed scale to verify precipitation estimation techniques.

The hydrological simulations are performed using a modified version of the University of Oklahoma's *r.water.fea* model (Vieux & Gauer, 1994; Vieux, 2001). The model is a distributed parameter, deterministic model that formulates runoff generation and routing using conservation equations on an event basis. The parameters used in the model are distributed in space and are derived from ancillary data that are then adjusted. The Ordered Physics-based Parameter Adjustment (OPPA) calibration method described by Vieux & Moreda (2002) employs scalars to adjust parameter maps. A majority of the ancillary data are freely available for most basins in the US in a geographical information system (GIS) format.

Observations of streamflow are needed for the evaluation component of this study. Hourly measurements of discharge at the outlet of the Blue River basin were obtained from the US Geological Survey (USGS) for the 12-day simulation period. The Blue River basin drains about 1200 km^2. The headwaters of the basin are about 80 km away from the nearest weather radar (KTLX located near Oklahoma City), while the basin outlet is over 200 km away.

METHODOLOGY

The ensemble methodology discussed herein inputs precipitation estimates to a distributed hydrological model to do the following:

(a) evaluate the relative performance of different precipitation inputs based on the observed and simulated basin hydrological response and
(b) assess the uncertainty caused by model parameterization by computing 90% simulation bounds for each input separately.

While the method is equally applicable to different sources of QPE from radar, satellite, raingauge, or combinations, the method is applied here to radar along with a scalar multiplier characteristic of a mean-field bias. The probabilistic approach yields an ensemble of results by examining the entire parameter space, thus avoiding the need to recalibrate the model for each input.

The ensemble procedure can accommodate parameters or inputs that have known ranges and/or distributions. If no information is known *a priori* about the parameter distributions, then a uniform distribution can and will be assumed. For application on the Blue River basin, parameter values for saturated hydraulic conductivity, Manning's roughness coefficient and the degree of initial soil saturation are distributed spatially. Thus, in order to allow the parameters to vary, a multiplier (the same type of scalar used in OPPA) is applied to each of the distributed parameter maps, adjusting the magnitude while preserving the spatial variation. The multipliers for saturated hydraulic conductivity and Manning's roughness coefficient vary from 50%, 75%, 100%, 133%, to 200% of their calibrated values. The initial degree of soil saturation varies from 30% (dry) to 70% (moist) in increments of 10%. The radar-derived precipitation estimates are also perturbed from 50%, 75%, 100%, 133%, to 200% of their given, deterministic values. For purposes of demonstrating the proposed evaluation methodology, the precipitation inputs are treated as if they are from different precipitation algorithms after the multipliers have been applied. Several different inputs that are truly independent could be utilized here instead of applying a bias factor to one algorithm as is done in this example. A number of hydrological simulations are then performed by uniform random sampling across the entire parameter space.

At this point, $5 \times 5 \times 5$ (= 125) hydrological simulations have been performed using all unique combinations parameter values within their specified ranges for each of the five different precipitation inputs. The next step requires the computation of observed hydrological variables. We have chosen to analyse the total volume of water discharged at the basin outlet as objective functions expressed as depth over the watershed, the time at which the maximum discharge was realized, and the peak discharge throughout the 12-day event. The likelihood measure chosen to determine the accuracy of each simulation is the difference between the simulated and observed variables mentioned above. The errors are computed for each simulation performed and are normalized such that the sum of all errors for a given variable equals one. This normalization satisfies the conditions required for a discrete probability distribution. Histograms may now be plotted for the different precipitation inputs perturbed by a scalar multiplier. Future tests will involve precipitation inputs independently derived using various multisensor sources. For the purpose of this demonstration, the precipitation input is perturbed using scalar multipliers from 50% to 200% of the initial value. As shown in the next section, the computed error histograms reveal the impact of altering the precipitation inputs on all three objective variables. Moreover, the probability densities permit the calculation of 90% simulation bounds and thus provide a measure of uncertainty.

RESULTS

The histograms for the simulated errors in the hydrological variables for the five different precipitation inputs are shown in Figs 1–3. The observed basin-averaged flow depth throughout the event was 34.0 mm, the maximum discharge was observed to be 206.7 $m^3 s^{-1}$, and the time of peak flow occurred at 5520 min. The histograms of the error in simulated basin-averaged flow depth and maximum discharge shown in Figs 1 and 2 indicate the expected result that increasing the magnitude of the rainfall inputs results in over-predictions of the two simulated outputs. Nonlinearity of the runoff response is evident. More importantly, the approach demonstrated is able to identify an optimal QPE calibration factor relying on streamgauge observations. The most probable rainfall multiplier that minimizes volume error is 0.75. Because the NWS has

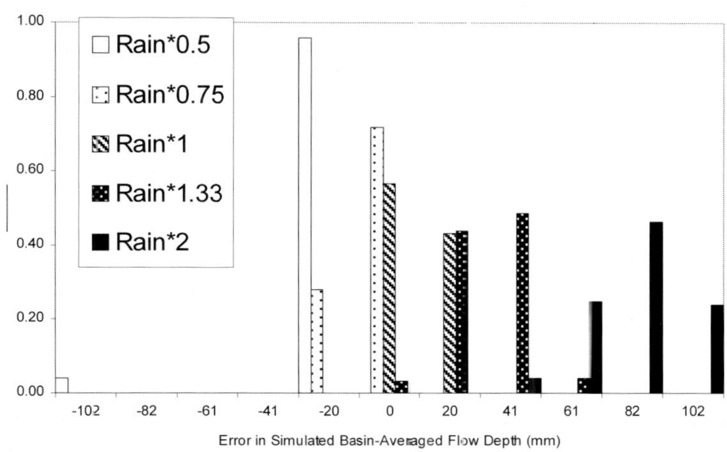

Fig. 1 Relative frequency histogram of the error between simulated minus observed basin-averaged flow depth (in mm). The patterned bars correspond to the different rainfall inputs varied from 50% (rain*0.5) to 200% (rain*2) of their default values. Observed basin-averaged flow depth was 34.0 mm.

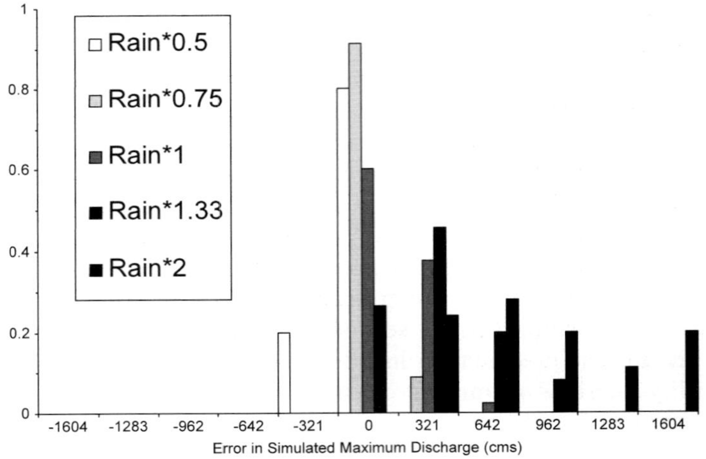

Fig. 2 Error computed as simulated minus observed maximum discharge ($m^3\ s^{-1}$). Observed maximum discharge was 206.7 $m^3\ s^{-1}$.

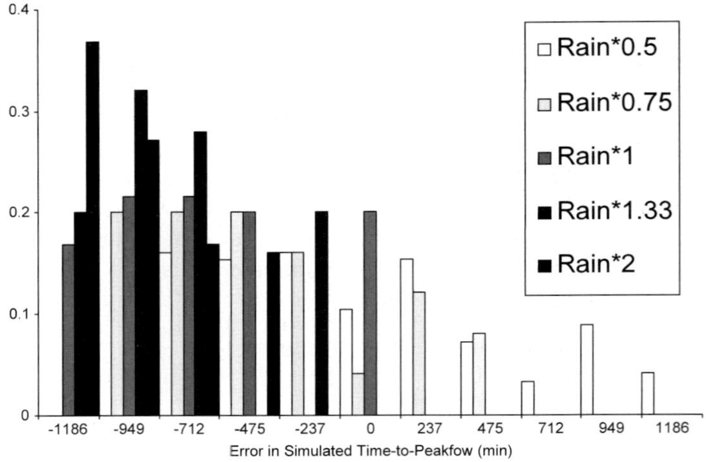

Fig. 3 Error between simulated minus observed time of maximum discharge (min). Time of maximum discharge occurred at 5520 min.

already adjusted the Stage III QPE using raingauges, the expected QPE multiplier is 1.0 rather than 0.75. One explanation is that the Stage III QPE was not tuned locally or specific to the Blue basin. Further work is needed for identification of the statistical significance of the different probability distributions, and separation of the error associated with model parameters from QPE input errors.

The error histogram for the time at which the maximum discharge was realized is shown in Fig. 3. In this case, increasing the rainfall amounts from their unperturbed values causes errors in the time-to-peak in the opposite direction relative to the total volume of flow and maximum discharge (Figs 1 and 2). Physically, soil saturation occurs earlier in the simulation and results in runoff arriving at the basin outlet earlier than observations suggest. In this case, a majority of the simulations are shifted to the left of the bin containing zero error regardless of the scalar used to adjust the rainfall inputs. This is an informative result to the hydrological modeller because it indicates that the model has a propensity to simulate the maximum discharge earlier than the observed time-to-peak of 5520 min. The proposed methodology has thus pointed out a possible structural error in the model due to a mis-specification of the channel cross-sectional representation or hydraulics. Future studies are planned to identify and resolve the causes for this early forecast, or it will need to be considered in the ensemble in addition to the uncertainty in the initial conditions and model parameters.

Results are summarized in terms of the 90% error bounds of the three objective functions for each of the rainfall scalars as shown in Figs 4–6. The estimates of

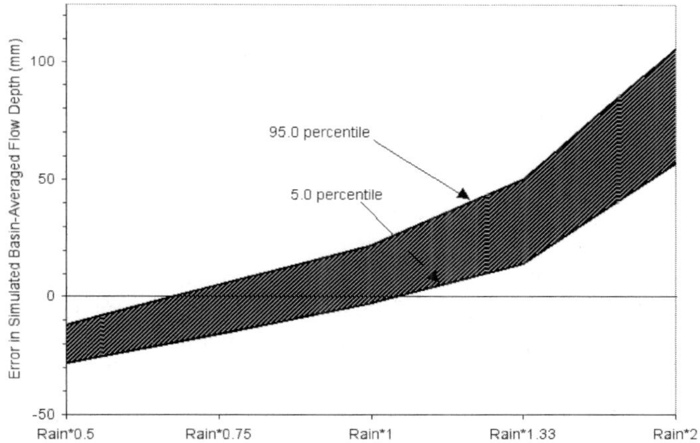

Fig. 4 Simulation bounds for the 90% error in simulated basin-averaged flow depth (mm) for the five different precipitation inputs. Hatched regions indicate simulation limits caused by uncertainty in model parameters.

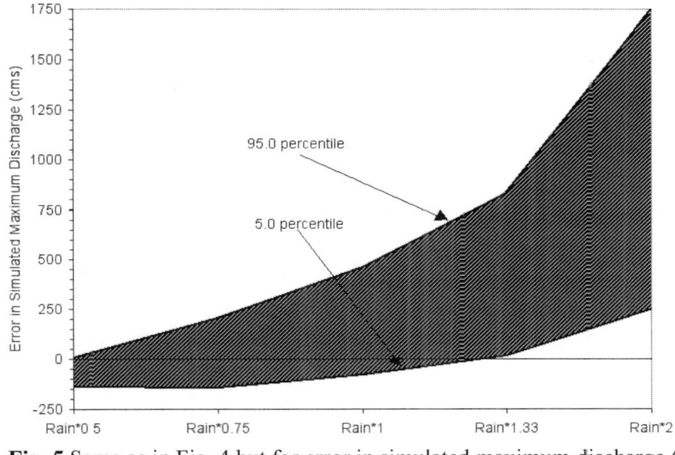

Fig. 5 Same as in Fig. 4 but for error in simulated maximum discharge ($m^3 s^{-1}$).

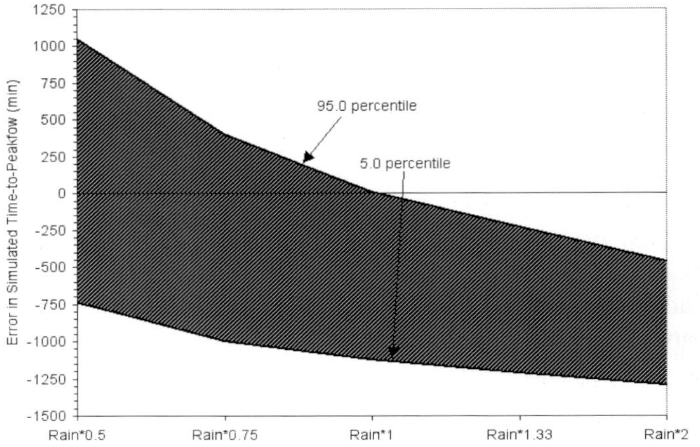

Fig. 6 Same as in Fig. 4 but for error in simulated time of maximum discharge (min).

are derived from the probability distributions shown in Figs 1–3 and should be interpreted as the 90% simulation limits caused by model parameterization alone. Volume has the narrowest range, and is relatively constant in difference between the 5- and 95-percentile indicating that scalar estimation for rainfall is best estimated using this objective function, which is consistent with expectations of a rainfall–runoff model. The imbalance in Fig. 6 between positive and negative error, with a propensity for negative (early) time-to-peak, is indicated, and is consistent with the histogram in Fig. 3. From this we can conclude that model structure or parameters representing channel hydraulics require improvement.

SUMMARY

Using streamflow and a rainfall–runoff model of a watershed, development, evaluation, and validation of new QPE algorithms can now be accomplished through comparison of simulated to observed discharge volumes. A method has been developed that can evaluate independent QPE algorithms using a hydrological model, and avoid the need to recalibrate the model to each input. The QPE validation is accomplished at the scale of a watershed that is commensurate with the goal of rainfall–runoff prediction. This method can also be used to identify the optimal bias adjustment independently from radar to gauge bias adjustment. From a statistical analysis of ensemble model simulations, sensitivity to QPE inputs and model parameters can be separated.

Acknowledgements Partial funding for this research was provided under NOAA-OU Cooperative Agreement no. NA17RJ1227. Support from the US Department of Education's Graduate Assistantship in Areas of National Need is also gratefully acknowledged.

REFERENCES

Beven, K. & Binley, A. (1992) The future of distributed models: model calibration and uncertainty prediction. *Hydrol. Processes* **6**, 279–298.

Droegemeier, K. K., Smith, J.D., Businger, S., Doswell, C., III, Doyle, J. Duffy, C., Foufoula-Georgiou, E., Graziano, T., James, L.D., Krajewski, V., LeMone, M., Lettenmaier, D., Mass, C., Pielke, R., Sr, Ray, P., Rutledge, S., Schaake, J. & Zipser, E. (2000) Hydrological aspects of weather prediction and flood warnings. Report on the ninth prospectus development team of the US Weather Research Program. *Bull. Am. Met. Soc.* **81**, 2665–2680.

Entekahbi, D., Anderson, M. P., Avissar, R., Bales, R., Hornberger, G. M., Nuttle, W. K., Parlange, M. B., Peters-Lidard, C., Potter, K. W., Roads, J. O., Wilson, J. L. & Wood, E. F. (2002) Report of a Workshop on Predictability and Limits to prediction in Hydrological Systems. Committee on Hydrological Science, National Research Council, National Academy Press, Washington, DC.

Freer, J., Beven, K. & Ambroise, B. (1996) Bayesian estimation of uncertainty in runoff prediction and the value of data: an application of the GLUE approach. *Water Resour. Res.* **32**, 2161–2173.

Gourley, J. J., Maddox, R. A, Howard, K. W. & Burgess, D. W. (2002) An exploratory multisensor technique for quantitative estimation of stratiform rainfall. *J. Hydromet.* **3**, 166–180.

Vieux, B. E. (2001) *Distributed Hydrological Modelling Using GIS*. Water and Science Technology Series, no. 38. Kluwer, Dordrecht, The Netherlands.

Vieux, B. E. & Gauer, N. (1994) Finite element modelling of storm water runoff using GRASS GIS. *Microcomp. Civ. Engng* **9**, 263–270.

Vieux, B. E. & Moreda, F. (2003) Ordered physics-based parameter adjustment of a distributed model. In: *Advances in Calibration of Watershed Models* (ed. by Q. Duan, H. V. Gupta, S. Sorooshian,, A. N. Rousseau & R. Turcotte), 267–281. American Geophysical Union, Washington, DC, USA.

Weather Radar Information and Distributed Hydrological Modelling (Proceedings of symposium HS03 held during IUGG2003 at Sapporo, July 2003). IAHS Publ. no. 282, 2003.

Real-time assimilation of radar-based precipitation data and streamflow observations into a distributed hydrological model

DONG-JUN SEO, VICTOR KOREN & NEFTALI CAJINA
Hydrology Laboratory, Office of Hydrological Development, National Weather Service, 1325 East-West Highway, Silver Spring, Maryland 20910, USA.
dongjun.seo@noaa.gov

Abstract Along with uncertainty in quantitative precipitation forecast (QPF), uncertainties in quantitative precipitation estimate (QPE) and in soil moisture states represent arguably the biggest sources of error in real-time hydrological prediction. To deal with these uncertainties, some form of state updating is employed in practice, that assimilates real-time streamflow observations. The purpose of this work is to explore real-time assimilation of radar-based QPE and streamflow observations into a distributed hydrological model via variational assimilation (VAR). The particular VAR formulation examined in this work deals with the inverse problem of estimating grid cell-specific initial model soil moisture states and spatially-uniform (but time-varying) multiplicative adjustment factors to QPE and potential evaporation (PE) based on real-time observations of streamflow, radar-based QPE and climatological estimates of PE.

Key words data assimilation; distributed hydrological modelling; radar rainfall; real-time forecasting; streamflow

INTRODUCTION

To deal with various sources of error in the initial and boundary conditions, and in model parameters and structure, some form of state updating is necessary in operational forecasting that makes use of real-time streamflow observations. Compared to state updating of lumped models that input raingauge-based precipitation estimates, that of distributed models, driven in particular by radar-based precipitation estimates, presents at least two additional difficulties: (a) the dimensionality of the estimation problem is very large; and (b) the error structure in radar-based precipitation estimates is not very simple. In theory, state updating of distributed models may be carried out using, e.g. variational assimilation (VAR) (Jazwinski, 1970), distributed-parameter Kalman filter (Tzafestas, 1978) and (distributed-parameter) ensemble Kalman filter (Evensen, 1994) techniques. In the least squares sense, they all attempt to essentially solve the same minimization problem and, if the model dynamics is linear, are essentially equivalent (Li & Navon, 2001). As such, we are not so much interested here in the particular choice of the assimilation technique as in formulating the observation models and prescribing the error statistics. These issues are common to both VAR and Kalman filtering, and are of fundamental interest in state updating of distributed models.

DESCRIPTION OF THE ASSIMILATION PROCEDURE

We cast state updating of distributed models as the following static estimation problem:

$$\boldsymbol{Z} = H(\boldsymbol{X}) + \boldsymbol{V} \tag{1}$$

where $\boldsymbol{Z}$, $\boldsymbol{X}$, and $\boldsymbol{V}$ denote the observation, state (or control) and error vectors, $\boldsymbol{Z} = [\boldsymbol{Z}_Q, \boldsymbol{Z}_P, \boldsymbol{Z}_E, \boldsymbol{Z}_B]$, $\boldsymbol{X} = [\boldsymbol{X}_0, \boldsymbol{X}_P, \boldsymbol{X}_E]$ and $\boldsymbol{V} = [\boldsymbol{V}_Q, \boldsymbol{V}_P, \boldsymbol{V}_E]$, respectively, and $H(\ )$ denotes the structure function that relates the states to the observations via the model dynamics. In the observation vector $\boldsymbol{Z}$, $\boldsymbol{Z}_Q$ denotes the vector of hourly streamflow observations in the most recent J hours (referred to herein as the assimilation window): $\boldsymbol{Z}_Q = [z_{Q,k-J}, z_{Q,k-J+1}, \ldots, z_{Q,k}]$, where $z_{Q,k}$ denotes the streamflow observation at hour k; $\boldsymbol{Z}_P$ and $\boldsymbol{Z}_E$ denote the vectors of precipitation and potential evaporation (PE) estimates in the most recent I hours (referred to herein as the control horizon): $\boldsymbol{Z}_P = [z_{P,u,k-I}, z_{P,u,k-I+1}, \ldots, z_{P,u,k}; u = 1,\ldots, n]$ and $\boldsymbol{Z}_E = [z_{E,u,k-I}, z_{E,u,k-I+1}, \ldots, z_{E,u,k}; u = 1,\ldots, n]$, respectively, where $z_{P,u,k}$ and $z_{E,u,k}$ denote the precipitation and PE, respectively, estimates at location u (i.e. the grid cell centred at u) at hour k, and n denotes the number of grid cells in the basin, and $\boldsymbol{Z}_B$ denotes the vector of *a priori* estimates of the model states, $\boldsymbol{Z}_B = [\boldsymbol{Z}_{B,s1}, \boldsymbol{Z}_{B,s2}, \ldots, \boldsymbol{Z}_{B,s6}, \boldsymbol{Z}_{B,r1}, \boldsymbol{Z}_{B,r2}, \boldsymbol{Z}_{B,r3}]$, where the subscripts s and r signify that the variables are associated with soil moisture accounting and routing models, respectively. For example, $\boldsymbol{Z}_{B,s1}$ in $\boldsymbol{Z}_B$ denotes the vector of the first soil moisture state variable, $\boldsymbol{Z}_{B,s1} = [z_{B,s1,u,k-I}; u = 1,\ldots, n]$. The size of the grid cell used in the distributed model is approximately 4×4 km^2. Soil moisture accounting is performed via the Sacramento model (SAC-SMA; Burnash *et al.*, 1973), as applied to each grid cell without lateral exchange. The routing model is based on kinematic wave, and has hillslope and channel flow components (Koren *et al.*, 2002). The above hydrological and hydraulic models constitute the current version of the National Weather Service Hydrology Laboratory's (NWS/HL) Research Modelling System (RMS) (for details see Koren *et al.*, 2002). There are a total of eight model state variables, six in SAC-SMA and two in routing (one for hillslope and one for channel). Though channel routing has only one state variable (i.e. the wetted channel cross-section), we used three control variables per grid cell following the discretization of the channel reach within a grid cell into three segments as implemented in RMS. Hence, for a basin made of 100 grid cells, there are 1000 control variables associated with the model states. In the state vector $\boldsymbol{X}$ in equation (1), $\boldsymbol{X}_0$ denotes the initial conditions of the model states at the beginning of the control horizon, $\boldsymbol{X}_P$ and $\boldsymbol{X}_E$ denote the hourly multiplicative adjustment factors to the precipitation and PE estimates, respectively, within the control horizon. In the error vector $\boldsymbol{V}$ in equation (1), $\boldsymbol{V}_Q$ denotes the zero-mean error vector associated with observation and model simulation of hourly streamflow, and $\boldsymbol{V}_P$ and $\boldsymbol{V}_E$ denote the zero-mean error vectors associated with estimation of grid cell-specific hourly radar precipitation and PE, respectively.

From equation (1), we have the following least squares minimization problem subject to the model dynamics:

$$\text{Minimize } J = 1/2[\boldsymbol{Z} - H(\boldsymbol{X})]^T \boldsymbol{R}^{-1}[\boldsymbol{Z} - H(\boldsymbol{X})] \tag{2}$$

Noting independence among the zero-mean error vectors, we may rewrite equation (2) as:

$$\text{Minimize} \quad \begin{aligned} J = {} & 1/2[\boldsymbol{Z}_Q - G(\boldsymbol{X}_0, \boldsymbol{X}_P, \boldsymbol{X}_E)]^T R_Q^{-1}[\boldsymbol{Z}_Q - G(\boldsymbol{X}_0, \boldsymbol{X}_P, \boldsymbol{X}_E)] \\ & + 1/2[\boldsymbol{Z}_P - H_P \boldsymbol{X}_P]^T R_P^{-1}[\boldsymbol{Z}_P - H_P \boldsymbol{X}_P] \\ & + 1/2[\boldsymbol{Z}_E - H_E \boldsymbol{X}_E]^T R_E^{-1}[\boldsymbol{Z}_E - H_E \boldsymbol{X}_E] \\ & + 1/2[\boldsymbol{Z}_B - \boldsymbol{X}_0]^T B^{-1}[\boldsymbol{Z}_B - \boldsymbol{X}_0] \end{aligned} \tag{3}$$

subject to

$$\boldsymbol{X}_{k+1} = g(\boldsymbol{X}_k, \boldsymbol{X}_{Pk}, \boldsymbol{X}_{Ek}) \tag{4}$$

$$\boldsymbol{X}_{\min} \le \boldsymbol{X}_{k+1} \le \boldsymbol{X}_{\max} \tag{5}$$

where $G(\ )$ denotes the (forward) run of the soil moisture accounting and routing models from the beginning of the control horizon to the hours concurrent to the respective streamflow observations in $\boldsymbol{Z}_Q$, $g(\ )$ denotes the model dynamics over one time step, X_k denotes the model (soil moisture accounting and routing) state variables at time step k and $\boldsymbol{R}_Q$, $\boldsymbol{R}_P$, $\boldsymbol{R}_E$, and B denote the error covariance matrices associated with streamflow, precipitation, PE and the *a priori* model states at the beginning of the control horizon, respectively. It is worth noting that the above constrained minimization is essentially the same as the Kalman smoothing problem (see e.g. Schweppe, 1973) if one considers adjustment to precipitation (and to a lesser degree PE) as correcting the model errors in the linearized dynamics of equation (4): $\boldsymbol{R}_P$ may then be interpreted as the model error covariance Q in the usual Kalman filter notation.

There are four error covariance matrices to be specified in equation (3); $\boldsymbol{R}_Q$, $\boldsymbol{R}_P$, $\boldsymbol{R}_E$, and $\boldsymbol{B}$, on which performance of the assimilation procedure depends critically. Here, as a first step in the continuing investigation, we made a simplifying assumption that all the error covariance matrices are diagonal. For precipitation, it is probably too gross an assumption in that uniform multiplicative adjustment of precipitation estimates does not change the spatial correlation structure of the estimation error. Another approximation of significance made here is that because the true states were not known, the error variances (i.e. the diagonal entries) could only be specified very loosely. As a first guess, we used the sample variances of the estimates and model simulations of the states in place of the error variances.

The minimization problem in equations (3)–(5) is solved by the conjugate gradient method of Fletcher-Reeves-Polak-Ribiere (Press *et al.*, 1986), which worked well for state updating of lumped models (Seo *et al.*, 2002). In the minimization, the gradients were evaluated via the adjoint code of the forward model (i.e. the SAC-SMA and routing), generated automatically by the Tangent linear and Adjoint Model Compiler (TAMC; Giering, 1999). Significant modifications to the TAMC-generated code were necessary to improve computational efficiency. To test the procedure, we performed assimilation retrospectively (assuming clairvoyant QPF) for the basin near Watts, Oklahoma, USA. The basin, modelled extensively in the Distributed Model Intercomparison Project (http://www.nws.noaa.gov/oh/hrl/dmip), has an area of 1645 km^2 and a time-to-peak of approximately 20 h. Figure 1 shows an example of the assimilation result, where the vertical line indicates the time of the most recent streamflow observation at the basin outlet. Figure 2 is the same as Fig. 1, but 1 h later (i.e. an additional streamflow observation was available). The beneficial effect of state updating to outlet streamflow simulation, particularly on the rising limb of the hydrograph, is apparent.

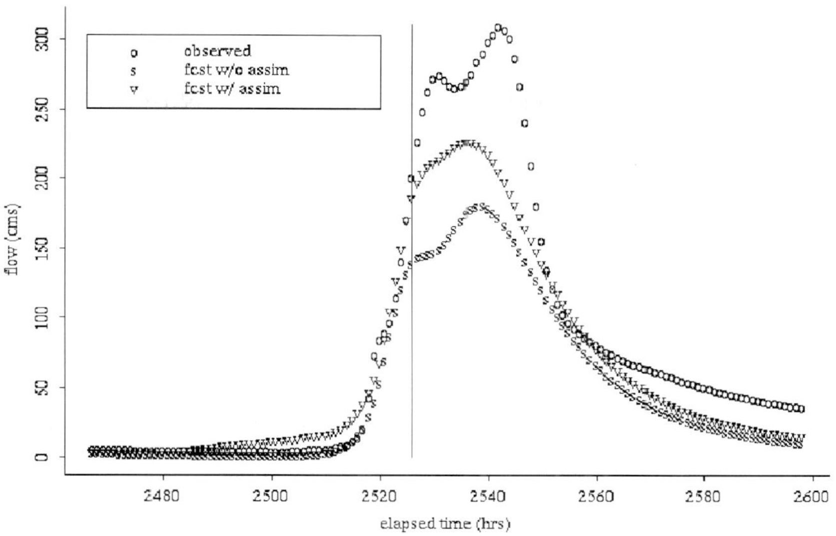

Fig. 1 An example of VAR-aided simulation as compared with simulation without state updating and the observed flow, where the vertical line indicates the time of the most recent streamflow observation at the basin outlet.

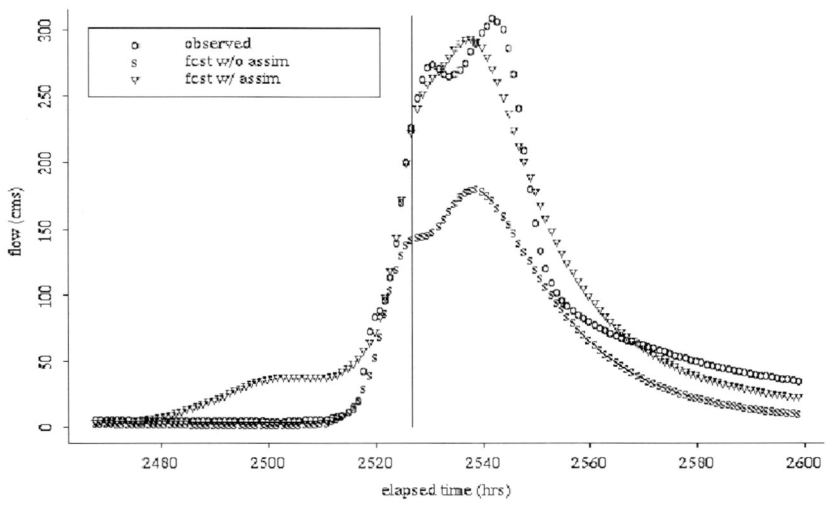

Fig. 2 An example of VAR-aided simulation, where an additional streamflow observation (one hour later) was available as compared with Fig. 1.

ISSUES

A number of questions arise from the above assimilation problem that are pertinent to state updating of distributed models in general. Here we identify several of them:

(a) How does the length of the control horizon affect the assimilation results? Presumably, the longer the control horizon is, the less sensitive the assimilation result is to the initial conditions at the beginning of the control horizon (Chao & Chang, 1992). Then, one may be able to specify the background error covariance matrix ***B*** only very approximately if a longer control horizon can be afforded computationally (e.g. always begin assimilation in an inter-storm period).

(b) How does the size of the assimilation window affect the results? Of further interest is the presumed dependence of sizing of the window to the model states: the size may depend on which part of the hydrograph the streamflow observations come from.

(c) How accurately should the error covariance in precipitation estimates be modelled? It is possible to model in $\boldsymbol{R}_P$ the space–time correlation structure of error in precipitation estimates. It is not clear, however, whether the marginal improvement may justify the far greater computational expense.

(d) Is bin-by-bin adjustment, as opposed to multiplicative adjustment over some large area, of model states really necessary? Given that soil moisture observations are not routinely available and that it is too difficult to relate model soil moisture states to (point) observations, one may consider adjusting the model states over some large area based on some predetermined scaling relationships. Such simplification will not only reduce the computational burden but may also improve the robustness of the procedure.

(e) How beneficial is the state updating based on streamflow observations at the basin outlet to streamflow prediction at interior locations? For state updating of distributed models to be operational viable, updating-aided streamflow prediction must consistently be at least as good as the prediction, without updating at all interior locations of the basin. We note here that work is under way to investigate these and other issues, and the results will be reported in the near future.

Acknowledgements This work is supported by the Advanced Hydrological Prediction Service (AHPS) of the National Weather Service.

REFERENCES

Burnash, R. J. C., Ferral, R. L. & McGuire, R. A. (1973) A generalized streamflow simulation system—conceptual modelling for digital computers. National Weather Service, NOAA, and the State of California Department of Water Resources, Technical Report Joint Federal-State River Forecast Center, Sacramento, California, USA.

Chao, W. C. & Chang, L.-P. (1992) Development of a four-dimensional variational assimilation analysis system using the adjoint method at GLA Part 1: Dynamics. *Mon. Weath. Rev.* **120**, 1661–1673.

Evensen, G. (1994) Sequential data assimilation with a nonlinear quasi-geostrophic model using Monte Carlo methods to forecast error statistics. *J. Geophys. Res.* **99**(C5), 10143–10162.

Giering, R. (1999) *Tangent Linear and Adjoint Model Compiler User's Manual*, TAMC Version 5.2 (available from http://www.autodiff.com/tamc).

Jazwinski, A. H. (1970) *Stochastic Processes and Filtering Theory*. Academic Press, San Diego, California, USA.

Koren, V., Smith, M., Reed, S., Zhang, Z. & Seo, D. J. (2002) Hydrology Laboratory Research Modelling System of the National Weather Service. *J. Hydrol.* (submitted).

Li, Z. & Navon, I. M. (2001) Optimality of variational data assimilation and its relationship with the Kalman filter and smoother. *Quart. J. Roy. Met. Soc.* **127**, 661–683.

Press, W. H., Flannerty, B. P., Teukolsky, S. A. & Vetterling, W. T. (1986) *Numerical Recipes*. Cambridge University Press, Cambridge, UK.

Schweppe, F. C. (1973) *Uncertain Dynamic Systems*. Prentice-Hall, Englewood Cliffs, New Jersey, USA.

Seo, D.-J., Koren., V. & Cajina, N. (2003) Real-time variational assimilation of hydrological and hydrometeorological data into operational hydrological forecasting. *J. Hydromet.* (submitted).

Tzafestas, S. G. (1978) Distributed parameter in state estimation. In: *Distributed Parameter Systems* (ed. by W. H. Ray & D. G. Lainiotis), Chapter 3. Dekker, New York, USA.

Weather Radar Information and Distributed Hydrological Modelling (Proceedings of symposium HS03 held during IUGG2003 at Sapporo, July 2003). IAHS Publ. no. 282, 2003.

Impact of on-line calibration of radar-measured rainfall in hydrological forecasting using a distributed hydrological model

MINJIAO LU & NORIO HAYAKAWA

Department of Civil and Environmental Engineering, Nagaoka University of Technology, Nagaoka, Niigata 940-2188, Japan

lu@nagaokaut.ac.jp

Abstract In order to find a more effective way to use both gauge- and radar-measured rainfall data in hydrological forecasting, a runoff analysis was carried out over the Uono River basin (355 km^2), a tributary of the Shinano River. A distributed hydrological model was constructed by using our distributed hydrological modelling system. The study basin was divided into small grid cells. This model computes runoff generated in all grid cells from rainfall data using the XinAnJiang model, and routes the runoff from these cells to the basin outlet through a channel network delineated from geographical information. In a discussion on estimation of radar constants, the necessity of on-line calibration of radar rainfall was shown, and an on-line calibration algorithm was developed. Comparing with hydrographs calculated from gauge rainfall, the hydrographs from calibrated radar rainfall showed significant improvement in both cases, peak flows calculated from gauge rainfall were overestimated or underestimated.

Key words distributed hydrological model; hydrological forecasting; on-line calibration; radar hydrology; radar-measured rainfall

INTRODUCTION

Weather radar has been considered to be a powerful tool to provide rainfall data to drive hydrological models, especially distributed models. The high spatial and temporal resolution, and large areal coverage makes it potentially possible to take into account the effects of spatial distribution of rainfall, and to make hydrological forecasting for small and/or ungauged catchments. Up to now, many hydrologists applied and tested the radar rainfall in various hydrological models (Fortin *et al.*, 1986; Kouwen, 1988; Kouwen & Garland, 1989; Lu *et al.*, 1989; Carpenter *et al.*, 2001; Borga, 2002). It has been recognized that radar rainfall uncertainty largely limits its applicability. Krajewski & Georgakakos (1985) showed that error of radar rainfall could be as large as about 50%. This uncertainty makes the spatial averaging necessary (Kouwen & Garland, 1989), and then hinders its application to hydrological forecasting for small and/or ungauged catchments where the advantage of radar rainfall data are considered most significant.

In our previous work (Lu *et al.*, 1996b), it was found that calibration of radar rainfall with gauge-measured basin-average rainfall improves the accuracy of hydrological forecasting. In this study, an investigation of the variability of radar constants was carried out. Their large variability and dependency on the objective functions of optimization methods, and then the necessity of on-line calibration are shown. The purpose of this work was to develop a methodology of on-line calibration of radar rainfall and complementary use of radar rainfall and gauge rainfall, and to show its impacts on hydrological forecasting. By using this newly developed method, calculated hydrographs are significantly improved for cases where peak flows calculated from gauge rainfall are overestimated or underestimated.

STUDY BASIN AND DATA

The study area is upstream part of the Uono River basin, a tributary of the Shinano River, the longest river in Japan. Its drainage area is 355 km^2, and the basin discharge is gauged at the outlet (Fig. 1). This basin is hilly with elevation ranging from 100 m to about 2000 m. Figure 1 also shows the polar coordinate system of the Yakushidake weather radar and ten raingauge stations among 16 stations used to calibrate the radar rainfall data. Our study basin was completely covered by this weather radar, which was installed exclusively for rainfall observation. The limit for quantitative observation of this radar is 120 km, and the spatial and temporal resolutions of this radar are 3 km and 5 min, respectively. Though 5-min radar rainfall data are available, the hourly rainfall data are used in this analysis because the conventional data of rainfall and stream flow are collected hourly.

In this basin, floods are mainly caused by frontal storm, thunderstorm, typhoon, cyclone and snowmelt. Heavy frontal storms with a thunderstorm can be observed frequently in summer. A total of 16 flood events shown in Table 1 are selected to

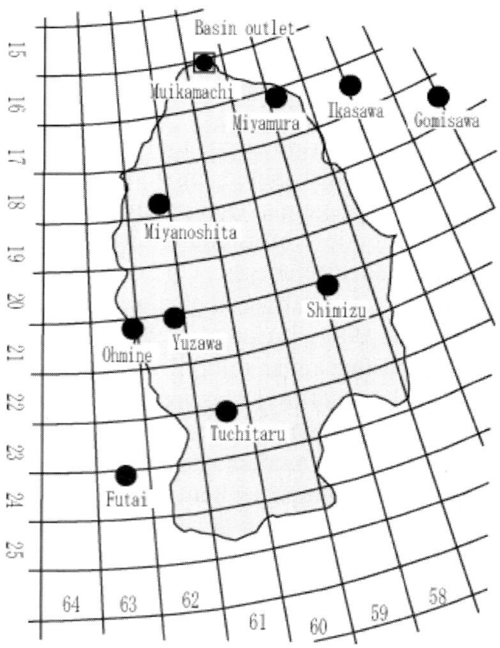

Fig. 1 Study basin.

Table 1 Selected flood events.

Flood no.	Period	Max. rainfall (mm h^{-1})	Peak flow ($m^3 s^{-1}$)	Time lag (h)	Cause
880824	1988/08/24–08/28	28.2	165.7	2	thunderstorm
880902	1988/09/02–09/04	13.0	195.0	4	thunderstorm
890525	1989/05/25–05/27	10.9	256.5	5	cyclone/front
890806	1989/08/06–08/07	10.6	286.5	4	typhoon
890826	1989/08/26–08/28	12.0	160.0	4	thunderstorm
890904	1989/09/04–09/06	12.1	201.6	4	cyclone/front
890918	1989/09/18–09/21	11.4	222.8	3	typhoon
900919	1990/09/19–09/21	13.9	387.1	2	typhoon
901007	1990/10/07–10/09	13.2	279.9	5	typhoon
901026	1990/10/26–10/28	8.2	195.2	3	cyclone/front
910830	1991/08/30–09/01	21.9	714.7	2	typhoon
911012	1991/10/12–10/14	5.2	33.5	5	typhoon
930826	1993/08/26–08/27	9.0	224.8	4	typhoon
940929	1994/09/29–10/01	9.4	381.5	3	typhoon
950708	1995/07/08–07/24	15.5	189.4	4	meiyu front
950915	1995/09/15–09/18	9.5	158.6	5	typhoon

investigate the possibility of using radar-measured rainfall in hydrological forecasting. Among them, three are caused by thunderstorms, nine by typhoons, three by cyclones and/or fronts, and one by a meiyu front. All causes of flood events in this area, except snowmelt, are considered.

DISTRIBUTED HYDROLOGICAL MODEL

A distributed hydrological model developed by Lu *et al.* (1996a) was used. The study basin was then divided into small grid cells. From field surveys and the result of distributed snowmelt analysis, we found that the hydrological processes can be classified into two groups: the processes within a grid cell and those between grid cells when the grid spacing is as large as several hundred metres. The processes within a grid cell can be simulated grid cell by grid cell, and the interaction between grid cells is assumed not to be dominant. The movement of water between grid cells is simulated by our runoff routing model. The runoff of a grid cell is concentrated to its centre and considered to flow to one of its eight neighbours, forming the steepest slope. The flow path between these two grid points (the centre of a grid cell) is modelled as an open channel with a wide rectangular cross-section. The runoff together with inflow from upstream channels is routed to the downstream end of this channel by using a routing scheme. In this model, all channels are arranged into an optimal routing sequence according to the structure of the channel network, which keeps the downstream channel to be routed after its upstream ones (Lu *et al.*, 1996a).

The study basin, the Uono River basin, is divided into 36 233 grid cells of 100 m × 100 m (slightly larger than officially published drainage area 355 km^2). From geographical information, a channel network that links all grid cells to the basin outlet is automatically delineated. At each grid point, the input data are assigned from available data sources, for example gauge rainfall and radar rainfall. Using the assigned rainfall data, runoff is calculated at each grid cell by using a conceptual rainfall–runoff model, the XinAnJiang model (Zhao, 1992), and then routed to the basin outlet through the channel network. In routing computation of each river channel, the kinematic wave approximation is used that considers the hilly topography of this basin. All model parameters are calibrated in previous studies.

ESTIMATION OF RADAR CONSTANTS

The radar rainfall uses the so-called radar equation in which the rainfall intensity is related to the radar reflectivity in the form of a power law given as follows:

$$Z = BR^{\beta} \tag{1}$$

where B and β are radar constants, which can be determined theoretically under some assumptions. For the Yakushidake radar, $B = 650$ and $\beta = 1.5$ is provided by a hardware company and used operationally (due to a hardware reason, Z and then B are multiplied by a factor of four). However, in most applications of radar rainfall, these constants are fitted with the observed data. Before running the distributed model, we tried to fit these constants for each flood event and for all events by using the following three methods:

(a) Optimize $\ln B$ and β by minimizing the objective function:

$$S(B,\beta) = \frac{1}{N}\sum_s\sum_t(\ln Z_s(t) - \ln B - \beta\ln R_s(t))^2$$
$$Z_s(t) = \frac{1}{12}\sum_{k=0}^{11} Z_s(t-k/12) \tag{2}$$

where $R_s(t)$ is hourly rainfall in the tth hour observed at the sth gauge station, and $Z_s(t)$ is hourly mean radar reflectivity corresponding to the sth station, N is the number of all effective data. In this study, all data of 16 stations are used.

(b) Optimize a and b by minimizing the objective function:

$$S(a,b) = \frac{1}{N}\sum_s\sum_t(\ln R_s(t) - a\ln Z_s(t) - b)^2 \tag{3}$$

and then obtain radar constants by $\beta = 1/a$ and $B = \exp(-b\beta)$;

(c) Optimize B and β directly by minimizing the objective function:

$$S(B,\beta) = \frac{1}{N}\sum_s\sum_t\left| R_s(t) - \frac{1}{12}\sum_{k=0}^{11}\left(\frac{Z_s(t-k/12)}{B}\right)^{1/\beta}\right| \tag{4}$$

In methods (a) and (b) (referred to as EST(Z) and EST(r)), radar constants can be explicitly obtained from observed gauge rainfall and radar reflectivity data. They, especially EST(Z), are mostly used in estimation of radar constants. Method (c) (referred to as OPT) is obviously a nonlinear optimization problem. Method (c) is

Table 2 Estimated radar constants.

Flood	EST(r)		EST(Z)		OPT	
	B_r	β_r	B_z	β_z	B_o	β_o
880824	693.00	2.074	1893.59	1.133	4062.42	1.004
880902	1631.80	2.077	3433.41	1.119	2644.18	1.389
890525	266.45	3.663	956.34	1.724	1879.76	1.523
890806	58.28	3.925	692.99	1.319	1841.73	1.150
890826	238.72	4.031	696.69	1.793	2297.18	1.294
890904	238.30	4.191	574.00	1.960	399.41	2.215
890918	188.94	3.703	724.23	1.750	200.32	2.757
900919	240.45	3.302	1766.07	1.149	288.92	2.480
901007	86.30	5.350	340.62	1.733	649.64	1.920
901026	409.65	3.101	1096.93	1.802	170.87	2.871
910830	99.22	3.662	766.67	1.475	1120.33	1.617
911012	5.52	7.901	447.29	1.439	0.02	10.376
930826	137.72	2.848	963.67	1.125	13.90	3.726
940929	204.21	4.652	443.17	2.190	2450.67	1.300
950708	292.311	2.708	626.63	1.428	434.59	1.690
950915	272.93	3.372	588.22	1.742	389.36	2.371
All	189.46	3.552	695.26	1.567	442.67	2.158

designed considering minimization of the error of rainfall itself is more important than that of ln(Z) or ln(R) in hydrological forecasting. Here, the quasi Newton method (Meza, 1994) is used to solve the nonlinear optimization problem.

Table 2 shows the estimated radar constants. It is shown that the radar constants vary largely from event to event, and from method to method. For flood no. 911012, methods minimizing the error of R or ln(R) gave extremely small B and extremely large β because the range of value of R is too small to perform such optimization. It is expected that data covering a wider range of R and Z will improve the estimation of radar constants. However, even using all data of 16 flood events, three methods gave largely different radar constants. Figure 2 shows the value of the objective function used in OPT and the estimated radar constants. In this figure, radar constants derived from EST(r) and EST(Z) are also plotted. From a viewpoint of water balance, it is very hard to say the results of EST(r) and EST(Z) are optimal. Figure 3 shows the accumulated basin-average rainfall of floods no. 880824 and no. 890806. Besides gauge-measured rainfall data, rainfall estimation using operational radar constants (referred to as OPE) and radar constants estimated by EST(r), EST(Z), and OPT are also plotted. Including these two flood events, EST(Z) overestimates and EST(r) underestimates the basin-average rainfall for all 16 flood events. OPT gave values closest to the gauge measured basin-average rainfall. This implies that selection of objective function is also an important issue in application of radar rainfall.

ON-LINE CALIBRATION OF RADAR-MEASURED RAINFALL

Because of the large variability of radar constants, we decided to use operational radar constants to estimate radar rainfall, and then make on-line calibration. For each grid

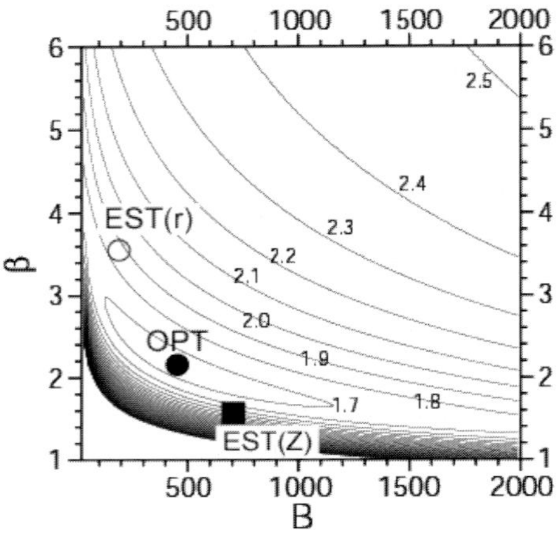

Fig. 2 Mean absolute error of radar rainfall.

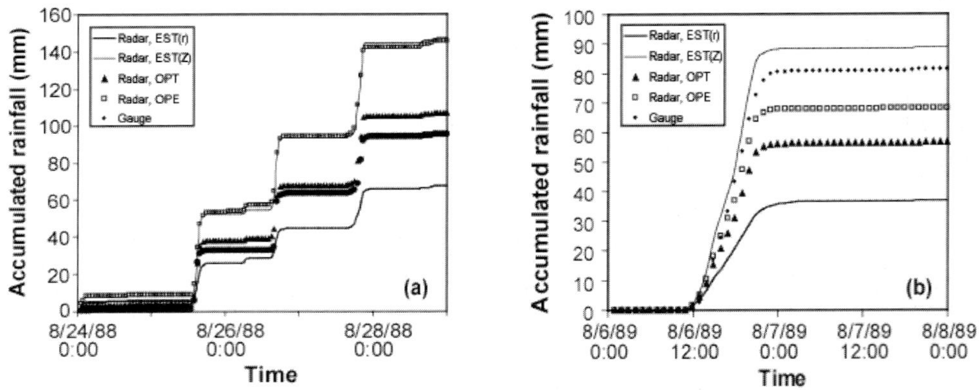

Fig. 3 Accumulated basin-average rainfall: (a) flood no. 880824; (b) flood no. 890806.

cell (i, j), the composite rainfall $r(i, j, t)$ is assigned as follows:

$$r(i, j, t) = \exp(-l / L) r_c(i, j, t) + (1 - \exp(-l / L)) r_r(i, j, t) \tag{5}$$

$$r_c(i, j, t) = G(i, j) r_r(i, j, t) = \frac{\sum_{m=0}^{\infty} k^m r_g(i, j, t - m)}{\sum_{m=0}^{\infty} k^m r_r(i, j, t - m)} r_r(i, j, t) \tag{6}$$

where t represents time, l is distance from grid cell (i,j) to nearest gauge station, $r_g(i,j,t)$ is rainfall data at the nearest gauge station, $r_r(i,j,t)$ is radar rainfall at the radar grid cell covering grid cell (i,j), k and L are parameters. $r_c(i,j,t)$ is radar rainfall calibrated by using gauge rainfall at the nearest gauge station. When the value of k increases, the antecedent rainfall will have more contribution on the calibration, and ratio $G(i,j)$ will become more stable. In a sensitivity analysis, it is found that k does not largely affect the calibrated rainfall, which is quite near to the gauge rainfall. In this study, $k = 0.5$ is used. As shown in equation (5), composite rainfall $r(i,j,t)$ consists of two parts, calibrated radar rainfall and raw radar rainfall. The nearer the grid cell (i,j) to a gauge station, the more the calibrated radar rainfall contributes. This allows us to use more information from gauge rainfall for a basin with a dense network of gauge stations, and more information from radar rainfall for a basin with a sparse network of gauge stations.

Using composite rainfall, hydrographs at the basin outlet are derived from our distributed hydrological model. Figure 4 shows the results for floods no. 880824 and no. 890806. In this figure, hydrographs derived by using gauge rainfall, radar rainfall, composite rainfall with different L are plotted. For flood no. 880824 (Fig. 4(a)), the third peak of discharge reproduced by gauge rainfall and radar rainfall is

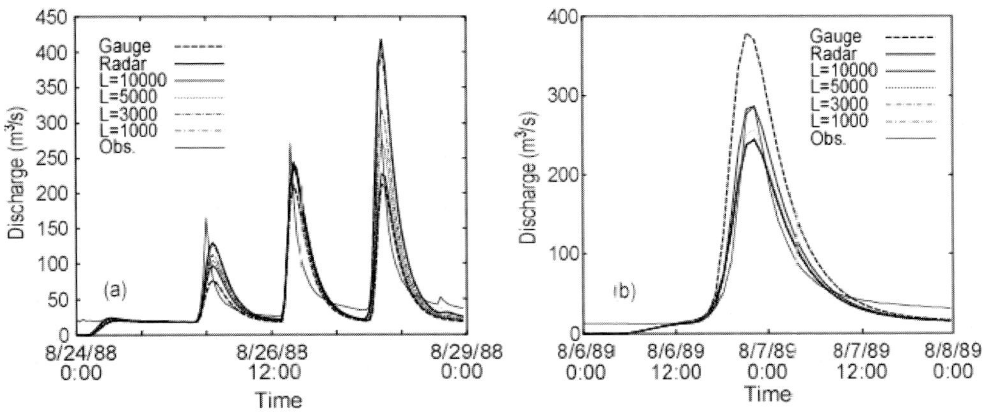

Fig. 4 Observed and calculated hydrographs: (a) flood no. 880824; (b) flood no. 890806.

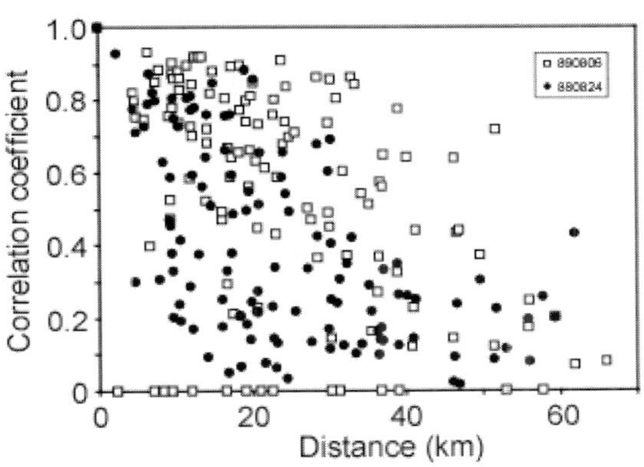

Fig. 5 Relationship between correlation coefficient of rainfall and distance of any two raingauges.

underestimated and overestimated, respectively, and the use of composite rainfall with L = 1000–3000 m significantly improves the representation of peak flow. In contrast, for flood no. 890806 (Fig. 4(b)) the peak flow reproduced by gauge rainfall and radar rainfall is overestimated and underestimated, respectively, and the use of composite rainfall with L = 10 000 m improves the representation of the peak flow. Parameter L seems to have a relationship with the spatial scale of the storm. Flood no. 880824 is caused by a thunderstorm with smaller spatial scale while flood no. 890806 is caused by a typhoon with larger scale. Figure 5 shows the relationship between the rainfall correlation coefficient and distance of any two gauge stations. Obviously, the correlation coefficient of rainfall decreases more slowly when the distance becomes larger for flood no. 890806, due to its larger spatial scale.

CONCLUSIONS

Three methods are used to estimate radar constants for each flood event. It is shown that the estimated radar constants change from event to event, and from method to method. Even using all data of 16 flood events, the radar constants estimated by the three methods are quite different. Radar constants estimated by minimizing RMS error of ln(Z) are the nearest to the operational radar constants. However, they differ from those estimated by minimizing RMS error of ln(R) or mean absolute error of rainfall. From the viewpoint of water balance, minimization of error of rainfall itself is an important issue. The mean absolute error shows that the method minimizing RMS error of ln(R) or RMS error of ln(Z) can hardly be considered as optimal.

Because of the large variability of the estimated radar constants, we decided to use operational radar constants and to make complementary use of gauge rainfall and radar rainfall. A new methodology of on-line calibration was developed. It was shown that significant improvements are achieved by using this method in cases where peak flows calculated from gauge rainfall are overestimated or underestimated. Also, the dependency of parameters of this method on the spatial scale of rainfall is clarified. Further work on this issue will be carried out to make this method applicable to operational hydrological forecasting.

Acknowledgements In this study, the Object-Oriented Nonlinear Optimization Library at the OPT++ developed in Sandia National Laboratories, Livermore, California, USA was used.

REFERENCES

Borga, M. (2002) Accuracy of radar rainfall estimates for streamflow simulation. *J. Hydrol.* **267**, 26–39.

Carpenter, T. M., Georgakakos, K. P. & Sperfslagea J. A. (2001) On the parametric and NEXRAD-radar sensitivities of a distributed hydrologic model suitable for operational use. *J. Hydrol.* **253**, 169–193.

Fortin, J. P., Villeneuve, J. P., Guilbot, A. & Seguin, B. (1986) Development of a modular hydrological forecasting model based on remotely sensed data, for interactive utilization on a microcomputer. In: *Hydrologic Applications of Space Technology* (ed. by A. I. Johnson) (Proc. Workshop, August 1985), 307–319. IAHS Publ. no. 160.

Kouwen, N. (1988) Watflood: a micro-computer based flood forecasting system based on real-time weather radar. *Can. Water Resour. J.* **13**(1), 62–77.

Kouwen, N. & Garland G. (1989) Resolution considerations in using radar rainfall data for flood forecasting. *Can. J. Civil Engng* **16**, 279–289.

Krajewski, W. F. & Georgakakos, K. P. (1985) Synthesis of radar rainfall data. *Water Resour. Res.* **21**, 746–768.

Lu, M., Koike, T. & Hayakawa, N. (1989) A rainfall–runoff model using distributed data of radar rain and altitude. *J. Hydraul. Coastal & Environ. Engng* **411**(11–12), 135–140.

Lu, M., Koike, T. & Hayakawa, N. (1996a) A distributed hydrological modelling system linking GIS and hydrological models. In: *Application of Geographic Information Systems in Hydrology and Water* (ed. by K. Kovar & H. P. Nachtnebel) (Proc. HydroGIS'96 Conf.), 141–148. IAHS Publ. no. 235.

Lu, M., Kioke, T. & Hayakawa, N. (1996b) Distributed XinAnJiang model using radar measured rainfall data. In: *Water Resources & Environmental Research: Towards the 21st Century* (Proc. Int. Conf.), **1**, 29–36.

Meza, J. C. (1994) OPT++: An object-oriented class library for nonlinear optimization. Sandia Tech. Report no. SAND94-8225, Sandia National Laboratories, Livermore, California, USA.

Zhao, R-J. (1992) The XinAnJiang model applied in China. *J. Hydrol.* **135**, 371–381.

Weather Radar Information and Distributed Hydrological Modelling (Proceedings of symposium HS03 held during IUGG2003 at Sapporo, July 2003). IAHS Publ. no. 282, 2003.

A new composite approach of physical and geostatistical aspects to groundwater modelling

TOSHIO HAMAGUCHI
Disaster Prevention Research Institute, Kyoto University, Uji, Kyoto 611-0011, Japan
hamaguti@urh.dpri.kyoto-u.ac.jp

Abstract This paper proposes a composite groundwater model with physical and geostatistical components to deal with spatial uncertainty in groundwater flow modelling. A hypothetical aquifer, assumed to be a true aquifer, is designed to test the ability of the composite model. The hypothetical aquifer composed of 20 zones with different hydraulic conductivity is represented as groundwater models with different zonation; those are models with one, two, four and 20 different hydraulic conductivity zones, respectively. The proposed scheme to reduce modelling errors is applied to each case, and the changes in the estimated water levels by applying the scheme are assessed. The results show that the composite estimates are reasonable and well in agreement with the true water level.

Key words geostatistical bias; groundwater; modelling error; spatial uncertainty; zonation

INTRODUCTION

Models in classical groundwater flow modelling are of the physically-based, deterministic type. The optimal parameters to be estimated can be obtained when the inverse problem is numerically solved. With such parameters, the calculated hydraulic heads are statistically best fits to the observed hydraulic, while the residuals between the two heads appear because the groundwater model in use is deterministic. They are consequently recognized as modelling errors, which cannot be sufficiently eliminated without a theoretical and sophisticated approach.

A universal and new strategy to completely eliminate modelling errors at the observed points is now needed. The strategy also should take into consideration space–time modelling errors. This paper describes a new theoretical approach to eliminate modelling errors and demonstrates its effectiveness through numerical studies in a hypothetical aquifer.

INTERPRETATION OF SPATIAL UNCERTAINTY IN MODELLING

Calibrations of model parameters are usually prerequisite to improve the fit between calculated and observed heads. In general, a difference between the above two heads is seen in actual problems. This difference derives from the physical modelling errors, which cannot be represented within the framework of physical modelling in space. For instance, when a true aquifer structure with inhomogeneous hydraulic conductivity is modelled as the one with uniform hydraulic conductivity, the best model with the most likely hydraulic conductivity is equivalently estimated against the true one. In contrast,

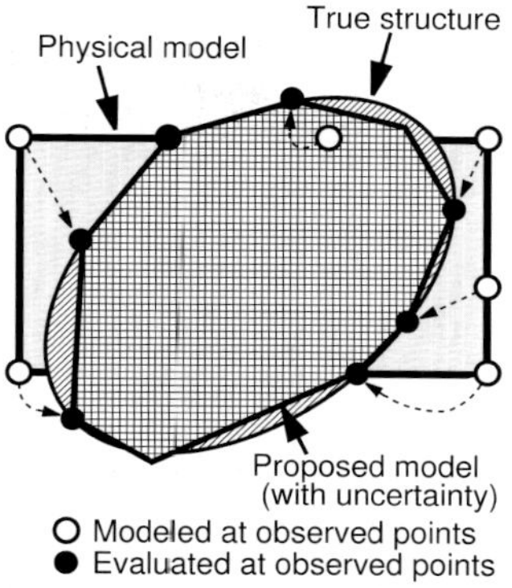

Fig. 1 A schematic view of the proposed model.

larger modelling errors appear in space. In this paper an approach is proposed for a geostatistical evaluation of spatial uncertainty, in particular, the homogeneous modelling error.

Figure 1 shows a schematic view of the relationships among a true structure of a natural phenomenon represented by the inclined ellipse, a corresponding model of the physically-based deterministic type represented by the rectangle, and a corresponding model proposed in this paper, represented by the polygon. Even though parameters in the physical model are identified in advance to fit the phenomenon, the differences between the natural structure and the physical model still remain at some points. A new approach to theoretically fill up the differences is developed and proposed in this paper. In the method, a spatial interpolation using the geostatistical technique of kriging is carried out to make the residuals at all observed points zero. Extensive improvements in the groundwater level estimates would be achieved by incorporating the geostatistical treatment.

MATHEMATICAL DESCRIPTION OF SPATIAL UNCERTAINTY

Spatial uncertainty due to a deterministic use of a numerical model mainly consists of following four kinds of errors the: (a) an error caused by representing a natural phenomenon using a physically-based model; (b) an error caused by representing a physical model using a numerical model; (c) an error caused by boundary conditions in space and time given to a numerical model; and (d) an error in system structure designing, program cording and numerical computing (Hamaguchi *et al.*, 1997a). The errors noted in (a)–(c) are generally defined as the modelling error denoted by $\eta(z,t)$, where z is a vector of space coordinates and t is a time variable. This paper focuses on the second modelling error, especially an error in zoning hydraulic conductivity.

The common assumption is that the trend and random variables are given by the formulation:

$$\phi(z,t) = m(z,t) + w(z,t) \tag{1}$$

where $\phi(z,t)$ denotes a true value, $m(z,t)$ and $w(z,t)$ are trend and random variables, respectively. It is assumed that spatial uncertainty and a random variable are static in time, and it can be rewritten as $\eta(z)$ and $w(z)$, respectively. In the absence of a drift $\hat{b}$, a constant expected value of $\eta(z)$ is proposed as:

$$\eta(z) = \hat{b} + w(z) \tag{2}$$

The specification of the drift indicates spatially uncertain but steady and stationary. The trend variable with the additional drift is then defined as:

$$m(z) = f(z,t) + \hat{b} \tag{3}$$

where $f(z,t)$ is a numerical solution of a physical model. A geostatistical approach of kriging in evaluating spatial uncertainty is proposed as the appropriate method for solving such a problem. The kriged value $w(z)$ can be then presented as:

$$w(z) = k(z)^T K^{-1} \hat{w} \tag{4}$$

with:

$$\hat{b} = \frac{\alpha^T K^{-1} \varphi}{\alpha^T K^{-1} \alpha}, \qquad \hat{w} = \varphi - \alpha \frac{\alpha^T K^{-1} \varphi}{\alpha^T K^{-1} \alpha} \tag{5}$$

where α denotes a constant vector of substituting a scalar 1 for all components, $\boldsymbol{\varphi}$ designates an observed vector, $k(z)$ describes a covariance vector of $w(z)$ between a given point to be kriged and a observed one, and K represents a covariance matrix of $w(z)$ between the observed points. The resulting evaluation $\eta(z)$ in this framework can be expressed as:

$$\eta(z) = \hat{b} + k(z)^{\mathrm{T}} K^{-1} \hat{w} \tag{6}$$

With the evaluated term $\eta(z)$, the solution to this estimation system $\phi^*(z,t)$ is finally obtained as:

$$\phi^*(z,t) = f(z,t) + \hat{b} + k(z)^{\mathrm{T}} K^{-1} \hat{w} \tag{7}$$

IDENTIFICATION OF EQUIVALENT HYDRAULIC CONDUCTIVITY

The unsteady two-dimensional unconfined groundwater flow in a non-uniform isotropic aquifer is described by:

$$\lambda \frac{\partial h}{\partial t} = \frac{\partial}{\partial x}\left\{k(h-s)\frac{\partial h}{\partial x}\right\} + \frac{\partial}{\partial y}\left\{k(h-s)\frac{\partial h}{\partial y}\right\} + \varepsilon \tag{8}$$

where λ is effective porosity, h is water head, k is hydraulic conductivity, s is elevation of bedrock, and ε is areal groundwater recharge rate. The groundwater model is numerically analysed by FEM.

The hypothetical example here is designed to satisfy three requirements: (a) sufficient simplicity to allow testing without excessive computation cost and interpretation

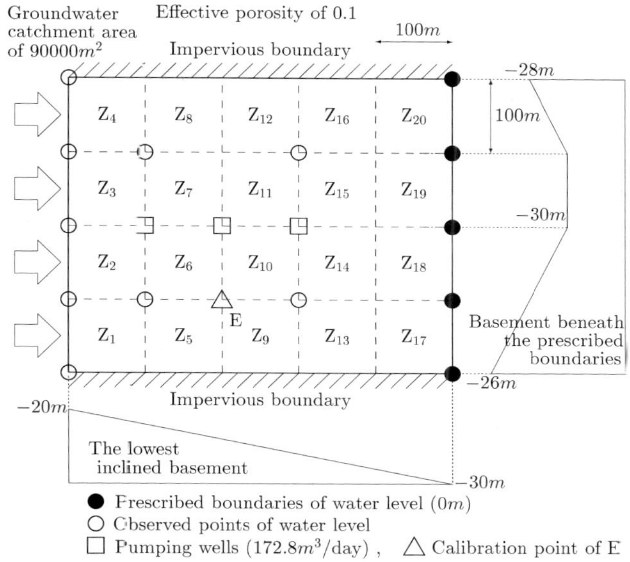

Fig. 2 Hypothetical aquifer modelled in the case studies.

Zone	Z_4	Z_8	Z_{12}	Z_{16}	Z_{20}
	Z_3	Z_7	Z_{11}	Z_{15}	Z_{19}
	Z_2	Z_6	Z_{10}	Z_{14}	Z_{18}
	Z_1	Z_5	Z_9	Z_{13}	Z_{17}
Exact	215.3322	255.5558	321.1791	400.0000	383.3276
	173.9983	200.0000	272.5400	329.4123	346.3049
	136.6654	171.4817	226.1393	273.9904	300.0000
	100.0000	146.4934	194.3980	235.9593	265.2360

Unit: m/day

Fig. 3 Original zonation of true hydraulic conductivity.

of the results; (b) sufficient complexity to allow testing various aspects of modelling errors; (c) a realistic representation of transient groundwater recharge by using the tank model. A rectangular domain of an unconfined aquifer having the area of 400 m × 500 m as shown in Fig. 2, was chosen for the case study. The domain has impervious boundaries at the top and bottom, a constant head boundary (0 m) at the right side, and a prescribed flow rate boundary based on the tank model at the left side. The groundwater recharge takes place at variable rates based on the tank model over the whole domain. A true aquifer has continuous properties in space, i.e. hydraulic conductivity, effective porosity, and so on. In preprocessing a numerical analysis by FEM, we need to discretize the two-dimensional domain of the aquifer by triangle or quadrangle elements. Representative model parameters on integrated average at each element are thus needed. The original aquifer in this study is subdivided into 20 constant-hydraulic conductivity zones with 4 × 5 square elements, as seen in Fig. 3. Here, the aquifer is represented by models with one constant-hydraulic conductivity zone (case 1), two zones (case 2), four zones (case 3), and 20 zones (case 4),

Case 1	221.5426				
Case 2	154.8059		335.7763		
Case 3	206.6947		354.0413		
	118.2425		293.4447		
Case 4	219.3408	263.1140	289.9044	382.6352	380.4207
	168.8113	224.7142	254.9413	348.1349	352.3683
	135.8332	185.8441	215.4022	273.9089	295.5448
	101.7502	144.2885	180.1621	237.3977	260.8399

Unit: m/day

Fig. 4 Results of identified hydraulic conductivity in each case.

respectively, as shown in Fig. 4. Three wells pump up water at a rate of 172.8 m^3 day^{-1} each to create strongly transient conditions. Head measurements are made at the nine observation wells.

The primary purpose of this study is to optimally describe a model with some equivalent hydraulic conductivity in an inhomogeneous aquifer. To identify the parameters from the observed head data, the extended Kalman filtering system in conjunction with FEM (Hasegawa *et al.*, 1994; Hamaguchi *et al.*, 1997b) is used in the framework of inverse analysis processing. The solutions of $\bar{k}_j$, the equivalent hydraulic conductivity in the *j*th zone, in each case are obtained by using the method with FEM computation. The results are arranged in Fig. 4.

The second purpose is to evaluate a geostatistical term added to the groundwater heads obtained by a groundwater model based on the proposed approach. The groundwater heads are numerically solved by using the models with equivalent hydraulic conductivity. After calculating the model heads, geostatistical estimates of the residuals of groundwater heads based on head measurements are obtained. The improved heads are obtained by adding the kriged residual heads with uncertainty to the modelled ones.

EFFECTIVENESS OF ADDITIONAL MODELLING ERRORS

To demonstrate the effectiveness of the improvement of the optimized groundwater model, the validity of its improvement is illustrated by the water level variation at point E (Fig. 2). Figure 5 shows the estimated water levels simulated by the physical models (marked by +, ×, Δ and ◊), those obtained by the proposed model with spatial uncertainty (marked by □, pentagon, ∇ and ○) in cases 1 to 4, and the true groundwater level drawn by the solid line. This figure reveals that the improvement of estimations by adding the modelling errors to the modelled solutions is greatly performed in all cases because all the geostatistically improved solutions are in better agreement with

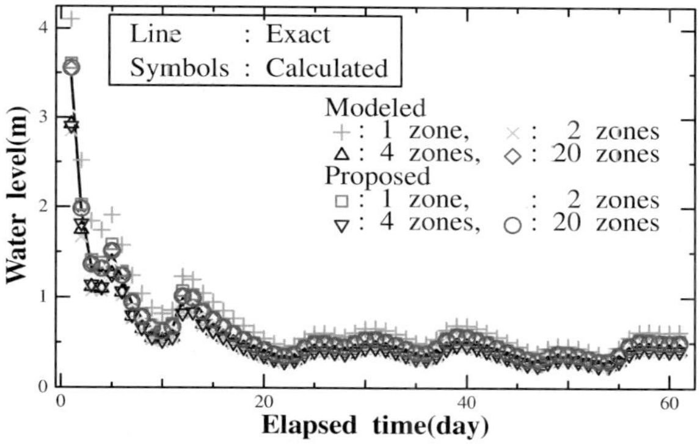

Fig. 5 Comparison of water level variations.

the true values than physically modelled ones. A striking feature of the fact in those studies is that the resulting variation in case 4 is the most accurate in estimating the true heads due to its similarity to the original zonation shown in Fig. 3. The improved variation in case 1 is in reasonable agreement with true variation, and it proves the proposed method effectively works to eliminate the modelling error.

EFFECTIVENESS OF A BIAS OF SPATIAL UNCERTAINTY

Introducing a drift term $\hat{b}$ in the proposed method is discussed in the section. To show the effectiveness of introducing $\hat{b}$, the index, a reciprocal number of a coefficient of variation

$$\frac{1}{|\delta_t|} = \frac{|\hat{b}|}{\sigma_t} \tag{9}$$

is used, where δ_t is a coefficient of variation, σ_t is a standard deviation of $w(z)$, and $|\bullet|$ indicates an absolute value of the variable. This index has two interesting properties in that a relative and absolute comparison among the variations at any time can be made, and that the effectiveness of $|\hat{b}|$ is increased in the geostatistical component when a value of $1/|\delta_t|$ becomes larger.

Figure 6 shows the temporal changes of $1/|\delta_t|$ in each case. The result of $1/|\delta_t|$ in case 1 keeps higher at any time than that in any other cases except case 4. The results in case 4 is higher than any other results, but this evaluation is practically overestimated because any values of σ_t are essentially smaller than the corresponding drifts. This means that the overall value of $\hat{b}$ computed in case 1 is most effectively worked to eliminate the model error corresponding to spatial uncertainty. As mentioned above, it is proved that a drift term of $\hat{b}$ is effective in decreasing the model error in the process of geostatistically describing spatial uncertainty.

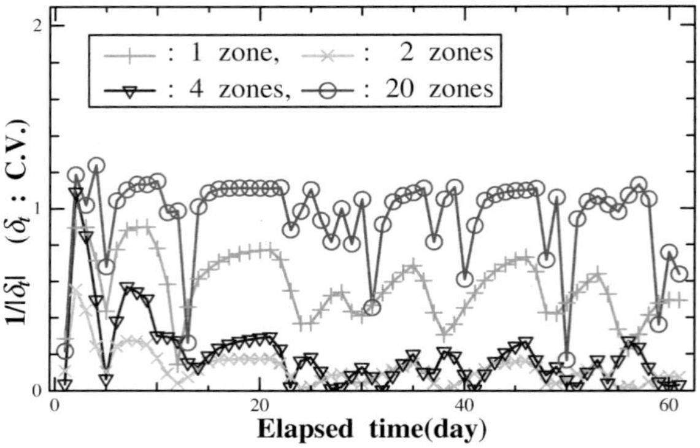

Fig. 6 Effectiveness of a drift term.

CONCLUSIONS

The proposed approach for groundwater modelling statistically assesses spatial uncertainty, which is added to the physically described spatial variability. The advantage of this scheme is the ability to fill up the differences between simulated and observed values. The study demonstrated four hypothetical cases, where a geostatistical calibration was applied successfully to model the aquifer behaviour under unsteady conditions. From these case studies, it was shown that drift term information was greatly helpful and effective in improving the physical model. The major conclusion is that a successful complement to spatial uncertainty due to modelling errors is performed for groundwater flow modelling.

REFERENCES

Hasegawa, T., Murakami, A. & Hamaguchi, T. (1994) Identification of parameters of groundwater flow model by extended Kalman filter and determination of pumping rate considering the charge of water level (in Japanese). *Proc. Japan Soc. Civil Engng* **505**, 297–305.

Hamaguchi, T., Murakami, A. & Hasegawa, T. (1997a): Inverse analysis method of two-dimensional groundwater model with prescribed boundaries including stochastic uncertainty (in Japanese). In: *Application to Inverse Analyses and Construction Assessment Relating to Geotechnical Engineering, JGS* (Proc. of Symp.), 111–114.

Hamaguchi, T., Murakami, A. & Hasegawa, T. (1997b) Groundwater model in consideration of moving boundaries in a two-dimensional analysis and its application to an inverse analysis (in Japanese). *Proc. Japan Soc. Civil Engng* **568,** 133–145.

3 Assessment of Performance in Distributed Hydrological Modelling

Distributed hydrological modelling: what accuracy is achievable?

BAXTER E. VIEUX
School of Civil Engineering and Environmental Science, University of Oklahoma, 202 West Boyd Street, Norman, Oklahoma 73019, USA
bvieux@ou.edu

FEKADU G. MOREDA
National Weather Service, National Atmospheric Administration, 1325 East-West Highway, W/OHD1, Silver Spring, Maryland 20910, USA

Abstract Distributed parameter watershed models that are physics-based offer distinct advantages over conceptual rainfall–runoff models. This paper describes the accuracy of simulated runoff achievable for the Illinois River (2300 km^2). Parameter values estimated from physical properties of the watershed are adjusted to reproduce system behaviour. Spatial heterogeneity is accounted for at the watershed scale by maps of parameters derived from digital maps of soil properties, land use/cover, and topography. Inputs come from radar or multisensor combinations of radar, satellite, and raingauge. Distributed hydrological models based on conservation laws have identifiable optimal values and expected behaviour and interaction during calibration. Predicted runoff accuracy for this subset is RMSE of 3.65 mm (±1.8 mm) with the best runoff accuracy achieved to within 1.6% or 0.3 mm. For this storm subset, consistent high accuracy is achieved (<10% by volume) with an important distinction that smaller absolute differences result for larger events.

Key words Distributed Modelling Intercomparison Project; floods; hydrological modelling; hydrology; rainfall runoff

INTRODUCTION

The goal of physics-based distributed models is to represent physical processes deterministically using conservation equations at a computational element scale that in aggregate reproduces the system behaviour. One advantage, among others, is that shorter historical records are required to calibrate a physics-based distributed model (PBD) compared to long records required by conceptual rainfall–runoff models (CRR). The spatial heterogeneity of watershed characteristics is represented by PBD models using such geospatial data as digital elevation models (DEMs), land use/cover, soils, or other digital maps or databases. Such models include those proposed by Abbott *et al.* (1986a,b); Zhang & Cundy (1989), James & Kim (1990), Julien & Saghafian (1991), and Vieux & Gauer (1994). These models seek to preserve the spatial heterogeneity at the watershed scale but are lumped at the scale of the computational element, which may be a finite difference grid, stream tube, or finite element. Scientifically, PBD models are useful for investigating and testing catchments to test new hydrological hypotheses. Practically, PBD models are becoming used in operational settings (Vieux *et al.*, 2003). Unlike CRR models, PBD models have identifiable optimal parameter sets and predictable parameter interaction.

The US National Weather Service (NWS) has recently promoted distributed hydrological modelling as a major new thrust for improving flood forecasting. The Hydrology Laboratory (HL) of the NWS has just completed the first phase of the Distributed Model Intercomparison Project (DMIP). Participants download data sets and run their models in continuous or event-based simulation modes. Model simulations are compared to observed streamflow data, as well as, with simulations generated from lumped and semi-distributed applications of the Sacramento Soil Moisture Accounting (SAC-SMA) model. Using the results provided by participants of DMIP, the NWS anticipates the development of an HL distributed model that incorporates distributed modelling principles. DMIP results will be published in the *Journal of Hydrology* in the near future.

MODEL DESCRIPTION

The model used herein was first developed by Vieux (1988) and applied to a small agricultural watershed without channel routing, which was later added to account for larger watersheds with defined channels. The solution using linear, one-dimensional (1-D) elements presented by Vieux *et al.* (1990) uses a single chain of finite elements for solving overland flow. Vieux & Gauer (1994) extended this finite element solution to a network of elements representing a watershed domain with channels within a GIS. The resulting model was the distributed hydrological model *r.water.fea*, developed in 1993 for the US Army Corps of Engineers, Construction Engineering Research Laboratory, Champaign, Illinois (USA-CERL). The initial development of the model is a part of the public domain GIS called GRASS (Geographic Resource Analysis Support System). Several derivative models now exist. The *r.water.fea* model was ported from Unix to a Windows version that runs as an ArcView extension, called *arc.water.fea* (Vieux, 2001). The *r.water.fea* model accesses a map database for various parameters controlling the hydrological process. A newly developed model called V*flo*™ implements the finite element approach, and is written in Java to take advantage of secure client-server configurations for real-time operational applications (Vieux *et al.*, 2003; Vieux & Vieux, 2002).

A mathematical analogy is derived by ignoring insignificant differential terms. The kinematic wave analogy (KWA) for overland flow is a simplification of the conservation of mass and momentum equations wherein the principle gradient is the land surface slope. The 1-D continuity for overland flow resulting from rainfall excess is expressed by:

$$\frac{\partial h}{\partial t}+\frac{\partial (uh)}{\partial x}=R-I \tag{1}$$

where R is rainfall rate; I is infiltration rate; h is flow depth; and u is overland flow velocity. In the KWA, bed slope is equated with the friction gradient. In open channel hydraulics, this amounts to the *uniform flow* assumption. Using this fact together with an appropriate relation between velocity, u, and flow depth, h, such as the Manning equation, we obtain:

$$u = \frac{S_O^{1/2}}{n} h^{2/3} \tag{2}$$

where S_o is the bed slope or principal land surface slope; and n is the hydraulic roughness. Though not presented here, infiltration rate, I, is modelled using the Green and Ampt equation, of which saturated hydraulic conductivity, k, has a significant control. Two of the most important parameters governing response are the saturated hydraulic conductivity, k, and the roughness n. These parameters are represented as grid maps of the same resolution as the DEM. The hydraulic conductivity controls the total amount of water that will be split into the surface runoff and subsurface flow, whereas the hydraulic roughness mainly affects the peak flow and the time to peak. Model response is controlled using scalars that multiply the distributed maps of parameters. Combining equations (1) and (2) with the addition of parameter controls results in:

$$\frac{\partial h}{\partial t} + \beta \frac{s^{1/2}}{n} \frac{\partial h^{5/3}}{\partial x} = \gamma R - \alpha I \tag{3}$$

where the three scalars α, γ, and β control infiltration rate, I, rainfall rate R, and hydraulic roughness n, respectively; h is the flow depth and s is the principal land-surface slope at the centre of each grid cell. The slope and hydraulic roughness are spatially variable, while rainfall, infiltration, and flow depth are spatially and temporally variable. If we consider the rainfall as perfectly known, with no bias, then γ is equal to one. These scalars are identifiable and have defined optimal values. The inverse solution of equation (3) and the optimal values obtained through the adjoint formulation is described in Vieux *et al.* (1998) and White *et al.* (2001a,b). Model calibration is achieved through adjustment of these scalars, whether by manual or automatic means, using the OPPA methodology described by Vieux & Moreda (2002).

Finite element solution of equation (3) is accomplished by the Galerkin formulation using 1-D elements arranged in the direction of principal slope. Finite element solution details may be found in Vieux & Gauer (1994), and a more recent summary in Vieux (2001). The finite difference solution solves the resulting ordinary differential equations in time using an explicit time stepping scheme subject to the Courant-Fredericks-Levy condition.

STUDY AREA

The Illinois River watershed straddles the Oklahoma/Arkansas border, with approximately 54% of the 2400 km^2 draining to Tahlequah located in Oklahoma (Fig. 1). The basin is located in Delaware, Adair, Cherokee, and Sequoyah counties in northeastern Oklahoma. The average annual flow of the Illinois River as it enters Oklahoma near Watts is 20 m^3 s^{-1}, which increases at Tahlequah to 29 m^3 s^{-1}, after which it flows into Lake Tenkiller. Agriculture dominates the economy of the region. Forest or pasture land uses are located in upland areas, whereas cropland is concentrated in more fertile lowlands. Digital geospatial data representing the watershed are described as follows.

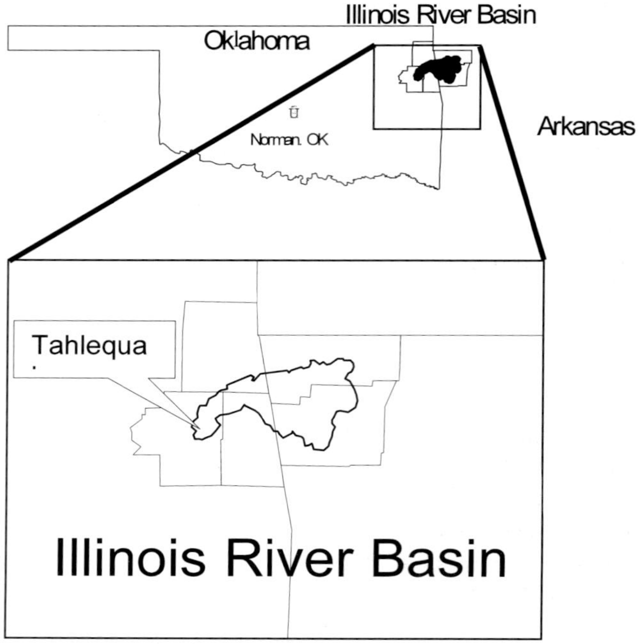

Fig. 1 Location of the Illinois River Basin in relation to Arkansas and Oklahoma.

Digital Elevation Model A 30-m resolution DEM is re-sampled to obtain a 960 m resolution map of elevation. This DEM map is used to delineate the watershed into three sub-basins. Delineation of the watershed can be done using the GRASS, r.watershed command or the ArcView 3.2 hydrological model extension. The *arc.water.fea* extension performs delineation and data extraction necessary for running a version of the *r.water.fea* model that has been exported to a PC such that simulation takes place on the PC in a GIS environment.

Channel characteristics Determining the exact characteristics of a channel requires extensive surveying or calibration to match hydrograph shape and peak. Trapezoidal cross-sections are interpreted from surveyed cross sections and stream gauging rating curves to obtain average values for the stream channels.

Rainfall Hourly rainfall maps as input are obtained from Stage III NEXRAD observations at 4 × 4 km resolution. Eight storms (see Table 1) are presented herein as described by Vieux & Moreda (2002) and in Vieux (2001). Each 960-m resolution grid cell receives rainfall from the overlying 4 × 4 km radar pixel. Other studies and comparison of raingauge accumulations show a mean field bias averaged over eight storms to be within 10%. No attempt to adjust the Stage-3 radar estimates are made considering that *ad hoc* adjustments to agree with raingauge accumulations are already introduced by the NWS River Forecast Centre, for which no adjustment record is generally available.

Table 1 Eight storm event summary of rainfall, simulated and observed runoff and associated errors.

Event	Rainfall (mm)	Simulated (mm)	Observed (mm)	Diff. (mm)	Abs diff. (mm)	Error (%)	(Diff.)^2
4 March 1995	30.4	4.3	9.7	-5.4	5.4	-55.7	29.20
10 May 1996	32.5	14.4	12.2	2.2	2.2	18.0	4.80
25 November 1996	53.3	23.5	29.6	-6.1	6.1	-20.5	37.20
8 June 1995	68.8	21.5	23.0	-1.5	1.5	-6.5	2.30
13 January 1995	71.4	22.9	26.4	-3.5	3.5	-13.3	12.30
4 November 1996	75.7	18.5	21.0	-2.5	2.5	-11.9	6.30
19 February 1997	76.1	29.3	33.1	-3.8	3.8	-11.5	14.40
20 April 1996	80.4	18.5	18.2	0.3	0.3	1.6	0.10
	61.1	**19.1**	**21.7**	**-2.5**	**3.2**	**-12.5**	***RMSE* = 3.65**

Discharge The simulated hydrographs for each storm are compared to the USGS observed hydrographs at USGS streamgauge 07196500, Illinois River near Tahlequah, Oklahoma (see Fig. 1). Since the model simulates only the surface flow component, the baseflow separation is accomplished using the straight-line method. Baseflow components are normally less than 10% of the flow during these events. The 50% chance exceedence flow rate is approximately 30 cm.

Infiltration and roughness parameters The Green and Ampt soil parameter maps are calculated using the Brooks and Corey relationships from Rawls *et al.* (1983a,b) to estimate the infiltration parameters based on soil properties such as clay, sand content, bulk density, and pore size distribution. These soil properties are then used to estimate the Green and Ampt infiltration parameters. The soil properties for soils in the Illinois River basin are obtained from the USDA Natural Resources Conservation Service for the State of Oklahoma from county-level soil surveys. The eight storms selected had relatively dry antecedent conditions except for the 25 November–December 1996 and 4 March 1995 events. Wet conditions were assigned a nominal 70% degree of saturation, whereas dry conditions were assigned 30%. The Manning parameter maps are obtained by relating land use/cover (LULC) maps to the corresponding roughness coefficients estimated according to the dominant LULC classification.

MODEL CALIBRATION

Calibration proceeding along the lines of OPPA, adjusts initial parameter maps by a scalar in order to minimize an objective function. Hydraulic conductivity is adjusted to achieve a volume match followed by adjustment of overland hydraulic roughness affecting peak discharge and timing. Subsequent adjustment of channel hydraulics principally affecting the time-to-peak agreement with observed. Similarly, hydraulic roughness is derived from land use/cover maps. Scalars are multiplied by these parameter maps to adjust the value in each grid cell while preserving the spatial heterogeneity. The sequence of adjustment is to first minimize the objective function for volume, and then peak, obtaining an overall *optimal* parameter set for the eight storms.

RESULTS AND DISCUSSION

Results for the eight storm events are presented in detail in Vieux (2001) and Vieux & Moreda (2002). A current effort to extend these results to 22 events is underway within the DMIP project. Initial results show that only minor adjustment to hydraulic conductivity resulted when additional storm events are added to the calibration suite. The results for the eight-storm suite are described as follows.

Hydraulic conductivity is used to adjust volume, after which, hydraulic roughness is used to adjust timing and peak. Several adjustments were made to ensure that timing controlled by channel hydraulics produced consistent results at an interior point located near Watts, Oklahoma. Calibration is performed for the suite of eight events by adjusting parameter maps by scalar multipliers (α and β in equation (3)) that minimize volume and peak discharge errors. Table 1 reports the results for each of the eight storms along with a summary (shown in bold on the bottom line of the table) for the suite of eight storms. The second column indicates the rainfall depth averaged over the watershed. The third and fourth columns are the runoff volumes in watershed depth (mm) for simulated and observed amounts, respectively. The fifth and sixth columns are the runoff differences and absolute runoff differences reported as mm of simulated and observed runoff. The seventh column is the simulated error reported as a percentage of the observed runoff. The final and eighth column is the square of the error, which when summed, is used to compute the root mean square error (*RMSE*) shown in the last line of the column.

Several interesting features emerge when considering the absolute difference between simulated and observed runoff. The absolute differences expressed as an equivalent watershed depth (mm) are all 6.1 mm or less, averaging 3.2 mm. The average rainfall is 61.1 mm. Using a criterion that simulated runoff should be within 10% of the average rainfall, then performance of the model for any one storm is 10% or better. If calibration were performed with any one storm, the simulated output would be no more than 10% of the input. Expressing the simulated runoff as a percentage error, the range of error is more variable. The 4 March 1995 event was the smallest event with 30.4 mm of rainfall.

The simulated runoff was 4.3 mm compared with 9.7 mm observed, or a 55.7% error. The model tends to predict runoff consistently to less than 6.1 mm, with an overall performance of *RMSE* of 3.65 mm (±1.8 mm). The relatively consistent prediction is evident in Fig. 2 where the rainfall and simulated runoff difference (Diff. in Table 1) are shown as vertical bars. It appears from this storm suite that there may be a tendency to more accurately predict with larger rainfall events. For example, the model predicted to within 1.6% or 0.3 mm for the 20 April 1996 event; the best accuracy achieved. While rainfall more than doubles from 30.4 mm to 80.4 mm, the runoff prediction errors remain consistently below 6.1 mm.

Considering the uncertainties in the radar rainfall inputs, the simulated volumes agree remarkably well with those observed. Removal of systematic error (bias) found in radar rainfall estimates is accomplished by adjusting the radar amounts to agree with raingauge accumulations. The adjustment procedure, called P1, performed by the NWS approaches bias correction by sampling the mosaic of hourly rainfall accumulation and comparing with raingauge amounts for the same time period. The radar/gauge ratio

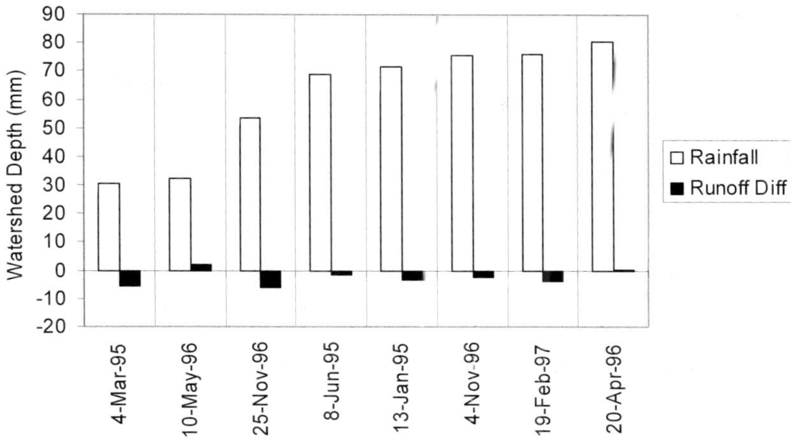

Fig. 2 Eight storm rainfall and simulated runoff differences expressed as an equivalent depth distributed over the watershed.

at each gauge location is then contoured with some smoothing and quality control on non-reporting gauges or other anomalies. Using these ratios, the radar is adjusted such that it agrees with raingauge accumulations. A limited comparison of storm totals for the eight storms revealed an average bias over the eight events to be less than 10%. Errors in runoff simulation and radar inputs for individual storm events are more or less than the average. Given this uncertainty in the input to the model, it is unlikely to obtain simulated volumes that are better than 10% error. The overall simulation accuracy for the eight-storm event suite has a regression coefficient of $R^2 = 0.87$ and a slope of 1.12, indicating a systematic bias towards underprediction. Figure 3 depicts a hydrograph occurring on 19–25 February 1997, selected from the 22 storm events simulated as a part of DMIP showing good agreement in this case. Because optimal parameters were selected that minimize error for the eight events, agreement can be improved on an individual event basis.

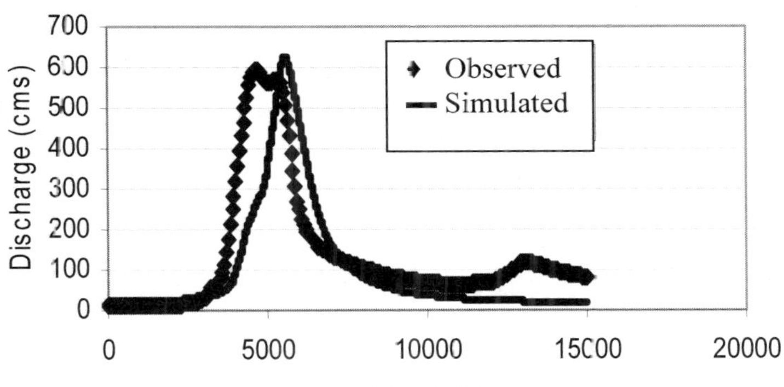

Fig. 3 Simulated and observed hydrograph selected from the 22 DMIP events (19–25 February 1997).

CONCLUSIONS

The results of this calibration are encouraging given that several storms during different seasons were well reproduced with high accuracy. Because of the consistent behaviour, the optimal parameter set would be chosen regardless of which storms were used. The accuracy achieved in runoff volume for all storms is 6.1 mm or less, with an *RMSE* of 3.65 mm. Additional events have been simulated for a total of 22 events in the DMIP experiment. Slight changes resulted, but did not materially change the calibration. Seasonality effects related to land use/cover, as well as radar rainfall bias or anomalies, decrease simulation accuracy. The optimal parameter set is stable and did not materially change as more events from different seasons or years are added. Further, calibration using any one of the events would have yielded about the same result. In no case did the simulated runoff exceed 10% of the rainfall input. An important behaviour of the model is that the absolute difference between simulated and observed runoff was the smallest for the larger events, which typically cause the most flooding.

REFERENCES

Abbott, M. B., Bathurst, J. C., Cunge, J. A., O'Connell, P. E & Rasmussen, J. (1986a) An introduction to European Hydrological System–Systéme Hydrologique Européen, "SHE", 1. History and philosophy of a physically-based distributed modelling system. *J. Hydrol.* **87**, 45–59.

Abbott, M. B., Bathurst, J. C., Cunge, J. A., O'Connell, P. E & Rasmussen, J. (1986b) An introduction to European Hydrological System–Systéme Hydrologique Européen, "SHE", 2. Structure of a physically-based distributed modelling system., *J. Hydrol.* **87**, 61–77.

James, W. P. & Kim, K. W. (1990) A distributed dynamic watershed model. *Water Resour. Bull.* **24**(4), 587–596.

Julien, P. Y. & Saghafian, B. (1991) CASC2D user manual—a two dimensional watershed rainfall–runoff model. Civil Engineering Report CER90-91PYJ-BS-12, Colorado State University, Fort Collins, Colorado, USA.

Rawls, W. J., Brakensiek, D. L. & Miller, N. (1983a) Predicting Green and Ampt infiltration parameters from soils data. *J. Hydraul. Div. ASCE* **109**(1), 62–70.

Rawls, W. J., Brakensiek, D. L., & Soni, B. (1983b) Agricultural management effects on soil water processes, Part I: Soil water retention and Green and Ampt infiltration parameters. *Trans. Am. Soc. Agric. Engrs*, **26**(6), 1747–1752.

Vieux, B. E. (1988) Finite-element analysis of hydrological response areas using geographical information system. PhD dissertation, Department of Agricultural Engineering, Michigan State University, East Lansing, Michigan, USA.

Vieux, B. E. (2001) *Distributed Hydrological Modelling Using GIS*. Water Science Technology Series, no. 38. Kluwer Doredrecht, The Netherlands.

Vieux, B. E. & Gauer, N. (1994) Finite element modelling of storm water runoff using GRASS GIS. *Microcomputers in Civil Engng* **9**(4), 263–270.

Vieux, B. E. & Moreda, F. G. (2003) Ordered physics-based parameter adjustment of a distributed model. In: *Advances in Calibration of Watershed Models* (ed. by Q. Duan, S. Sorooshian, H. V. Gupta, A. N. Rousseau & R. Turcotte) Chapter 20, Water Science and Application Series no. 6, 267–281. American Geophysical Union.

Vieux, B. E. & Vieux, J. E. (2002) V*flo*™: a real-time distributed hydrological model. *Proc. Second Federal Interagency Hydrological Modelling Conf.* (July 28–August 1, 2002, Las Vegas, Nevada, USA).

Vieux, B. E., Bralts, V. F., Segerlind L. J. & Wallace, R. B. (1990) Finite element watershed modelling: one-dimensional elements. *J. Water Resour. Plan. Manage. ASCE* **116**(6), 803–819.

Vieux, B. E., LeDimet, F. & Armand, D. (1998) Optimal control and adjoint methods applied to distributed hydrological model calibration. In: *IV World Congress on Computational Mechanics* (Proc. Int. Assoc. for Computational Mechanics, 29 June–2 July, Buenos Aires, Argentina).

Vieux, B. E., Vieux, J. E., Chen, C. & Howard, K. W. (2003) Operational deployment of a physics-based distributed rainfall–runoff model for flood forecasting in Taiwan. In: *Weather Radar Information and Distributed Hydrological Modelling* (Proc. Sapporo Symp., July 2003) (ed. by Y. Tachikawa, B. E. Vieux, K. P. Georgakakos & E. Nakakita), 251–257. IAHS Publ. no. 282.

White, L. W., Vieux, B. E. & Armand D. (2002a) Surface flow model: inverse problems and predictions. *Ad. in Water Resour.* **25**, 317–324.

White, L. W., Vieux, B. E. & Armand D. (2002b) Estimation of optimal parameters for a surface hydrology model. *Adv. Water Resour..* **26**, 337–348.

Zhang, W. & Cundy T. W. (1989) Modelling of two-dimensional overland flow. *Water Resour. Res.* **25**(9), 2019–2035.

Model behaviour of distributed hydrological modelling with different forcing data resolutions

ROSHAN K. SHRESTHA
Disaster Prevention Research Institute, Kyoto University, Uji, Kyoto 611-0011, Japan
roshan@rdp.dpri.kyoto-u.ac.jp

YASUTO TACHIKAWA & KAORU TAKARA
Disaster Prevention Research Institute, Kyoto University & CREST, JST, Uji, Kyoto 611-0011, Japan

Abstract Acquiring distributed data of particular grid resolution for a particular size of catchment is a matter of interest in distributed hydrological modelling. Using a macroscale distributed hydrological model in Huaihe River basin at Bengbu (132 000 km^2), Wangjiaba (29 800 km^2) and Suiping (2093 km^2) in Japan, the behaviour of the model's performance was observed using the concept of IC ratio (ratio of input grid area and catchments area). An experimental 10-min data setwas generated that refers to the 1.25-degree GAME Re-analysis data and 10-min HUBEX EEWB data. These data are averaged to create coarser distributed data before feeding it to the hydrological model. The change in behaviour of results displays a rapid improvement of performance with finer resolution. The rate of improvement saturates afterwards around an IC ratio range (1:10 ~ 1:20) suggesting the existence of a threshold IC ratio to obtain a satisfactory result. Higher sensitivity in model performance due to spatial resolution is observed in smaller catchments, but low sensitivity may appear if spatial variability is weak.

Key words: Huaihe River, Japan; IC ratio; macroscale distributed hydrological modelling

INTRODUCTION

Hydrological models, which can broadly be classified as either lumped-conceptual models or distributed physically-based models, have evolved to get a better understanding of surface runoff generated from complex watersheds. The lumped conceptual models are not capable of handling the spatial variability. Distributed models are developed, with a hope of obtaining a solution to the problems of lumped models, to represent the variability in physical watershed characteristics. For many of the problem areas the need for the distributed models reflects a demand for predictive capability on the effects of man-induced impacts (Refsgaard & Abott, 1996). The distributed models' strength to assess the effects of spatially variable impacts on hydrological responses has made it an essential tool for the analysis of the hydrological behaviour of a watershed with increased accuracy.

Development of hydrological simulations for gauged and ungauged basins frequently uses meteorological data inputs retrieved from global climatic scenarios, which are provided by General Circulation Models (GCMs). Unfortunately, the structure of GCMs is such that their space resolution is often too coarse and inadequate (Burlando, 2002) to describe variability in hydrological process components at the basin scale. Large efforts to produce finer resolution GCM data with greater accuracy have increased the possibility of extensive use of GCM data in distributed hydrological models.

The use of distributed models is complicated by the need to establish a suitable spatial scale to be used in characterizing watershed conditions such as topography, drainage density, degree of soil saturation and rainfall properties, or in a broad sense the input information to the model. The challenge in applying distributed models to large watersheds is to determine a scale at which spatial variability can be neglected, with average characteristics of a given area providing sufficient information for accurate modelling of basin runoff generation (Molnar, 2000) with respect to catchments characteristics.

Some criteria are felt necessary to determine an acceptable scale of distribution. Shrestha *et al.* (2002) illustrated that the results of distributed hydrological models are better while an IC ratio (ratio between input data resolution and catchments area) is more than 1:10. Results of a distributed model are widely believed to be affected by data resolution. This paper will illustrate the model behaviour of a macroscale distributed hydrological model to clarify the effect of forcing data resolutions. This experiment is conducted for three basins in China, namely Huaihe (132 350 km^2), Wangjiaba (29 844 km^2) and Suiping (2093 km^2).

DESCRIPTION OF MODEL AND DATA

A macroscale distributed hydrological model (Tachikawa *et al.*, 2000) is used to simulate the hydrological processes and obtain river discharge results. This model consists of a runoff process model based on Xinanjiang model (Zhao, 1992) and lumped stream kinematic-wave flow routing model for each grid cells to constitute an element model (see Fig. 1). Output from the runoff process model inside the grid cell is fed as lateral input to the routing model. These element models are interconnected as per the river network arrangement to construct the total simulation process of flow routing (see Fig. 2). The model parameters are calibrated on the basis of field observation records in the Shigan River basin (6000 km^2), a sub-basin inside Huaihe

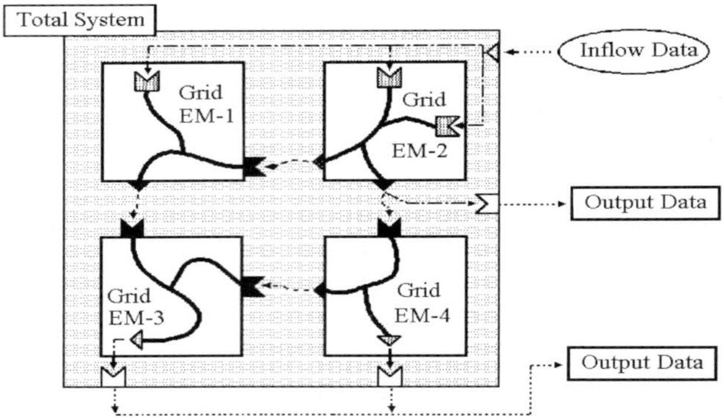

Fig. 1 Structure of an element model.

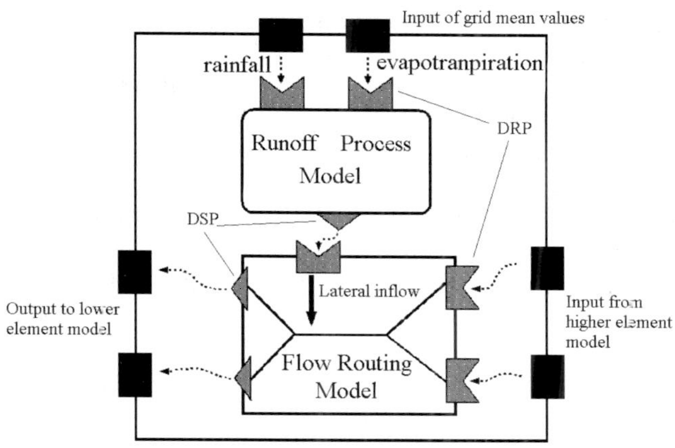

Fig. 2 Schematic total hydrological simulation system.

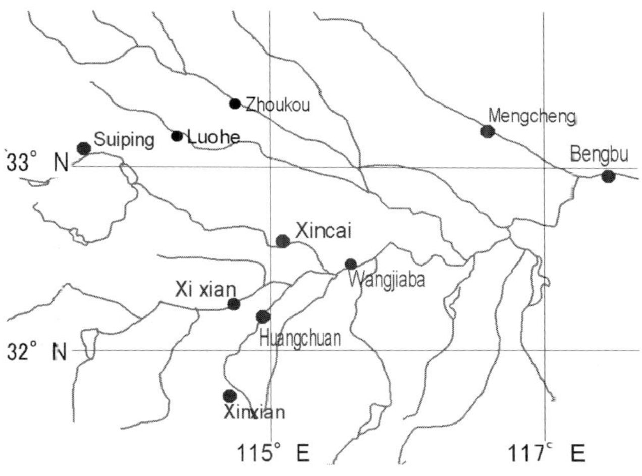

Fig. 3 River networks and location of basin confluences.

River basin (see Fig. 3), which is assumed to include the representative geomorphology of the entire basin.

A finest resolution of 10-min experimental data is created by preserving the spatial pattern of HUBEX-IOP EEWB data (Kozan *et al.*, 2001; termed EEWB data hereafter) and magnitude of GAME Re-analysis (Version 1.1) with 1.25 degree resolution (JMA, 2000; termed GAME data hereafter). EEWB data are basically ground-based data, which consider both distance and direction of each ground observation station to produce distributed 10-min grid data for the precipitation field. EEWB data also include evapotranspiration data simulated by a simple biosphere with urban canopy (SIBUC) model (Tanaka, 1998). The EEWB data set is the finest resolution data that is used for discharge simulation in the experimental basin (Tachikawa *et al.*, 2002) among the available distributed data until now. Therefore this data set is assumed to represent the spatial pattern better than any other data. On the other hand, GAME data are coarse resolution data produced by the 4DDA system of the Japan Meteorological Agency, which has produced good simulation results (Shrestha *et al.*, 2002); therefore its magnitude is believed to contain fewer errors than the actual data.

An arithmetical approach is used in the process of generating the 10-min experimental data to include the spatial variability of EEWB with respect to its averaged figure at the scale of GAME data. Thus the new data set is represented as:

$$P_{4j} = \begin{cases} P_{1j} + P_{2j} + P_{3j} & \text{if} \geq 0 \\ 0 & \text{otherwise} \end{cases} \quad \text{for time j} = 1, 2, 3\ldots \text{m} \tag{1}$$

$$P_{5j} = P_{4j} \frac{\sum P_{1j}}{\sum P_{4j}} \quad \text{for time j} = 1, 2, 3\ldots \text{m} \tag{2}$$

Here, P_{5j} is the newly created experimental 10-min data set that preserves the spatial pattern of EEWB data and the magnitude of GAME data; P_{1j} is GAME 1.25-degree data; P_{2j} is EEWB 10-min data; and P_{3j} is an average of EEWB 10-min data at 1.25-degree resolution and P_{4j} is intermediate data. The experimental 10-min data is used further to create 20-min, 30-min data, etc. by passing an averaging window of (2 × 2), (3 × 3), etc. up to 150-min (2.5 degrees), respectively. The total accumulated values of the input data are kept constant.

RESULTS AND DISCUSSION

Simulation results with experimental data, while compared against the observed discharge, are better obtained than the simulation results using either original EEWB or GAME data for all three experimental basins. The model performances are evaluated by using several indexes such as Pearson's Moment Correlation Coefficients (PMC coeff.), the Nash-Sutcliff coefficient (NSI coeff.), the Index of Agreement (IOA), root mean square error (RMSE) and covariance of simulated and observed discharges.

Distinct changes in the hydrograph are clearly visible for different values of spatial resolution data input. These changes indicate the influence of the spatial resolution with a clear sign of worse simulation results at coarser resolution (see Fig. 4). A point worth noting is that the coarser resolution data applied here are nothing more than the spatially averaged values from the same finer resolution data. Spatial averaging just smoothes the spatial variability at a coarser scale (Vieux, 1993) and keeps the same data pattern and the same accumulated values over the time series.

IC–RATIO: A COMPARISON INDEX

A poorer model performance with the coarser resolution forcing data than with the finer resolution is obvious; in addition, it is also expected that variable model performances are obtained with different test basins. The constitutive relations for different flow processes that are functions of the detailed geometry of flow pathways in different catchments are difficult to compare (Beven, 2002). Some indices, for

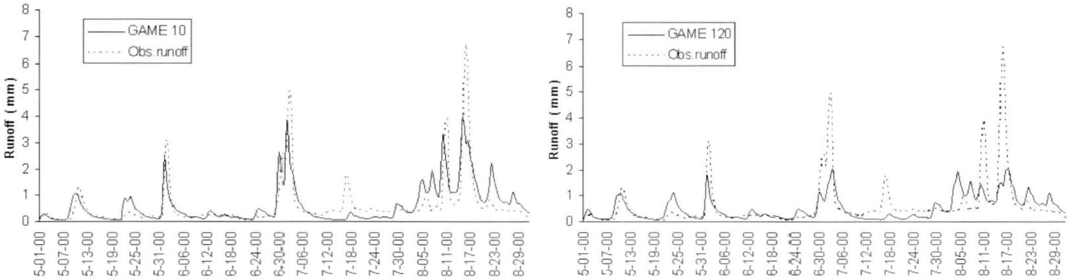

(a) Comparison of hydrograph at Suiping with 10 min and 2.0 degree data.

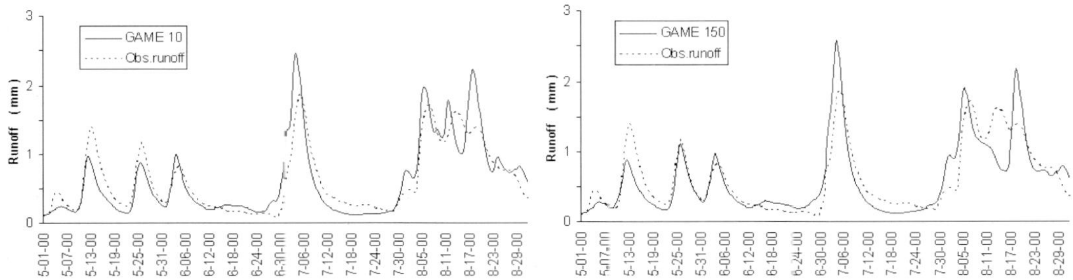

(b) Comparison of hydrograph at Wangjiaba with 10 min and 2.5 degree data.

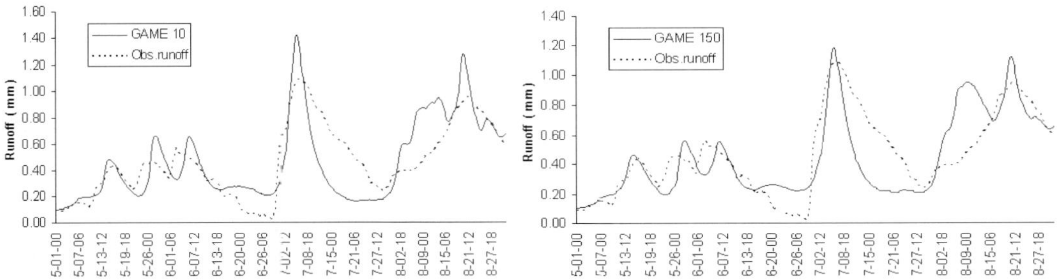

(c) Comparison of hydrograph at Bengbu with 10 min and 2.5 degree data.

Fig. 4 Comparison between the simulated discharges and observed discharges at: (a) Suiping; (b) Wangjiaba; and (c) Bengbu.

example average slope, flow lengths, watershed relief, etc. exist to represent the catchments features but they may not be compatible to test against forcing data resolutions and scale issues.

Spatial variability of data reduces with the coarseness of resolution if catchment size is the same. If larger and larger catchment size is considered, then the magnitudes of spatial distribution of hydrological value such as precipitation become larger too. At the same time the effect of channel routing process attenuates the discharge, which may have a similar effect of variability smoothening. Therefore, while checking the model performance and comparing the effect of data resolution, the scale of catchment size cannot be neglected. In this regard, the concept of ratio between the input grid resolution and catchments size, termed IC ratio (Shrestha *et al.*, 2002), may generally be a useful index for investigating the phenomenon observed due to effects of resolution changes.

The finest resolution in this experiment is 10 minute spatial distance (approximately 13.5 km). The input data for Suiping basin (2093 km^2) varies from 10 minutes to 2 degree resolution. Therefore the IC ratio value ranges from 1:11.77 to 1:0.08. The IC ratio values for Wangjiaba basin (29 844 km^2) and Bengbu basin (132350 km^2) vary from 1:167.8 to 1:0.75 and 1:744.5 to 1:3.3, respectively, within the framework of this experiment.

CHANGE IN MODEL PERFORMANCE

As the output discharge hydrograph changes with change in resolution of data, the model performances also change. Model performances are consistently better toward IC ratio 1:10 (called higher IC ratio values here) from IC ratio 1:0.08 (lower IC-Ratio) in Suiping Basin (see Fig. 5). Higher IC ratio corresponds to finer resolution and lower IC ratio corresponds to coarser resolution of input as catchment size remains constant. The trend of change in model performance with IC ratio shows the smaller basin has higher sensitivity in response to the forcing data resolutions. The rate of change in performance is higher in Wangjiaba (29 844 km^2) than that in Bengbu (132 350 km^2) and higher still in Suiping (2093 km^2) than that in Wangjiaba. This indicates a need for more attention to resolution issues for smaller basins than larger ones. Perhaps in larger basins, a larger heterogeneous part may be considered as homogenous parts, than in smaller basins.

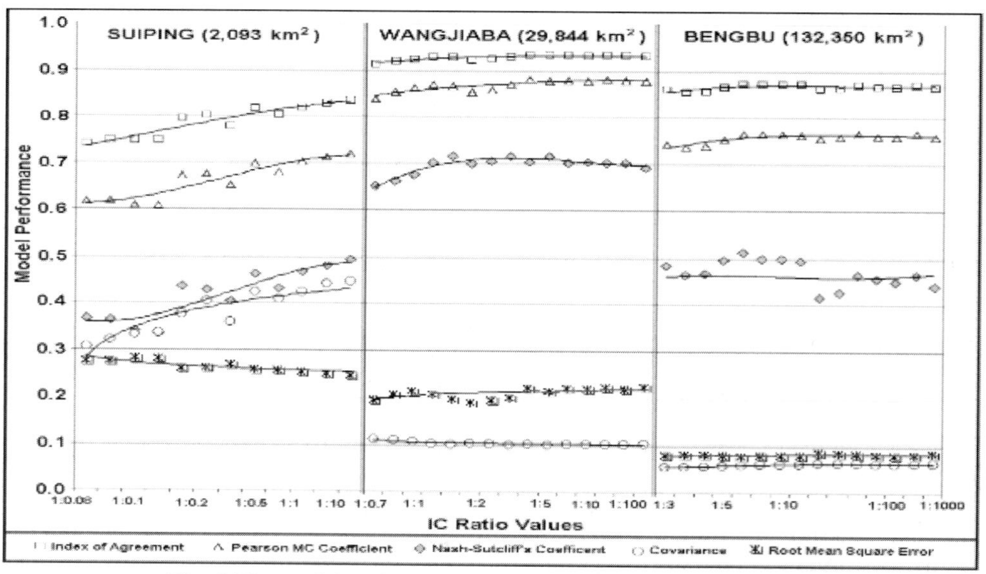

Fig. 5 The model performances of Suiping, Wangjiaba, and Bengbu at different IC ratios.

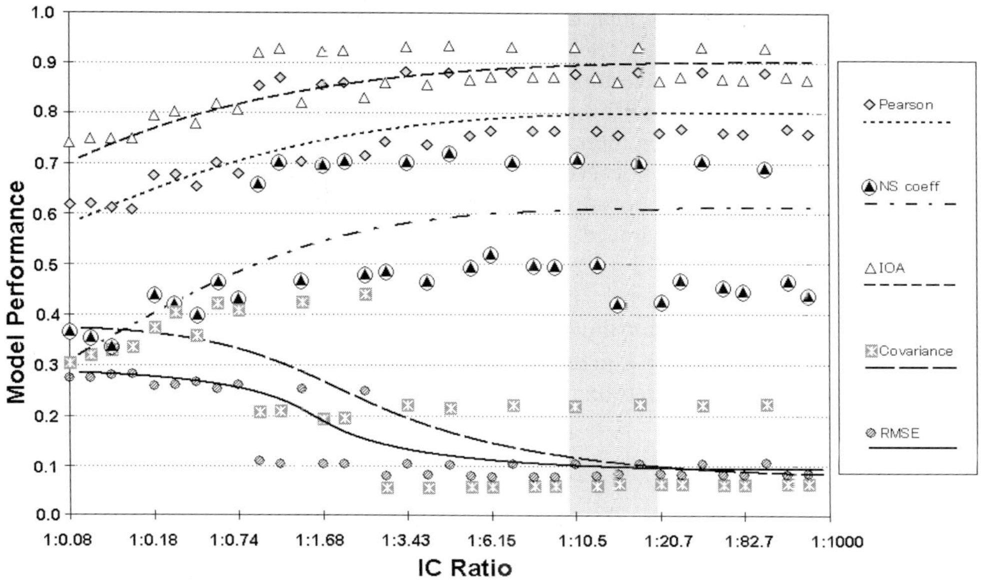

Fig. 6 Model performance *vs* IC ratio.

The rate of performance improvement is higher when the IC ratio is less than 1:10. It is an indication that less than 10 units of distribution inside a catchment may have a higher chance of obtaining better results by increasing the degree of distribution. When the IC ratio values reach higher than 1:20, the improvement rates of the model performances are not more significant but they tend to saturate (see Fig. 6). Conventionally, the more the degree of distribution was assumed to produce, the more accurate the result. It is an indication that there is some limit of improvement; at some point, no practical improvements can be obtained Looking from this aspect, the IC ratio range (1:10–1:20) may be referred to as a marginal value to obtain the simulation result at a satisfactory level.

CONCLUSION

The performance of a distributed hydrological model is generally believed to be better than the lumped-conceptual type of modelling. However, it greatly depends upon the quality of modelling that refers to the quality of modelling technique, model parameters and input data. An increased tendency to use GCM data and other distributed data in hydrological modelling has increased concerns on their accuracy and resolution issues. Observing the behaviour of the model performance different resolution data are analysed using a macroscale distributed hydrological model. The ratio of input data resolution catchment area, termed the IC ratio, can be an index to evaluate the suitability of data resolution in distributed hydrological modelling.

In this experiment, three catchments, ranging from small (2093 km^2) to large (132–350 km^2), have been taken to study the data resolution effects on the discharge simulation results. Smaller catchments are found to be more sensitive than the larger catchments, which indicates that the resolution issues need more attention in the smaller basins than in the larger ones.

The simulation results deteriorate sharply when the IC ratio is below 1:10. The improvement rate in model performance is small after increasing the IC ratio values above 1:20. A smaller improvement in model performance with higher IC ratio, using high resolution data, depicts no need of too high resolution data. A marginal IC ratio value of 1:10–1:20 seems to produce favourable performance of distributed hydrological modelling, viewing the need for efforts to process the high resolution data.

Acknowledgements The contribution of providing HUBEX-IOP EEWB data by Dr K. Tanaka, DPRI, Kyoto University and GAME Re-analysis data by the Japan Meteorological Agency are gratefully acknowledged.

REFERENCES

Beven, K. (2002) Towards an alternative blueprint for a physically based digitally simulated hydrological response modelling system. *Hydrol. Processes* **16**, 189–206.

Burlando, P. & Rosso R. (2002) Effects of transient climate change on basin hydrology. *Hydrol. Processes* **16**, 1151–1175.

Japan Meteorological Agency (JMA) (2000) GAME-IOP Reanalysis Data Version 1.1. downloaded from ftp://hydro.iis.u-tokyo.ac.jp/GAME/ GAIN/GAME-REANL/

Kozan, O., Tanaka, K. & Ikebuchi, S. (2001) Estimation of energy and water budget in the Huaihe River basin, China. *Game Publication* **28**, 32–35.

Molnar, D. K. & Julien, P. Y. (2000) Grid-size effects on surface runoff modelling. *J. Hydrol. Engng ASCE* **5**(1), 8–16.

Refsgaard, J. C. & Abott, M. B. (1996) The role of distributed hydrological modelling in water resources management. In: *Distributed Hydrological Modelling* (ed. by M. B. Abbott & J. C. Refsgaard). Kluwer Academic Publishers, Dordrecht, The Netherlands.

Shrestha, R., Tachikawa, Y. & Takara. K. (2002) Effect of forcing data resolution in river discharge simulation. *Ann. J. Hydraul. Engng JSCE* **46**, 139–144.

Tachikawa, Y., Takara, K., Ichikawa, Y. & Shiiba, M. (2000) Development of a macroscale distributed hydrological model using an object-oriented hydrological modelling system. Hydro-Informatics, CD-ROM, Cedar Rapids, Iowa, USA.

Tachikawa, Y., Takara, K., Tanaka, K., Suishu, T., Ichikawa, Y. & Shiiba, M. (2002) Simulation of river discharge in the Huaihe River basin in China. *J. Japan Soc. Hydrol. & Water Resour.* **15**(2), 139–151.

Tanaka, K.(1998) Activities in Huaihe River basin (HUBEX) study area, Asia-Pacific Network for Global Change Research. Report of 1997–1998, 69–91.

Vieux, B. E. (1993) DEM aggregation and smoothing effects on surface runoff modelling. *J. Comput. Civ. Engng ASCE, Special Issue on Geographic Information Analysis* **7**(3), 310–338.

Zhao, R. J. (1992) The Xinanjiang model applied in China. *J. Hydrol.* **135**, 371–381.

Investigating the effect of catchment characteristics on the response time scale using a distributed model and weather radar information

EFRAT MORIN

Department of Hydrology and Water Resources, University of Arizona, Tucson, Arizona 85721, USA

morine@hwr.arizona.edu

KONSTANTINE P. GEORGAKAKOS

Hydrological Research Center, 12780 High Bluff Drive, Suite 250, San Diego, California 92130 and *Scripps Institution of Oceanography, UCSD, La Jolla, California 92093, USA*

URI SHAMIR

Faculty of Civil Engineering and Stephen and Nancy Grand Water Research Institute, Technion, Haifa 32000, Israel

RAMI GARTI

Soil Erosion Research Station, Emek Hefer 40250, Israel

YEHOUDA ENZEL

Institute of Earth Sciences, The Hebrew University of Jerusalem, Jerusalem 91904, Israel

Abstract The response time scale (RTS) is a characteristic time scale of the catchment that represents the amount of smoothing performed by the hydrological system in transforming the rainfall input into runoff. Previous studies using catchments in Israel and Panama indicate that the RTS is stable for a given catchment and it depends on the catchment characteristics. In order to study the relationship between catchment characteristics and the RTS, a physically-based, non-calibrated distributed hydrological model is applied to a 24-km^2 rural catchment in Israel. The radar rainfall data are used to obtain the computed runoff hydrographs, and these computed data are then used to derive the RTS of the modelled catchment. The effect of catchment parameters (such as length and roughness of hillslopes and channels) on the RTS is examined by changing parameter values and deriving the RTS for each case. Special emphasis is placed in distinguishing the effects of hillslope *vs* channel processes. The results indicate that the effect of hillslope processes on the response time scale is greater than the effect of the channel processes for the study catchment.

Key words catchment response; characteristic time scale; distributed hydrological model; hydrological processes; small catchment; weather radar

INTRODUCTION

The hydrological response of a catchment is a composition of the response of different processes with a variety of characteristic time scales. Associating the catchment response, as a whole, with one representative characteristic scale is important in many aspects of hydrology, both theoretical, for better understanding the integration of different processes for the production of runoff and stream flow, and practical, for modelling and design purposes. In the current study we present a characteristic time scale for the catchment hydrological response and investigate the relationships of this time scale with the hydrological processes and the catchment characteristics.

THE RESPONSE TIME SCALE (RTS)

The RTS concept

Consider a general system that receives an input time series and generates an output time series. One can characterize the time scale of the system by measuring the amount of smoothing that is performed on the input in this transformation. A possible way to obtain this measure is to compare the input time series, smoothed over different time scales, with the corresponding output time series, and to select the time scale that generates a similar measure of smoothness for the two series. Thus, the characterization involves analysis of the two time series, without assuming a specific model for representing the system. For a system that represents the catchment hydrological response, application of this approach implies analysis of rainfall and runoff time series. This is the idea behind the concept of the "response time scale" (RTS) that was developed to characterize the time scale property of catchment hydrological response (Morin *et al.* 2001, 2002).

Identifying the RTS based on rainfall–runoff observations

Morin *et al.* (2002) presented an objective algorithm to determine the RTS of a catchment based on observed gauge or radar rainfall and observed outlet runoff for a single rainfall–runoff event or for a group of events. The algorithm was applied to five small catchments (10–150 km^2) in Israel and in Panama. Table 1 presents the estimated RTS for the analysed catchments. In general, all the RTS values found are reasonably stable and the algorithm was shown to be relatively robust to errors in rainfall and runoff data. It was shown that stable RTS values are more likely to be estimated for high flows with multiple peaks. These RTS values indicate that, as expected, the urban and arid catchments have faster hydrological response compared to the rural and forested catchments. The RTS algorithm successfully captures this inherent difference in catchment response.

Table 1 Catchment characteristics and RTS.

Catchment	Size (km^2)	Climate	Land use	RTS (min)	Uncertainty range (min)
Habel	24	Mediterranean	Rural	70–85	60–125
Raanana	10	Mediterranean	Urban	10–15	5–15
Evtach	43	Semiarid	Rural	125–130	90–215
Ramon	98	Arid	Natural	30–35	10–85
Rio Pequeni	133	Tropical	Forested	120–180	120–180

Relationships of the RTS with catchment characteristics

In order to study the relationships of the catchment characteristics to the RTS, a physically-based, non-calibrated distributed hydrological model was used. The rest of the current paper deals with this issue. The model allowed investigation of what hydrological processes primarily govern the RTS, where emphasis was placed on comparing hillslope *vs* channel processes. The model was applied to the Habel catchment, a 24-km^2 rural catchment with a Mediterranean climate. The only significant runoff-generation process in the catchment is surface runoff.

THE HYDROLOGICAL MODEL AND ITS APPLICATION

Model description

The model assumes that the catchment is composed of a sub-catchment network. Each sub-catchment has a book-like structure: two equal sloping plains (the hillslope) that drain into a sloping channel. The rainfall input to each sub-catchment represents the mean areal rainfall over the sub-catchment area. The rainfall intensities are estimated from weather radar reflectivity data using a power law relationship (Marshall & Palmer, 1948) with bias removal according to raingauge storm depth (Wilson & Brandes, 1979). The rainfall time resolution is 5 min.

An infiltration model used to calculate rainfall excess for each sub-catchment is based on Morin & Benyamini (1977) and on Morin & Kosovsky (1995). It assumes an exponential decay of infiltration capacity with accumulated rain depth and a surface storage. The model also allows the specification of the relative contributing area. The model was found suitable for application to the environment as the study catchment.

The rainfall excess generated over the hillslope is routed toward the channel using the kinematic flow equation. The kinematic flow equation is a simplification of the continuity and momentum equations combined with the Manning equation (see for example Eagleson, 1970). A numerical solution with an explicit, backward difference approximation scheme is used to solve the above equations for time varying rainfall excess. The initial condition is zero (dry) and the boundary condition is the rainfall excess. The runoff discharge at the hillslope bottom enters the channel as lateral flow.

The kinematic flow approximation is also used for channel flow with a similar numerical solution algorithm. The outflow from the sub-catchment is the runoff discharge at the channel's downstream end. The initial condition is zero. The boundary conditions are zero for first-order channels and the sum of outflow from the two contributing upstream channels for higher order channels.

Model application to the Habel catchment

The delineation of the Habel catchment into sub-catchments was done using geographical information system (GIS) algorithms applied on a 50 m digital elevation model (DEM). A 0.5 km^2 threshold area was used to define the channel network, and this resulted in 27 sub-catchments (Fig. 1). The geometrical parameters (area, length,

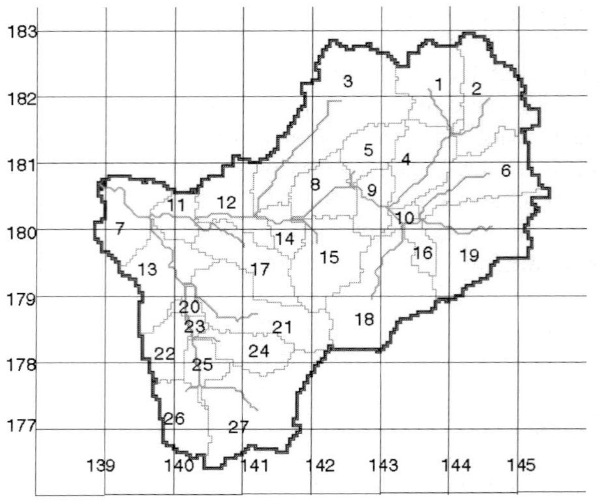

Fig. 1 Delineation of the Habel catchment from the DEM with a threshold area of 0.5 km^2. Numbers indicate basin ID from 1 to 27. The grids represents local coordinates system with 1 km spacing.

slope) of hillslopes and channels are identified using GIS. The infiltration model parameters were obtained from published and unpublished studies according to soil type and land use (Morin & Benyamini, 1977 and Morin, personal communication). Manning's roughness parameters were set based on published tables (for example, Chow *et al.* 1988) as 0.1 s m$^{1/3}$ and 0.035 s m$^{1/3}$ for hillslope flow routing and channel flow routing, respectively. All the model parameters were selected *a priori* and no calibration step was conducted.

RESULTS AND DISCUSSION

RTS of measured and computed outlet runoff

The RTS of each of the events and for the group of three events is given in Table 2. For the first two events the RTS based on the measured and computed runoff agree quite well with each other, while for the third event, the computed runoff hydrograph is too smooth relative to the measured hydrograph and that results in a higher RTS

Table 2 RTS of the analysed events based on measured and computed runoff.

Event	RTS based on measured hydrographs (min)	RTS based on computed hydrographs (min)
17–20 Jan 1996	85–125	65–125
22–23 Jan 1997	60–85	60–85
21–23 Feb 1997	85–115	120–185
All events	70–85	70–85
All events (uncertainty range)	60–125	60–225

value. The RTS for a group of events is a more stable value (Morin *et al.*, 2002). For the current case study the three-events RTS is 70–85 min (60–125 min range) based on the measured runoff and 70–85 (60–225 min range) based on the computed runoff. The fit of the RTS is good considering the model was not calibrated. It suggests that the model sufficiently represents the hydrological processes affecting the RTS. In the following analysis we use the model to identify these processes.

RTS of hydrological processes

The three main processes represented by the model are: (a) infiltration, (b) hillslope routing, and (c) channel routing. The amount of smoothness performed in each one of these processes can be estimated by applying the RTS algorithm to the inputs and outputs of the related model components for the different sub-catchments. Figure 2 demonstrates this for the first-order sub catchment number 19 for the February 1997 storm. The RTS of the infiltration process is estimated from the analysis of rainfall and rainfall excess, the RTS of the hillslope routing process is estimated by analysis of rainfall excess and hillslope outflow (lateral inflow into the channel) and the RTS of the channel routing process is estimated by analysis of lateral inflow and channel outlet flow. The RTS of the combined hydrological response is estimated from the analysis of the rainfall and channel outlet flow. For non-first-order sub-catchments the channel routing includes additional

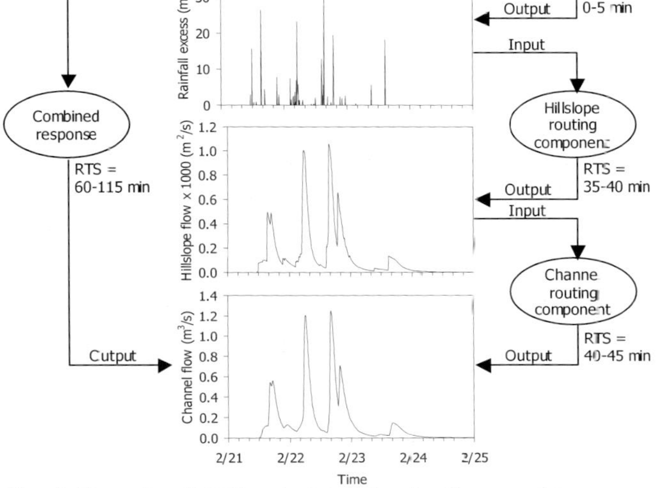

Fig. 2 Example of RTS calculated for the three model components and for the combined response for sub-catchment 19 and the February 1997 event.

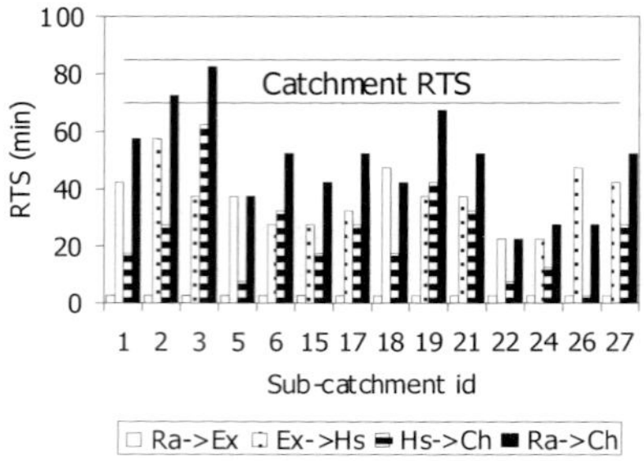

Fig. 3 RTS of first-order sub-catchments for each of the three model components and the combined response. The RTS of the whole catchment response is indicated.

input from the upstream channels. Figure 3 shows for first order sub-catchments the RTS of the three individual components and of the combined response. The RTS is computed for the three events group and the value shown in the figure is the mid-point of the resulted range. The RTS identified for the infiltration process is 0–5 min for all the sub-catchments, indicating no significant smoothness in this process according to the model (recall that the data time interval is 5 min). The RTS values of the hillslope routing and channel routing processes lie between 20 and 60 min and between 0 and 65 min, respectively. For most sub-catchments the RTS associated with the hillslope routing is larger than that associated with the channel routing, suggesting higher importance of the hillslope process for the first order sub-catchments.

We examine next the change in RTS moving to higher order sub-catchments, proceeding toward the catchment outlet. The RTS for each sub-catchment outlet point is shown in Fig. 4(a) as a function of the contributing area. It is calculated from the rain over the contributing area and the modelled runoff at the sub-catchment outlet. To associate the RTS with hillslope and channel routing processes we calculated for each sub-catchment the average flow length over hillslopes and in channels up to the examined outlet point. We assume that water that falls on a hillslope of a given sub-catchment flows on average for half the length of the hillslope, half the length of the local channel and then the full length of downstream channels. For any given outlet of a sub-catchment, the average hillslope flow length and the average channel flow length is calculated from all the sub-catchments that drain into this outlet and it is weighted by sub-catchment area. Figure 4(b) plots the change of hillslope and channel average flow length as a function of contributing area. The flow length on hillslopes is not accumulated downstream. It is relatively variable for small areas, first- or second-order sub-catchments, and it becomes constant toward the catchment outlet. The flow length in channels, on the other hand, is accumulated and is increasing toward the catchment outlet. The RTS at the small contributing areas (smaller than 10 km^2) seems to reflect the variability in the hillslope average flow lengths. At the larger contributing areas the RTS generally increases with contributing area similarly to the increase of the average channel flow length. Recall that flow velocity is considerably smaller over hillslopes than in channels, the overall change of RTS can be associated with average flow time. For typical velocities of 0.1 m s^{-1} and 2.2 m s^{-1} on hillslopes and in channels, respectively, the average flow length can be transformed into average flow time.

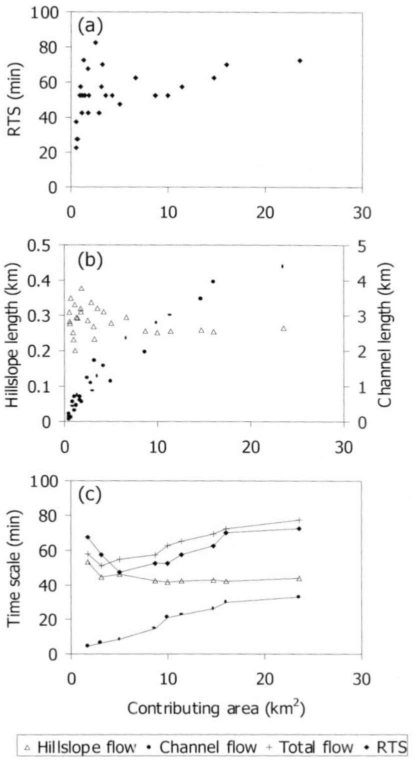

Fig. 4 (a) RTS of all the sub-catchments, (b) average flow length to sub-catchment outlets on hillslopes and in channels, and (c) estimated duration of flow to points along the main channel assuming velocities of 0.1 m s^{-1} over hillslope and 2.2 m s^{-1} in channel, and the RTS for the same points.

Figure 4(c) presents for points along the main channel (from sub-catchment 19 to the outlet) the average flow durations over hillslopes, in channels the total flow duration in hillslopes and channels, and the RTS of the same points. The RTS agrees quite well with the total flow duration.

The above results suggest that the RTS represents the average flow time to the point of interest. Because flow on hillslopes is slower than in channels, at the small catchment areas the hillslope routing process affects more the RTS, while at large catchment areas the time of flow in the channel becomes more dominant and has a larger effect on the RTS. The 24 km^2 studied catchment is still largely affected by the hillslope routing, but the channel effect is expected to be more dominant as we move downstream from the examined outlet point.

Effect of model parameters

The effect of hillslope *vs* channel processes on the RTS is examined also by changing related model parameters. Figure 5(a) presents the change of RTS when lengths of

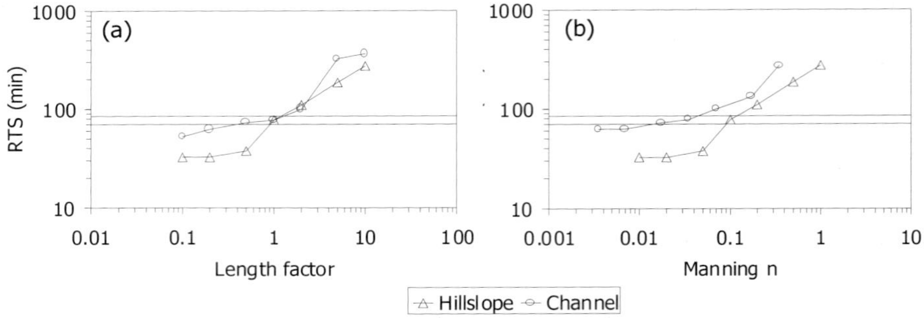

Fig. 5 Effect on the RTS of changing (a) length and (b) roughness parameter for hillslopes and channels. The RTS with the original parameters is indicated.

hillslopes and channels were multiplied by a factor between 0.1 and 10 and Fig. 5(b) presents the change in RTS with changes in roughness parameter. The RTS with the original parameters is indicated in the figures for comparison. We see that for the lower range of parameter values the RTS is more sensitive to changes in hillslope parameters than in channel parameters. For the higher range of parameter values this trend is not as clear and even shows larger effect of channel parameters for large increase of length. This result agrees with the above explanation associating the RTS with the average flow duration. Since originally the channel flow duration is already relatively small, reducing it more hardly changes the RTS. The hillslope flow duration however is significant and its reduction directly affects the RTS. When considerable slowing in channel response is implied the channel flow duration becomes long enough to be significant in its effect on the RTS.

CONCLUDING REMARKS

The response time scale represents a measure of the characteristic time of the catchment hydrological response. It quantifies the amount of smoothness performed on the input when transformed into output by identifying the equivalent averaging that has to be applied to the rainfall such that it will have the same smoothness as the runoff. In the current study a radar-based distributed hydrological model is used to study what controls the response time scale. The results suggest that this time scale represents the average flow duration to the examined concentration point. In small catchments this duration is more affected by hillslope routing processes while as the catchment increases in size channel routing processes become more dominant. This result still needs to be validated by modelling different catchments in a variety of geomorphologic and climatic regimes.

Acknowledgements This study was funded by grants from the United States–Israel Binational Science Foundation (Grant 97-00026), the International Arid Lands Consortium (IALC, 00R-11) and the United States–Israel Binational Agricultural Research and Development Fund (BARD, FI-303-2000). Additional support was provided by a cooperative agreement between HRC and the Hydrology Laboratory (HL) of the US National Weather Service, National Oceanic and Atmospheric Administration.

REFERENCES

Chow, V. T., Maidment, D. R. & Mays, L. W. (1988) *Applied Hydrology*. McGraw-Hill Book Company, Singapore.

Eagleson, P. S. (1970) *Dynamic Hydrology*. McGraw-Hill Book Company, New York, USA.

Marshall, J. S. & Palmer, W. M. (1948) The distribution of raindrops with size. *J. Meteorol.* **5**, 165–166.

Morin, E., Enzel, Y., Shamir, U. & Garti, R. (2001) The characteristic time scale for basin hydrological response using radar data. *J. Hydrol.* **252**, 85–99

Morin, E., Georgakakos, K. P., Shamir, U., Garti, R. & Enzel Y. (2002) Objective, observational-based, automatic, estimation of the catchment response time scale. *Water Resour. Res.* **38**(10), 1212–1227.

Morin, J. & Benyamini, Y. (1977) Rainfall infiltration into bare soils. *Water Resour. Res.* **13(5)**, 813–817.

Morin, J. & Kosovsky, A. (1995) The surface infiltration model. *J. Soil Water Conserv.* **50(5)**, 470–476.

Wilson, J. W. & Brandes, E. A. (1979) A radar measurement of rainfall—a summary. *Bull. Am. Met. Soc.* **60**, 1048–1058.

Weather Radar Information and Distributed Hydrological Modelling (Proceedings of symposium HS03 held during IUGG2003 at Sapporo, July 2003). IAHS Publ. no. 282, 2003.

Effect of spatial and temporal resolution of precipitation data on the accuracy of long-term runoff simulation

HIROSHI ISHIDAIRA, KUNIYOSHI TAKEUCHI, ZONGXUE XU
Graduate School of Engineering, Yamanashi University, Takeda 4-3-11, Kofu 400-8511, Japan
ishi@ccn.yamanashi.ac.jp

TIANQI AO
Public Works Research Institute, 1-6 Minamihara, Tsukuba 305-8516, Japan

JUN MAGOME, MAKOTO KUDO
Yamanashi University, Takeda 4-3-11, Kofu 400-8511, Japan

Abstract The purpose of this study is to investigate the effect of spatial and temporal resolution of precipitation data on the accuracy of long-term runoff simulation. The Radar-AMeDAS precipitation data are used to represent the spatial and temporal variability of precipitation. The precipitation is averaged over different spatial and temporal scales, and both original and averaged data are used as the input for a hydrological model. A rainfall–runoff simulation is carried out in the Fuji River basin by using the BTOPMC model. The difference between the simulated daily hydrographs using original precipitation and the averaged ones is investigated through the comparison of the Nash efficiency for each simulation. In addition, the same analyses are carried out for hydrographs at other time scales, and the influence of spatial and temporal resolution of precipitation data on the accuracy of the hydrological simulation is identified.

Key words distributed hydrological model; long-term runoff simulation; Radar-AMeDAS

INTRODUCTION

Precipitation is usually dramatically distributed with time and space within a basin. In a physically-based distributed model, precipitation data with rational resolution to represent the variability of precipitation needed for an accurate estimation of river discharge. Ichikawa *et al.* (2002) showed that the spatial variability of precipitation causes a significant difference in the shape of the flood hydrograph. However, it is not yet clear at which time and space scale runoff is well estimated with rational accuracy.

The objective of this study is to identify the effect of spatial and temporal resolution of precipitation data on the accuracy of long-term runoff simulation. Precipitation data with different spatial and temporal resolution are processed as input data for a distributed hydrological model, and rainfall–runoff simulations are conducted using these data. The simulated hydrographs are compared so as to identify the influence of the resolution of precipitation data on runoff simulation.

PRECIPITATION DATA

The Radar-AMeDAS precipitation data provided by the Japan Meteorological Agency (JMA) are used in this study. The data are based on the hourly radar gauged precipitation data calibrated by ground gauged precipitation. The spatial resolution of the data is 0.05° × 0.0625° in latitude/longitude (about 5 × 5 km). The original Radar-AMeDAS data are averaged over different spatial scales: 10 × 10 km, 20 × 20 km, 30 × 30 km area and the whole basin area. The temporal averaging is also performed for 1 day, 5 days, 10 days, and 30 days. The precipitation data used in this study are listed in Table 1. Figure 1 shows the precipitation data with different spatial resolution and averaging period.

DISTRIBUTED RAINFALL–RUNOFF MODEL

A rainfall–runoff simulation is carried out using the Block-wise TOPMODEL with Muskingum-Cunge method (BTOPMC). The BTOPMC is a grid-base distributed rainfall–runoff model developed by Ao *et al.* (1999a).

Table 1 Precipitation data with different time and space scales used in the study.

Data	Resolution	
	Spatial	Temporal
S05T00	5 km	
S10T00	10 km	
S20T00	20 km	1 hour
S30T00	30 km	
S60T00	Basin ave	
S05T01	5 km	
S10T01	10 km	
S20T01	20 km	1 day
S30T01	30 km	
S60T01	Basin ave	
S05T05	5 km	
S10T05	10 km	
S20T05	20 km	5 days
S30T05	30km	
S60T05	Basin ave	
S05T30	5 km	
S10T30	10km	
S20T30	20km	10 days
S30T30	30 km	
S60T30	Basin ave	
S05T30	5 km	
S10T30	10 km	
S20T30	20 km	30 days
S30T30	30 km	
S60T30	Basin ave	

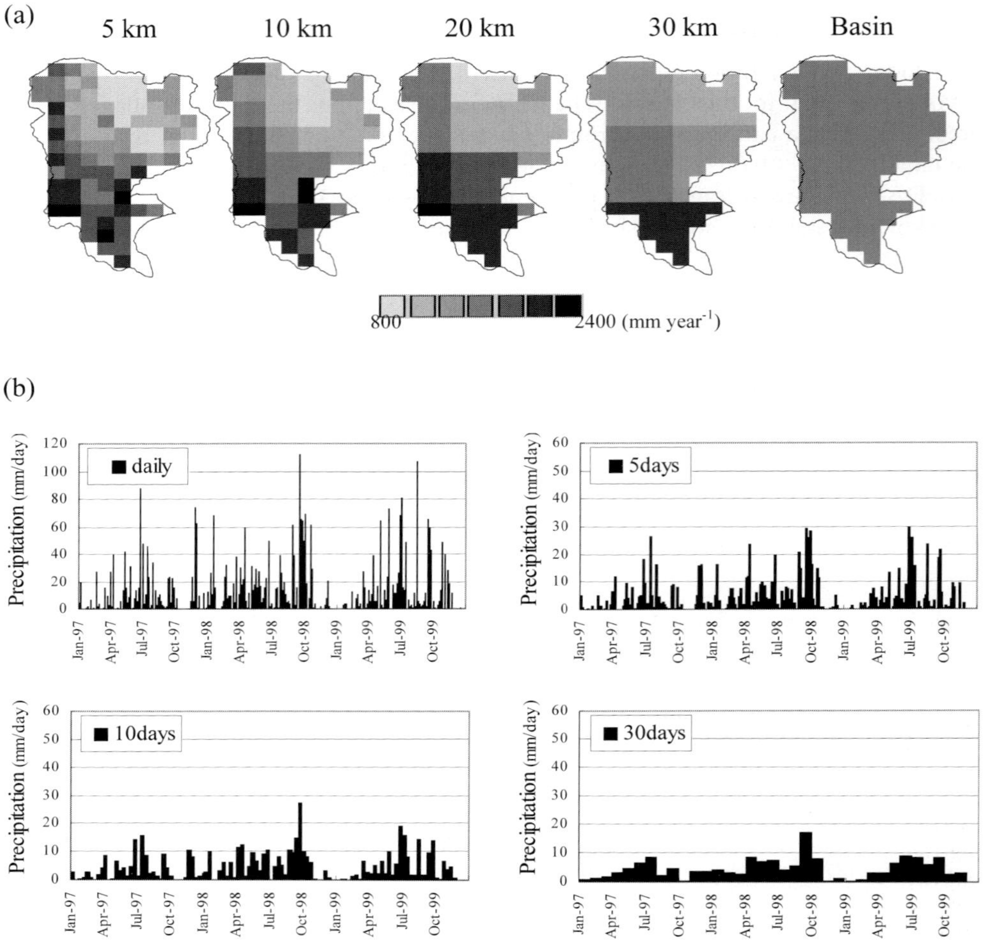

Fig. 1 (a) Spatial distribution of annual precipitation with different spatial resolution. (b) Precipitation data with different averaging period at the Fuji River basin, 1997–1999.

In this model, runoff generation at each grid cell is estimated on the basis of the TOPMODEL concept (Beven & Kirkby, 1979). A river basin is divided into blocks by referencing land cover, soil type and the basin scale, and model parameters are subsequently identified for each block rather than for the whole basin. The average saturation deficit at each time step and the average soil-topographic index are also thereby calculated for each block of the basin. The Muskingum-Cunge method is used for flow routing. This is useful for a large stream network since it can express the flood wave diffusion and propagation. The parameters for river reach (i.e. width and roughness coefficient of river segments) are distributed according to the topography (Ao *et al.*, 1999a; Lu *et al.*, 1989). For more details, see Takeuchi *et al.* (1999) and Ao *et al.* (1999b).

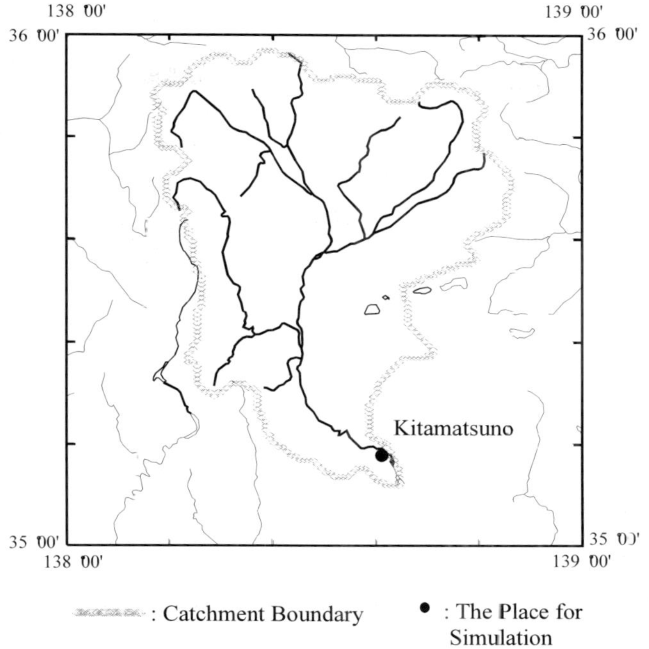

Fig. 2 The Fuji River basin.

Table 2 Model parameters of BTOPMC for the Fuji River basin.

	T_0 (m^2 h^{-1})	m (m)	Sr_{max} (m)	S_{bar0} (m)	n_0
Block 1	6.0	0.0235	0.005	0.1200	0.085
Block 2	12.0	0.0235	0.005	0.0500	0.088
Block 3	10.0	0.0235	0.005	0.1290	0.035
Block 4	12.0	0.0235	0.005	0.0500	0.088
Block 5	10.0	0.0235	0.005	0.0345	0.035
Block 6	10.0	0.0235	0.005	0.0345	0.035

T_0 : saturated hydraulic transmissivity of soil,
m: decay factor of T_0,
Sr_{max}: maximum capacity of root zone,
S_{bar0}: initial saturation deficit of soil,
n_0: Manning roughness coefficient at the outlet of the block.

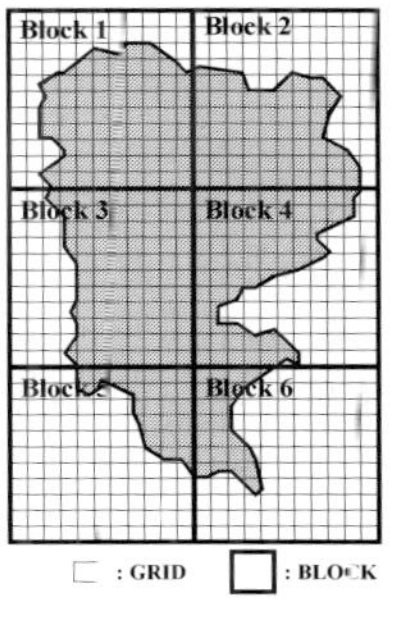

EVALUATION

Using the precipitation data listed in Table 1 and the distributed hydrological model presented above, a rainfall–runoff simulation is conducted in the Fuji River basin (Fig. 2).

The Fuji River with a length of 128 km and basin area of 3570 km^2 is located in the central Honshu Island of Japan. It originates from mountains with a height of 3000 m and flows down to the Pacific Ocean (RSC, 1997). The grid size of the model is 1 × 1 km, and the basin is divided into six blocks. The model parameters are calibrated by trial-and-error approach. The original Radar AMeDAS precipitation (S05T00 in Table 1) is used as the input data for the calibrations. The identified parameters are listed in Table 2. The simulation is carried out for three years (1997–1999).

The difference between the hydrographs of each simulation is evaluated through the Nash efficiency, E_f, defined as follows:

$$E_f = \left\{1 - \left(\sum_{i=1}^{n} (Q_r(i) - Q_c(i))^2 / \sum_{i=1}^{n} \left(Q_r(i) - \overline{Q_r}\right)^2\right)\right\} \times 100 \quad (\%) \tag{1}$$

where Q_r(i) is referred discharge, $Q_c(i)$ is the compared discharge, $\overline{Q_r}$ is the average of $Q_r(i)$, n is the number of data. A higher value of E_f indicates good agreement between $Q_r(i)$ and $Q_c(i)$, and E_f = 100% means the simulated hydrographs agree with the observed ones in terms of the variance.

Table 3 Results of case studies for Experiments 1 and 2.

Case	Precipitation:: Data	Spatial	Temporal	Evaluation Time (T_e)	Nash efficiency (E_f)
Experiment 1: C1-1*	S05T00	5 km			100.0
C1-2	S10T00	10 km			99.6
C1-3	S20T00	20 km	1 hour	1 day	98.6
C1-4	S30T00	30 km			95.7
C1-5	S60T00	Basin ave			94.7
Experiment 2: C2-1*	S05T01	5 km			100.0
C2-2	S10T01	10 km			99.7
C2-3	S20T01	20 km	1 day	1 day	98.9
C2-4	S30T01	30 km			97.1
C2-5	S60T01	Basin ave			96.4
C2-6	S05T05	5 km			80.3
C2-7	S10T05	10 km			77.2
C2-8	S20T05	20 km	5 days	5 days	75.3
C2-9	S30T05	30 km			70.8
C2-10	S60T05	Basin ave			71.7
C2-11	S05T10	5 km			61.1
C2-12	S10T10	10 km			57.1
C2-13	S20T10	20 km	10 days	10 days	55.1
C2-14	S30T10	30 km			50.5
C2-15	S60T10	Basin ave			52.3
C2-16	S05T30	5 km			13.2
C2-17	S10T30	10 km			6.9
C2-18	S20T30	20 km	30 days	30 days	4.9
C2-19	S30T30	30 km			–2.4
C2-20	S60T30	Basin ave			–1.8

*The estimated discharge from this data is used as reference discharge $Q_r(i)$.

CASE STUDIES

In this study, two different kinds of experiments are designed. Experiment 1 is designed to understand the effect of spatial scale of precipitation data on the runoff. The estimated discharge from the original precipitation data (5 × 5 km, hourly) is used as the reference discharge $Q_r(i)$, and the Nash efficiency E_f is calculated for each hydrograph simulated from the spatially-averaged precipitation. The evaluation time T_e, the time scale for calculation of Nash efficiency E_f is taken as one day. Experiment 2 is designed to show the effect of both spatial and temporal scale of precipitation data on runoff. The calculated discharge simulated by using daily precipitation with original spatial resolution (5 × 5 km) is used as the reference discharge $Q_r(i)$, and the Nash efficiency E_f is calculated for each hydrograph simulated by using the spatially- and temporally-averaged precipitation. The evaluation time T_e is the same with the temporal averaging time for precipitation. For example, the Nash efficiency is calculated by using 10 days averaged discharge when precipitation data are averaged over 10 days.

RESULTS AND DISCUSSION

Figure 3 shows the simulated hydrograph using the precipitation data at different spatial scales. The differences between each hydrograph are not significant except the peak discharge. The change in E_f is approximately 5% even when precipitation is averaged over the basin. In this application, spatial resolution of precipitation does not significantly influence the accuracy of long-term runoff simulation. It is noted that the

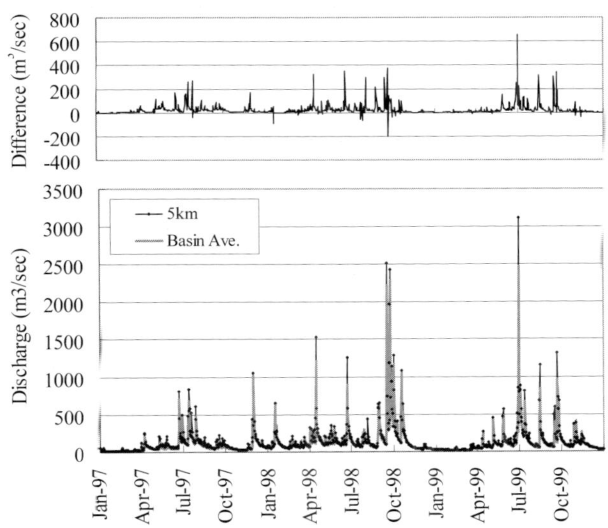

Fig. 3 Simulated hydrograph using precipitation with different spatial scales. In the figure, 5 km denote the simulated discharge using original Radar-AMeDAS data, Basin Ave. denote the simulated discharge using basin averaged precipitation, and the difference between these hydrographs are shown.

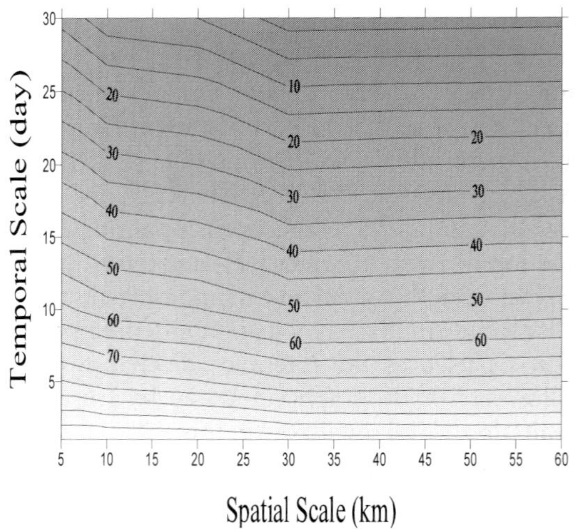

Fig. 4 Change in the Nash efficiency due to spatial and temporal averaging of the precipitation. The spatial scale for "basin average" is represented as 60 km.

results show less influence of the spatial distribution of the precipitation on long-term runoff simulation if the areally-averaged precipitation is accurately observed in the basin. The adequate resolution to detect the basin average precipitation is still problematic.

Figure 4 shows the results of Experiment 2, where the changes in E_f due to spatial and temporal averaging of the precipitation is displayed. It can be seen that the temporal averaging will cause significant difference for long-term hydrological simulation compared with spatial averaging. For example, the difference between daily precipitation with resolution of 5 km and overall basin is 4.6%. On the other hand, the difference between E_f using daily precipitation and monthly averaged precipitation with 5 km resolution is more than 85%. One of the reasons for this large discrepancy is considered to be the change in simulated evaporation. In the case of using temporally-averaged precipitation, the soil layer will be kept in wet conditions due to the longer precipitation duration, and this will cause increasing evaporation from the land surface. For the long-term runoff simulation, the calculation scheme of evaporation within the model should also be changed according to the temporal scale of precipitation. In addition, if the precipitation data are temporally averaged, direct runoff will decrease due to the smaller precipitation rate. This will result in a change in partitioning of the runoff into surface flow and baseflow at each grid cell, and the shape of the resulting hydrograph.

CONCLUSIONS

The influence of spatial and temporal resolution of precipitation data on the accuracy of long-term hydrological simulation is investigated. The result is as follows: the spatial resolution of precipitation dose not show significant influence on the accuracy of long-term runoff simulation, when the areally-averaged precipitation is accurately observed for the basin. However, the adequate resolution to observe the basin-average precipitation is still problematic.

The temporal resolution of precipitation data is sensitive for the long-term hydrological simulation.

Presently, the study of the Prediction in Ungauged Basins (PUB) is planned. PUB is an IAHS decadal initiative to provide hydrological data in ungauged or information-poor basins. It is a scientific endeavor to assemble, promote and build the science and technology capacity to predict and estimate the hydrological phenomena without depending on calibration data. The runoff simulation by using physically-based distributed hydrological models is one of the key components of PUB. For the hydrological simulation in PUB, it is necessary to clarify at which time and space scale runoff is well estimated with rational accuracy. In this study, the impact of spatial and temporal scale of precipitation is investigated. Further studies would be needed for different scales of basins.

REFERENCES

Ao, T. Q., Ishidaira, H. & Takeuchi. K. (1999a) Study of distributed runoff simulation model based on block type TOPMODEL and Muskingum-Cunge method. *Ann. J. Hydraul. Engng JSCE* **43**, 7–12.(In Japanese).

Ao, T. Q., Takeuchi, K., Ishidaira, H., Kazuhiko, F. & Kaneki, M. (1999b) The Naka River floods analyses by BTOPMC method. In: Proc. Int. Symp. Floods and Droughts (October 1999, Nanjing), 414–420. China.

Beven, K. J. & Kirkby, M. J. (1979) A physically based, variable contributing area model of hydrology. *Hydrol. Sci. Bull.* **24**(1), 43–69.

Ichikawa, Y., Tachikawa Y., Hori, T., Takara, K. & Shiiba, M. (2002) Investigation on the scale of rainfall spatial variability to be considered in runoff simulation. *Ann. J. Hydraul. Engng JSCE* **46**, 133–138 (In Japanese).

Lu, M., Koike, T. & Hayakawa, N. (1989) A rainfall–runoff model using distributed data of radar rain and altitude (in Japanese). *Proc. JSCE* **411**(11-12), 135–140.

RSC (1995) *Catalogue of Rivers for Southeast Asia and the Pacific*, vol. I (ed. by K. Takeuchi, A. W. Jayawardena & Y. Takahasi). UNESCO IHP Regional Steering Committee for Southeast Asia and the Pacific (RSC).

Takeuchi, K., Ao, T. Q. & Ishidaira, H. (1999) Introduction of block-wise use of TOPMODEL and Muskingum-Cunge method for the hydro-environmental simulation of a large ungauged basin. *Hydrol. Sci. J.* **44**(4), 633–646.

Weather Radar Information and Distributed Hydrological Modelling (Proceedings of symposium HS03 held during IUGG2003 at Sapporo, July 2003). IAHS Publ. no. 282, 2003.

Effect of precipitation spatial distribution on the hydrological response in the upper Tone River of Japan

DAWEN YANG, TOSHIO KOIKE & HIROSHI TANIZAWA
Department of Civil Engineering, University of Tokyo, Bunkyo-ku, Tokyo 113-8656, Japan
dyang@hydra.t.u-tokyo.ac.jp

Abstract Besides the land surface heterogeneity, spatial distributions of precipitation are another key factor for hydrological simulations. With the development of weather radar techniques it is possible to incorporate radar measurements with distributed models for investigating hydrological applicability, especially for flood forecasting. A grid-based distributed hydrological model was used to evaluate the effect of spatial distributions of precipitation on the hydrological response in the Aimata catchment of the upper Tone River basin in Japan. The precipitation data in 2001, obtained by gauge and radar measurements, were used as input data to the hydrological model. As a result of the two input data sources, the long-term hydrological responses and simulation of flood event are compared and discussed.

Key words distributed hydrological model; hydrological response; radar precipitation; Tone River

INTRODUCTION

With the progress in atmospheric sciences, hydrology can greatly update the modelling approach and then upgrade the simulation accuracy. Apart from conventional raingauge measurements, operational weather radars and numerical weather prediction models offer new opportunities in hydrology to improve flood forecasting and the prediction of water resources. For practical applications of water resource management and flood forecasting, lumped models are commonly employed. Current lumped models use averages for meteorological inputs and watershed characteristics and such are prone to calibration and extrapolation difficulties in the ungauged or poorly gauged basins. The spatial pattern of precipitation input cannot be translated into the simulated hydrograph by the lumped models. However, distributed models incorporate the temporal and spatial variability of watershed and meteorological inputs and allow better representations of the hydrological processes compared with other model types (Yang *et al.*, 2002). Developments of the distributed hydrological models have emphasized capturing the local unique contributions due to different land cover, soil property and topography, and the sensitivity analysis of surface spatial heterogeneity (Yang *et al.*, 2001a). At the same time, many studies have addressed the hydrological model response to precipitation inputs of various spatial and temporal resolutions (Finnerty *et al.*, 1997; Cranmer *et al.*, 2001). These studies have mainly focused on calculation of area mean precipitation and evaluation of the lumped models. Recently, a few studies evaluated the performance of several distributed hydrological models by utilizing spatial distribution of precipitation (Cranmer *et al.*, 2001; Taschner *et al.*, 2001; Anderson *et al.*, 2002). Incorporation of distributed inputs of precipitation was expected to improve the simulation of hydrographs, but some research showed no significant improvement. It is still necessary to examine the performance of different distributed models incorporating different data sources of spatial distributed precipitation.

Now the operational weather radars cover most of the land area of Japan and radar interpreted precipitation is available. Several research projects have been initiated for studying the applicability of radar precipitation data for flood forecasting and water resources management. The main objective of this study was to identify the effect on hydrological responses of the spatial distribution of precipitation and to examine the applicability of radar precipitation data for flood forecasting. For this purpose, a grid-based distributed model has been applied to the upper Tone River of Japan. The precipitation data of 2001, obtained by gauge measurement and radar measurement, were used as the input data. The results corresponding to the two input data sources are compared for long-term hydrological responses and flood events.

METHODOLOGY

Study area

The Tone River is located in northeast Tokyo, Japan. The upper Tone River basin is a high mountainous region. The long-term mean annual precipitation in this region is near 2000 mm. This region suffers heavy snowfall, commonly during the winter (December–February) and typhoons, sometimes in late summer (August and September). The snowmelt runoff is the major water resource used to supply Tokyo, and typhoons also cause high flood risks to Tokyo. Therefore, the upper Tone River has been selected as one of the study areas in the research project of radar precipitation application. As the first step of this research programme, the Aimata catchment, a headwater area of the upper Tone River, was studied. The study area is the same drainage area of the Aimata Dam, by which a reservoir with total storage of 25 million $\times 10^6$ m^3 has been formed. The Aimata catchment has a drainage area of 110.8 km^2, of mainly high mountains covered by forests. The elevation varies from about 400 m to 2000 m, and the depth of topsoil is very shallow.

The geographical information used in this research includes topography, land cover, soil and geology maps. The digital data were obtained from the Japanese Geographical Survey Institute. The topography is simulated by a grid-based DEM (digital elevation model) of 50-m resolution with the available finest data. The digital land cover map is available in 100-m resolution, and has been grouped into four categories including forest (85%), bare soil covered by sparse grass (14%), built-up land (<1%) and water body (<1%). The soil map is available in 1000-m resolution, and it is re-sampled into three types. Combining with the geology map the topsoil depth has been estimated to be from 0.8 to 5 m, and around 1 m in most areas. The vertical distribution of saturated soil hydraulic conductivity is assumed to decrease exponentially from K_0 near the soil surface to K_g of groundwater in a depth of 5 m (the maximum depth of topsoil). The value of K_0 is estimated from the soil data, and the hydraulic conductivity (K_g) of groundwater in unconfined aquifer is a calibrated

Table 1 Soil parameters used in the hydrological model.

Soil type		High permeable soil	Black soil	Forest soil
Coverage (%)		34	14	52
Saturated hydraulic conductivity (mm h^{-1})	Near surface	50	15	25
	Groundwater	1.0	1.0	1.0
Volumetric moisture content of topsoil	Saturated	0.51	0.51	0.51
	Residual	0.17	0.17	0.17
Parameter in Van Genuchten's expression (1980) of soil water retention curve and hydraulic conductivity *	α	0.01695	0.01695	0.01695
	n	1.4132	1.4132	1.4132

*The unit for matric pressure is cm water.

parameter according to the baseflow. Table 1 shows the soil-water parameters used in the hydrological simulations.

Precipitation

Conventional meteorological measurement Conventional meteorological measurements are available from the Automated Meteorological Data Acquisition System (AMeDAS) by the Meteorological Agency. The AMeDAS data include precipitation, temperature, wind speed and sunshine hours, and are available in hourly temporal resolution. By checking the data availability for the Aimata catchment, there is no AMeDAS gauge located inside of this catchment. The nearest AMeDAS gauge (around 10 km from the catchment centre) was selected and the potential evaporation was calculated using the Priestley-Taylor method (Brutsaert, 1982). Because there is only one gauge measurement that can be used for the hydrological simulation in the Aimata catchment, no interpolation has been used to distribute the point data. The values of rainfall and potential evaporation are specified uniformly (as the same as the point value) in the discrete catchment at the 100-m grid size.

Radar measured rainfall The radar interpreted rainfall is provided in 5-min temporal resolution and 1-km spatial resolution for 2001 by the Foundation of Rivers & Basin Integrated Communications (FRICS) of Japan. These rainfall data have been corrected according to the surface gauge measurements. A simple disaggregation is performed, duplicating each grid value to a 100-m grid. The hourly precipitation data are created by aggregating the 5-min rainfall.

Comparison between the gauge and radar measurements Because the accuracies of snowfall measurement by both radar and gauge are poor, only the no-snow season (i.e. from June to November) was examined in this study. Firstly, the daily catchment mean rainfall was compared between the gauge and radar measurements. The total amount of the gauge measured rainfall from 1 June to 30 November 2001 was 1248 mm, which is considered as uniform over the catchment. The rainfall measured by radar shows high spatial variability ranging from 1228 to 2639 mm, and has catchment mean value of 1655 mm in the same time period.

Distributed hydrological model

The grid-based distributed hydrological model used in this study was modified from the geomorphology-based hydrological model (GBHM) (Yang *et al.*, 2002). As mentioned above, the finest DEM is available in 50-m spatial resolution. For accurate simulation of the topography, the hydrological model should also use the same resolution. Unfortunately, grid-based distributed hydrological models usually cannot run at the high spatial resolution for practical applications due to heavy computation and limited land surface information. Considering the capacity of the present commonly used PCs Windows system, this model uses a grid size of 100 m as the same resolution as the land cover map. Since the grid-based DEM represents the topography of a grid by a single slope and average elevation, aggregating DEM from 50-m resolution to 100-m will decrease the gradient and increase the length of the hillslopes. A sub-grid parameterization scheme is necessary when the coarse grid size is used.

Sub-grid parameterization Distinguishing characteristics of the methodology on the sub-grid parameterization is representing sub-grid variability in topography by the geomorphological properties. In the geomorphology-based hydrological model (GBHM) developed by Yang *et al.* (2002), the geomorphological property of river–hillslope formation was used to represent catchment topography. In a similar way, it is assumed that a grid comprises a set of hillslopes located along the rivers. Within a grid, the hillslopes located in a grid are viewed as geometrically similar. The same hillslope model in the GBHM was used, which is a rectangular inclined plane. Length of hillslope is calculated from the 50-m DEM. Slope angle is taken by the mean slope of all sub-grids in the 100-m DEM. The impermeable bedrock slope is assumed to be parallel to the surface. Land cover and soil type are uniform in each grid.

Hillslope hydrological model A physically-based model is used to simulate hillslope hydrological response, in which the vertical plane is divided into several layers, including canopy, soil surface, unsaturated zone and groundwater aquifer (Yang *et al.*, 2002). Canopy interception depends on the LAI (leaf area index) and differs over time. The actual interception is determined according to the rainfall intensity and deficit of canopy storage (Yang *et al.*, 1998). The actual evapotranspiration is estimated from the potential evaporation considering individual component from canopy water storage, root zone, and soil surface. The Richards equation is employed to describe soil water movement in the unsaturated zone. The surface runoff is the infiltration and saturation excesses, which flow through the hillslope into the river and is modelled by the kinematic wave method. The subsurface flow is generated by gravity movement of soil moisture above the field capacity along the hillslope. The groundwater aquifer is treated as individual storage corresponding to each grid. The exchange between saturated zone and the river is considered as steady flow and calculated by Darcy's law (Yang *et al.*, 1998).

The runoff output from one grid is the sum of hillslope responses (both surface and subsurface runoff) within the same grid. The state variable, soil moisture content, of a grid is the average soil moisture of hillslopes. The vertical flux, actual evapotranspiration, of a grid is the total evapotranspiration simulated from the hillslopes.

River network routing In order to simulate flow routing in the river networks, it is necessary to number the river segments (links) and define the flow sequences. Here the Pfafstetter basin numbering scheme (Yang *et al.*, 2000, 2001b) has been applied. It allows a catchment to be divided into a number of small sub-catchments systematically. The size of the final sub-catchments depends on the simulation requirement and the DEM resolution. In the present application to the Aimata catchment, a total of 81 sub-catchments have been derived in an average area of 1.4 km^2. Within a derived minimum sub-basin, the river networks are simplified using the main stream. The flow sequences of the river networks are defined by the codes of the divided sub-catchments. River routing is modelled by the kinematic wave approach. The lateral inflow to a flow interval is the runoff generated from the grids at the same flow distance in the same sub-catchment. Once the water is routing in the river network it is still keeping a dynamic connection with the groundwater, the exchanges between river and groundwater has been described in the hillslope hydrological model.

Model application

Since the precipitation input from both radar and gauge measurements are available hourly, the hydrological simulations have been carried out at an hourly time step. Inflow of the Aimata reservoir is available in hourly temporal resolution from the Ministry of Land, Infrastructure and Transportation (MLIT), and is used for evaluating the model performance. Due to poor accuracy of snowfall measurement, this hydrological simulation focuses on the no-snow season (i.e. from June to November). After the snowmelt season most areas of the catchment are very wet, therefore initial conditions over the study area are specified with soil moisture at field capacity in the root zone and saturated soil moisture in the deeper layers, and full groundwater. The groundwater hydraulic conductivity is calibrated by checking the base discharges in a long-term (June–November) simulation. The calibrated value of the groundwater hydraulic conductivity is shown in Table 1.

RESULTS AND DISCUSSION

Rainfall input effect to the hydrological responses

Hydrological simulations are performed employing different rainfall data obtained by gauge and radar measurements. Figure 1 compares the daily hydrographs simulated using the two data sets. The simulation overestimates peaks for both gauge and radar data. Results given by gauge rainfall give higher simulation of the large peaks, and the radar data mainly overestimate the small peaks. In general it is difficult to evaluate which data source is better from the long-term simulation of daily hydrographs. From the water balance in Table 2, it is known that the gauge rainfall produces slight underestimation of total runoff, but the radar precipitation produces near 40% over-estimates of total runoff. It can be concluded that the gauge gives better measurement of total rainfall in the long-term than the radar does.

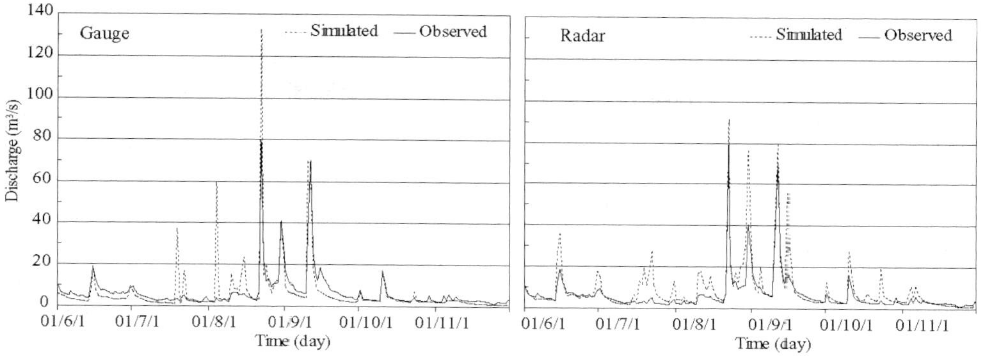

Fig. 1 Comparison of daily hydrographs simulated using different precipitation data.

Table 2 Comparison of long-term hydrological simulations by different data sources (simulated period: 1 June–30 November).

Data source	Precipitation (mm)	Simulated component (mm): Evapotranspiration	Runoff	Observed runoff (mm)	Total runoff error (%)
Gauge data	1248	414	900	941	–4.4
Radar data	1655	431	1289	941	+37.0

For evaluating the effect of rainfall input on the simulation of flood, the hourly hydrograph obtained by the same simulation as the above is examined for a flood event with the highest peak (21–23 August). Figure 2 shows the comparison of the simulated and observed hydrographs for a 72-h period centred to the peak, and the characteristics are summarized in Table 3. The radar rainfall gives much better simulation of this flood event than the gauge rainfall. The difference of total rainfall amounts between gauge and radar measurements is only about 20%. The maximum rainfall intensities

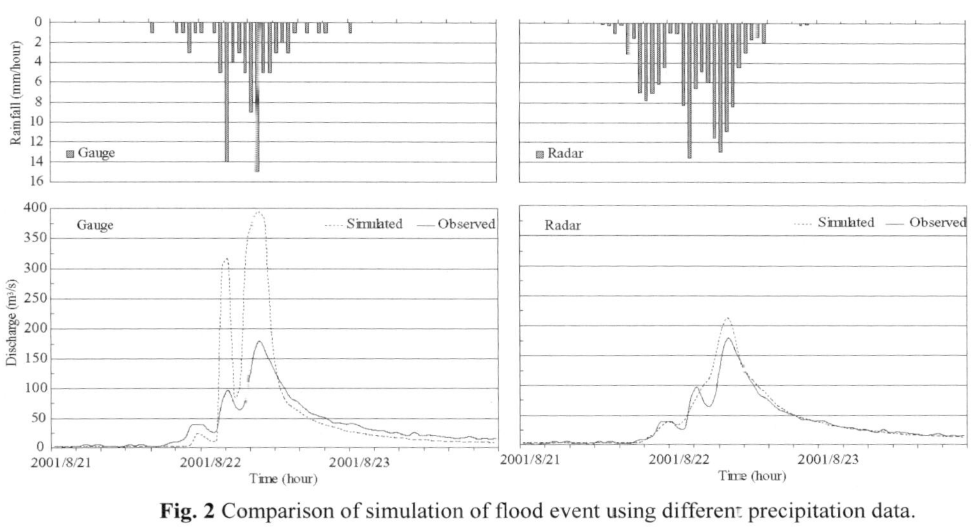

Fig. 2 Comparison of simulation of flood event using different precipitation data.

Table 3 Comparison of flood event (21–23 August) by different data sources.

Data source	Precipitation (mm)	Deviation to the measured hydrograph			Efficiency ε *
		Peak flow (%)	Peak time (hr)	Peak volume (%)	
Gauge data	159	+220	±0.0	+137	-1.6
Radar data	136	+1.2	±0.0	+1.1	0.9

* The efficiency criterion $\varepsilon = 1 - \frac{\sum_{i=1}^{72}(Q_{s_i} - Q_{o_i})^2}{\sum_{i=1}^{72}(Q_{o_i} - \overline{Q_{o_i}})^2}$ (where Q_s and Q_o are hourly discharges

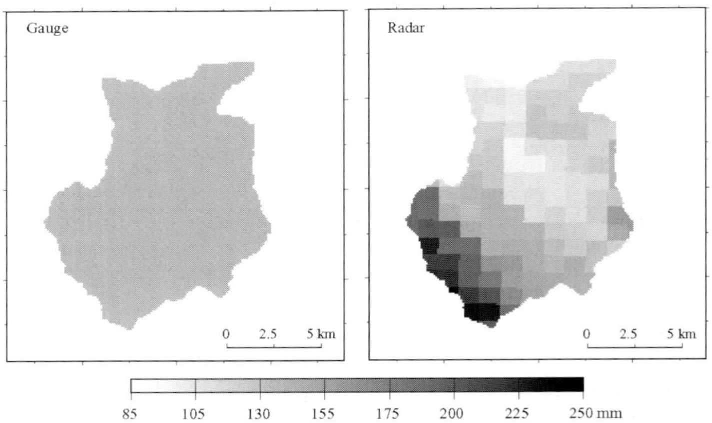

Fig. 3 Spatial distribution of rainfall during the flood event (21–23 August).

from both the gauge and radar are not much different too, but the peak flow simulated by the gauge rainfall is about two times higher than the peak flow simulated by the radar rainfall. This difference is caused mainly by the spatial distribution of rainfall shown in Fig. 3.

Flood sensitivity to the spatial distribution of rainfall

The difference of flood hydrographs between the gauge and radar simulations has shown the importance of rainfall spatial distribution. A simple sensitivity test has been carried out in the study. To account for possible movements of heavy rainfall in space, 2-km adjacency shifts of the rainfall field to four directions are performed and the sensitivities of flood responses are examined. Figure 4 shows the consequent results to the four shifted rainfall fields. When the rainfall field shifts 2 km towards the east and north from the original field, the peak flow occurs about one hour earlier and become near 50% higher than the original case. However, the peak flows slow down slightly and become a little smaller when the rainfall field shifts to western and southern directions. The results again enforce our understanding of the importance of spatial distribution of precipitation in hydrological simulations, especially for flood forecasts.

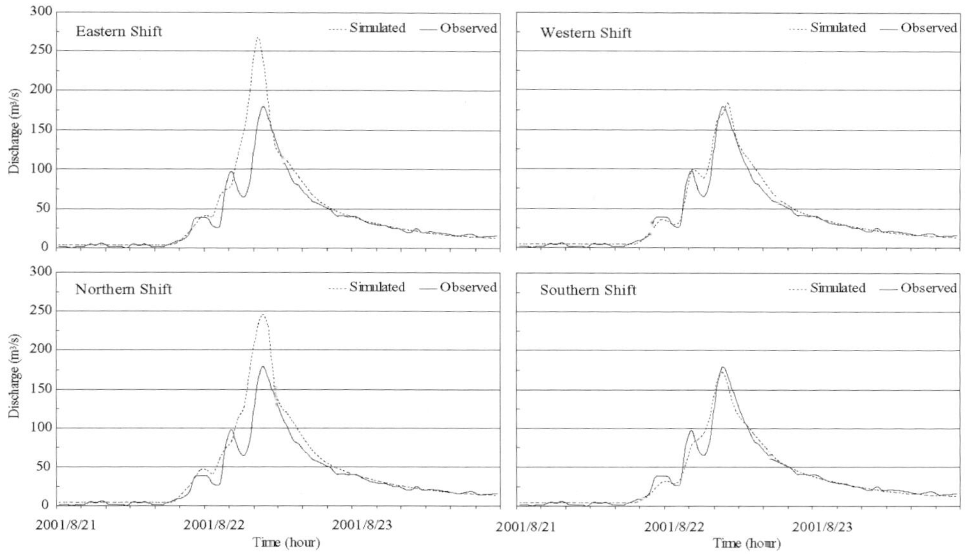

Fig. 4 Flood sensitivity to the spatial distribution of rainfall (2 km shift from the original data).

CONCLUSIONS

A grid-based distributed model has been applied to investigate the effect of rainfall spatial distribution to catchment hydrological responses. Results have shown that the spatial pattern of rainfall can cause significant changes to the catchment hydrological response at high temporal resolution, and this is especially important for flood simulation. A comparison between the gauge and radar measurements of precipitation indicates that the gauge offers a better measurement of long term (no snowfall season) total amount than the radar. However, the radar could measure the spatial pattern of precipitation, which is a key factor for flood simulation. It has been shown that the distributed hydrological model has the capability of incorporating the spatial distribution of precipitation into hydrological simulation. The radar rainfall measurement and distributed model have potentially important applications to flood forecasting in applied hydrology.

Acknowledgements The authors would like to thank the Reservoir Management Office of the Upper Tone River, MILT, for providing the river discharge data, and the Foundation of Rivers & Basin Integrated Communications (FRICS) of Japan for providing the radar precipitation data.

REFERENCES

Andersen, J., Dybkjaer, G., Jensen, K. H., Refsgaard, J. C. & Rasmussen K. (2002) Use of remotely sensed precipitation and leaf area index in a distributed hydrological model. *J. Hydrol.* **264**, 34–50.

Brutsaert, W. (1982) *Evaporation into the Atmosphere*. D. Reidel, Dordrecht, The Netherlands.

Cranmer, A. J., Kouwen, N. & Mousavi, S. F. (2001) Proving WATFLOOD: modelling the non-linearities of hydrologic response to storm intensities. *Can. J. Civ. Engng* **28**, 837–855.

Finnerty, B. D., Smith, M. B., Seo, D. J., Koren, V. & Moglen, G. E. (1997) Space-time scale sensitivity of the Sacramento model to radar-gage precipitation inputs. *J. Hydrol.* **203**, 21–38.

Taschner, S., Ludwig, R. & Mauser, W. (2001) Multi-scenario flood modelling in a mountain watershed using data from a NWP model, rain radar and raingauges. *Phys. Chem. Earth (B)*, **26**, 509–515.

Van Genuchten, M. (1980) A closed-form equation for predicting the hydraulic conductivity of unsaturated soil. *J. Soil Sci. Soc. Am.* **32**, 329–334.

Yang, D., Herath, S. & Musiake, K. (1998) Development of a geomorphology-based hydrological model for large catchments. *Ann. J. Hydraul. Engng JSCE* **42**, 169–174.

Yang, D., Musiake, K., Kanae, S. & Oki, T. (2000) Use of the Pfafstetter basin numbering system in hydrological modelling. In: *Japan Society of Hydrology and Water Resources* (Proc. Ann. Conf.), 200–201. Shiga/Kyoto, JSHW, Japan.

Yang, D., Herath, S. & Musiake, K. (2001a) Spatial resolution sensitivity of catchment geomorphologic properties and the effect on hydrological simulation. *Hydrol. Processes* **15**, 2085–2099.

Yang, D., Kanae S., Oki, T. & Musiake, K. (2001b) Expanding the distributed hydrological modelling to continental scale. In: *Soil–Vegetation–Atmosphere Transfer Schemes and Large-Scale Hydrological Models* (Proc. Maastricht Symp., July 2001) (ed. by A. J. Dolman, A. J. Hall, M. L. Kavvas, T. Oki & J. W. Pomeroy), 125–134. IAHS Publ. no. 270.

Yang, D., Oki, T. Herath, S. & Musiake K. (2002) A geomorphology-based hydrological model and its applications. In: *Mathematical Models of Small Watershed Hydrology and Applications* (ed. by V. P. Singh & D. K. Frevert). Water Resources Publications, Littleton, Colorado, USA. Chapter 9, 259–300.

Grid-based dam inflow prediction on a large river basin using Thiessen polygon and spatially-distributed rainfall data

SEONG JOON KIM, HYUNG JOONG KWON, IN KYUN JUNG & GEUN AE PARK

Department of Rural Engineering, Konkuk University, #1 Hwayang-dong, Gwangjin-gu, Seoul 143-701, Republic of Korea

kimsj@konkuk.ac.kr

Abstract The surface runoff and streamflow responses between Thiessen average rainfall controlled by rainfall stations (T) and spatially distributed rainfall that is cell by cell units (R) would be different from each other. To identify the effects quantitatively, the grid-based kinematic wave storm runoff model (Kim, 1998; Kim *et al.*, 1998) was adopted. The model was tested for the above rainfall conditions on Yeoncheondam watershed (1875 km^2) in South Korea. There were big temporal and spatial surface runoff/discharge differences between T and R when the rainfall is moving in a north to south direction rather than moving in an east to west direction because the watershed width is relatively narrow in shape.

Key words grid-based; kinematic wave; spatial rainfall; storm runoff; Thiessen

INTRODUCTION

Rainfall is a key input for all hydrological/water quality (H/WQ) models because it activates flow and mass transport. Chaubey *et al.* (1999) mentioned that large uncertainty in the model outputs could be expected if the spatial variability of rainfall is not properly taken into account.

The distributed storm runoff model, KIMSTORM (Kim, 1998; Kim *et al.*, 1998) described here was developed to predict temporal and spatial distributions of overland flow, subsurface flow, and streamflow in a watershed. The model coded in C language uses several GRASS (US Army CERL, 1993) grid data and the results are generated as ASCII-formatted map data, and displayed on GRASS–GIS. There are two grid-based hydrological models available with the GRASS distribution. *r.water.fea* (Vieux *et al.*, 1990) is an interactive program that allows the user to simulate stormwater runoff analysis using the finite element numerical technique. *r.hydro.CASC2D* (Saghafian, 1993) is a physically-based, distributed, raster hydrological model which simulates the hydrological response of a watershed subject to a given rainfall field. These models are fully embedded in GRASS. On the other hand, KIMSTORM just uses model input and generates model output as GRASS format.

The objective of this research is to study the effect on runoff when using point-averaged rainfall by Thiessen polygon and spatially-distributed rainfall interpolated by using point value rainfall using KIMSTORM. This study attempts to quantify the dam inflow due to rainfall spatial variability considering moving direction when applied to a large watershed (1875 km^2) with a loose network of four raingauges. The study watershed is in a special situation because part of the watershed belongs to North Korea, thus the hydrological data have not been collected in good condition.

MODEL (KIMSTORM) DESCRIPTION

The watershed is divided into regular cells and the schematic representation of water balance components in a cell is shown in Fig. 1. We can choose a 3 × 3 grid filter to determine lateral flow direction. The flow direction from the centre grid cell is determined by identifying the steepest slope to the eight neighboring grid cells using elevation data. The direction of the steepest descent downhill, numbering 1–8 to the centre grid cell, eventually generates flow direction. The water of the grid cell flows in the steepest direction. Single output flowpath is chosen in this study, but the inflow to a grid cell can come from multiple sources. To deal with multiple lateral inflows, the eight neighbouring grid cells are treated as inflows only if the flow direction number among them matches the already determined flow direction grid. If flows drain into a stream, then they are treated as lateral flow to the stream.

The kinematic wave equation for saturation overland flow/streamflow is obtained by the Manning equation:

$$Q = \alpha R^{m1} A^{m2} \quad (1)$$

where Q is discharge ($m^3\ s^{-1}$); $\alpha = n^{-1} W^{-2/3} \tan^{1/2}\beta$ for overland flow, $\alpha = n^{-1}\tan^{1/2}\beta$ for streamflow; n is Manning's roughness coefficient; W is cell width orthogonal to streamline (m), β is cell slope (degree), $R = H$ for overland flow as a shallow sheet flow, $\gamma \cdot A^{1/2}$ for streamflow; H is flow depth (m); $\gamma = 0.354$ for rectangular channel (Moore & Foster, 1990; Moore & Burch, 1986); A is the cross-sectional flow area (m^2); m1 = 0 for overland flow, 2/3 for streamflow; and m2 = 5/3 for overland flow, 1.0 for streamflow. The lateral saturated subsurface flow equation approximated by the kinematic assumption (Beven, 1982; Sloan & Moore, 1984) was adopted:

$$Q_{sub} = K_s A \sin\beta \quad (2)$$

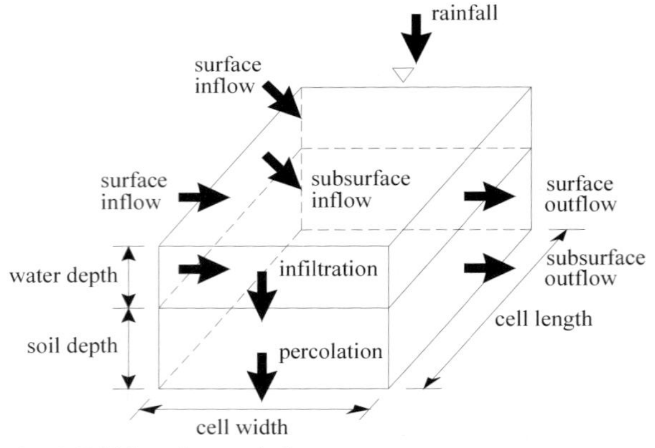

Fig. 1 Grid-based water balance components.

where Q_{sub} is subsurface discharge; K_s is saturated hydraulic conductivity (m s^{-1}). The Huggins and Monke infiltration equation (Beasley *et al.*, 1980) was adopted, and the percolation rate was applied when the soil moisture content is above the field capacity:

$$f = f_c + f_i (SM_r / PO_e)^b \tag{3}$$

where f is infiltration rate (mm h^{-1}); f_c is final infiltration rate (mm h^{-1}); f_i is initial infiltration rate (mm h^{-1}); SM_r is available storage capacity (m^3 m^{-3}); PO_e is effective porosity (m^3 m^{-3}); and b is a constant coefficient. The spatial distribution of initial subsurface flow depths in a watershed can be obtained from soil information maps describing porosity, field capacity and initial soil moisture conditions. The initial flow depth for each cell can be calculated by:

$$\begin{aligned} H_i &= D_c (SM_i - F_c)/(PO_e - F_c) && F_c < SM_i < PO_e \\ &= D_c && SM_i > PO_e \\ &= 0 && SM_i \le F_c \end{aligned} \tag{4}$$

where H_i is initial flow depth in cell (m); D_c is soil depth above the impeding layer (m); SM_i is initial soil moisture content (m^3 m^{-3}); and F_c is field capacity (m^3 m^{-3}). Water balance in a cell is sequentially calculated beginning from the uppermost left cell to the lowest right cell. The calculated outflow delivered to the neighbouring cell by flow direction is stored and used as inflows of the cell at the next time step. The water balance equation for overland flow and streamflow is:

$$\frac{dS_i}{dt} = P_i - F_i + \sum Q_{in.i} - Q_{out.i} \quad \text{for overland flow} \tag{5}$$

$$\frac{dS_i}{dt} = P_i - F_i + \sum Q_{in.i} + \sum Q_{sub.in.i} - Q_{out.i} \quad \text{for streamflow}$$

where i is cell address; S_i is cell storage (m^3); P_i is rainfall (m^3 s^{-}); F_i is infiltration (m^3), $Q_{in.i}$ is inflows to the cell (m^3 s^{-1}); $Q_{out.i}$ is outflow from the cell (m^3 s^{-1}); $Q_{sub.in.i}$ is subsurface inflows to the cell (m^3 s^{-1}); and t is time (s). The soil moisture routing equation is:

$$\frac{dSM_i}{dt} = F_i + \sum Q_{sub.in.i} - Q_{sub.out.i} - DP_i \tag{6}$$

where SM_i is soil moisture content in the cell (m^3); $Q_{sub.out.i}$ is subsurface outflow from the cell (m^3 s^{-1}); and DP_i is deep percolation to the groundwater (m^3).

A schematic flow diagram of the model is shown in Fig. 2. As input data for the model, six GRASS regular grid maps that are elevation, stream and flow direction, land use, soil, Thiessen or spatial rainfall data are converted into ASCII-formatted map data using the GRASS command r.out.ascii. Streamflow at the watershed outlet, and ASCII-formatted discharge map, flow depth map, and soil moisture map were generated for 1-h interval while the calculation time step is 1 min. The ASCII-formatted map data were converted into GRASS maps using the GRASS command r.in.ascii.

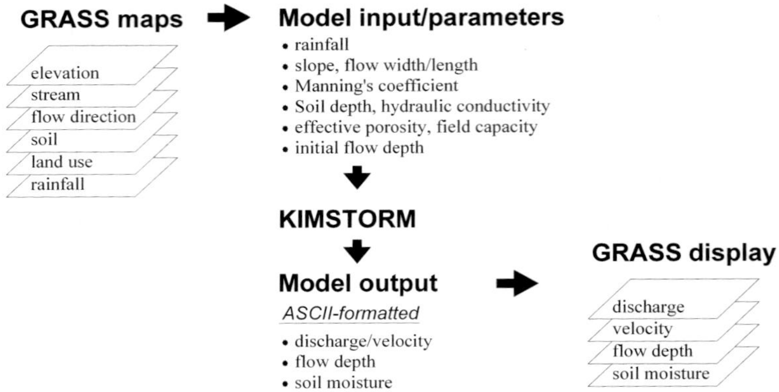

Fig. 2 Schematic diagram of the model.

MODEL APPLICATION

Watershed, soils and land-use data

The model was tested on Yeoncheondam watershed located in the north of South Korea. About one third of the watershed belongs to North Korea. The latitude ranges from 37°45′ to 38°33′. The watershed area is 1875 km^2 and the elevation ranges from 73 to 1171 m. Watershed location, elevation, and the streams are shown in Fig. 3. The subsurface unsaturated layer of most soils is permeable with soil depths ranging from 50 to 150 cm. Clay loam dominates covering 56.7% of the watershed. As soil information of North Korea was not available, it was assumed as clay loam. More than 52.3% is forested and 30.2% of lowland is paddy fields. The remaining area is dry field farming (11.3%) and farm village (0.6%) scattered between forest and paddy.

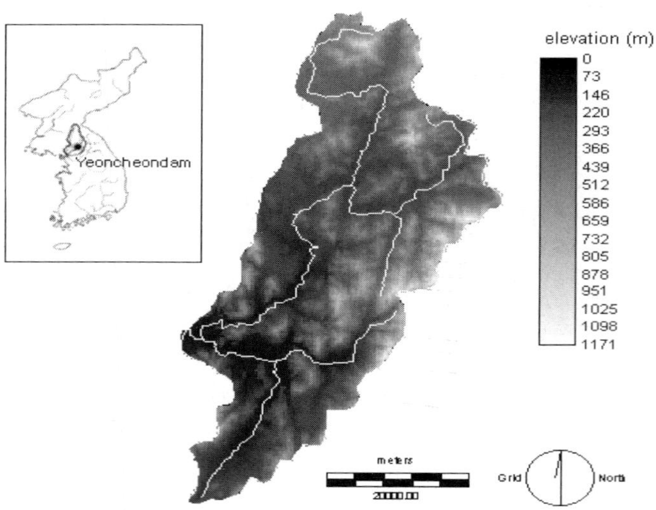

Fig. 3 Watershed location, main streams and topography.

Map data preparation

Elevation data was rasterized from the vector map of 1:5000 scale supplied by the Korea National Geography Institute. The flow direction map and flow accumulation map were generated using elevation data with the GRASS r.watershed command. Soil data were rasterized from the vector map of 1:50 000 scale supplied by the Korea Rural Development Administration. The land-use map was prepared using Landsat TM image of 2 June 1992 supplied by RESTEC (Remote Sensing Technology of Japan). Land cover analysis was achieved through the maximum likelihood classification with overall accuracy of 95.6%. The Thiessen polygon network was rasterized from the point vector data of geometric coordinates of four rainfall gauges. The Spatial rainfall map was generated from the valued point vector files using distance-weighted average interpolation. These maps were prepared to have a 400 × 400 m cell size.

Dam inflow prediction by Thiessen polygon network and spatially distributed rainfall data

Simulation by T (Thiessen polygon rainfall data) and R (spatially-distributed rainfall data) was conducted in four moving directions [NS (north to south), SN (south to north), EW (east to west), and WE (west to east)] with 2 h duration of arbitrary rainfall. Table 1 shows the applied rainfall conditions. Figures 4 and 5 show the predicted dam inflows by NS/SN direction, and EW/WE direction for T and R, respectively. Table 2 summarizes the predicted statistics of each case.

Table 1 Applied 2 h rainfall characteristics for each condition.

Rainfall type		Thiessen					Spatial distribution	Remarks
		Kim-hwa	Dong-song	Yeon-cheon	Po-cheon	Total/Mean		
Area (km^2)		839.9	522.3	123.3	389.2	1874.7	1874.7	2 h rainfall
Rainfall Intensity ($mm\ h^{-1}$)	ⓐ	90	70	40	15	65.6	72.9	ⓐ+ⓑ: NS
	ⓑ	70	90	120	145	94.4	86.5	ⓑ+ⓐ: SN
	ⓒ	90	50	10	45	64.3	68.3	ⓒ+ⓓ: EW
	ⓓ	70	100	130	110	90.6	90.0	ⓓ+ⓒ: WE

Table 2 Summary of predicted dam inflows by NS/SN direction, and EW/WE direction for T and R.

	NS:		SN:		EW:		WE:	
	R	T	R	T	R	T	R	T
Peak discharge ($m^3\ s^{-1}$)	1216.1	3572.1	1605.0	3550.2	3536.6	3307.1	3470.8	3350.1
Time to peak (h)	12	7	8	7	9	9	8	8
Total discharge ($10^6\ m^3$)	100.7	207.4	125.4	207.8	204.5	198.2	204.9	199.0

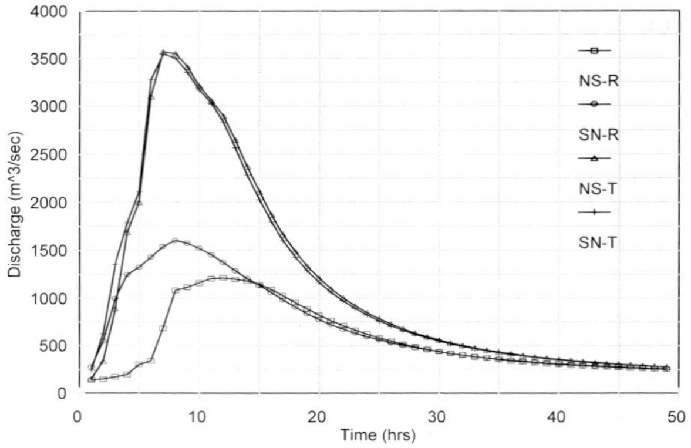

Fig. 4 Predicted dam inflows by NS and SN direction for T and R.

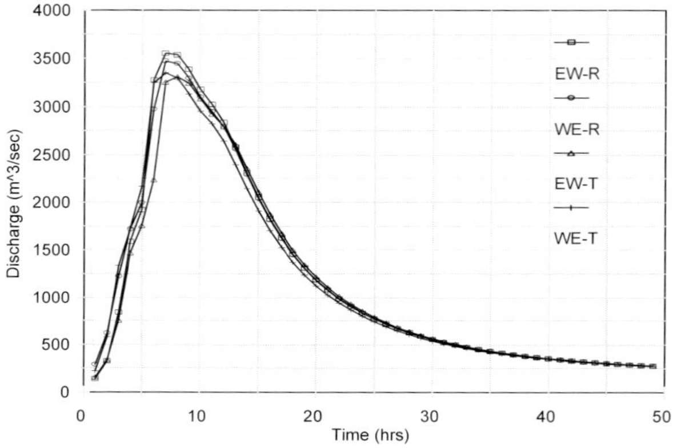

Fig. 5 Predicted dam inflows by EW and WE direction for T and R.

Runoff volume and peak flow rate of NS-R were 48.6% and 34.1% compared with those of NS-T, and the time to peak flow was delayed by 5 h. The runoff response of NS-R was slower than that of NS-T. This is because the first 1 h rainfall in the downstream areas of the watershed is relatively low to contribute as a streamflow and the produced runoff at the upstream areas needs time to reach to the watershed outlet. Runoff volume and peak flow rate of SN-R were 60.3% and 45.2% comparing with those of SN-T, and the time to peak flow was delayed by 1 h. The simulated results of under-produced runoff volume and peak flow by R application indicate the reflection of the spatial variations of rainfall to runoff. The results also suggest that the rainfall stations of this watershed are not sufficient to represent the areal average rainfall of the watershed as a well-designed Thiessen network. The simulated results by EW/WE direction for T and R showed little difference in the hydrograph. This is because the watershed is elongated to the north and south, and the rainfall by T and R did not give considerable difference from each other.

SUMMARY AND CONCLUSIONS

An analysis of the effect of rainfall by Thiessen polygon and spatial distribution map on runoff by using distributed storm runoff model (Kim, 1998; Kim *et al.*, 1998) was accomplished. The model was applied to a 1,875 km^2 watershed with a loose network of four raingauges. It revealed that the hydrograph generation is strongly related to the spatial variations in the rainfall, moving direction of the rainfall, and the shape of a watershed. Simulated results imply that the watershed needs more raingauge stations or other rainfall detection methods such as radar techniques to represent spatial variations of rainfall for the whole watershed. We can infer a result that if the watershed has a dense network of raingauges enough to represent spatial variations of rainfall, then the runoff difference by Thiessen polygon and spatially distributed rainfall will decrease.

REFERENCES

Beasley, D. B., Huggins, L. F. & Monke, E. J. (1980) ANSWERS: a model for watershed planning. *Trans. Am. Soc. Agric. Engrs* **23**(4), 938–944.

Beven, K. J. (1982) On subsurface stormflow: predictions with simple kinematic theory for saturated and unsaturated flows. *Water Resour. Res.* **18**, 1627–1633.

Chaubey, I., Haan, C. T., Salisbury, J. M. & Grunwald, S. (1999) Quantifying model output uncertainty due to spatial variability of rainfall. *Trans. Am. Soc. Agric. Engrs* **35**(5), 1113–1123.

Kim, Seong J. (1998) Grid-based KIneMatic Wave STOrm Runoff Model (KIMSTORM) I. Theory and model. *J. Korea Water Res. Assoc.* **30**(3), 303–308.

Kim, Seong J., Chae, H. S. & Shin, S. C. (1998) Grid-based KIneMatic Wave STOrm Runoff Model (KIMSTORM), II. Application—applied to Yoncheon Dam watershed. *J. Korea Water Res. Assoc.* **30**(3), 309–315.

Moore, I. D. & Burch, G. J. (1986) Sediment transport capacity of sheet and rill flow: application of unit stream power theory. *Water Resour. Res.* **22**, 1350–1360.

Moore, I. D. & Foster, G. R. (1990) Hydraulics and overland flow. In: *Process Studies in Hillslope Hydrology* (ed. by M. G. Anderson & T. P. Burt), 215–254. John Wiley, New York, USA.

Saghafian, B. (1993) Implementation of a distributed hydrological model within Geographic Resources Analysis Support System (GRASS). In: *Integrating Environmental Models and GIS* (Proc. Second Int. Conf.). Breckenridge, Colorado, USA.

Sloan, P. G. & Moore, I. D. (1984) Modeling subsurface stormflow on steeply sloping forested watersheds. *Water Resour. Res.* **20**, 1815–1822.

US Army CERL (1993) *GRASS 4.1 User's Manual*. Champaign, Illinois, USA.

Vieux, B. E., Bralts, V. F., Segerlind, L. J. & Wallace. R. B. (1990) Finite element watershed modeling: one-dimensional elements. *J. Water Resour. Plan. Manage.* **116**(6), 803–819.

Application of radar-measured rain data in hydrological processes modelling during the intensified observation period of HUBEX

LILIANG REN, CHUNHONG LI & MEIRONG WANG

Department of Hydrology and Water Resources, Hohai University, No. 1 Xikang Road, Nanjing 210098, China

rll@hhu.edu.cn

ABSTRACT Raster flow vectors and watershed delineation were generated from the DEM for the upper area of Huangnizhuang station in the Shihe catchment intensified observation field for the HUBEX/GAME project. The Xinanjiang model was then applied for runoff production in each grid element where rain data measured by weather radar at Fuyang station were used as the input of the hydrological model after correction. Those elements are connected by flow vectors with the outlet of drainage catchment, in which runoff is routed by the Muskingum method from each grid element to the outlet. The Nash-Sutcliffe model efficiency coefficient is up to 92.41% from 31 May to 3 August 1998. The index of model efficiency improvement is 51.22% if compared with the case where raingauge data were used. This shows that radar-measured data are superior to raingauge data as the input of hydrological modelling.

Key words catchment; digital elevation model; hydrological processes; radar; raster; Xinanjiang model

INTRODUCTION

Topography is of great importance to the description, quantification, and interpretation of land surface processes, such as water and energy fluxes, evapotranspiration and soil erosion (Beven, 1989; Moore *et al.*, 1991). Topography varies from one location to another and topographical variability affects the apparent slope, flow path, land cover and soil formation. Moreover, topography is easily available earth surface information. Extracting topographical information for a watershed by traditional and manual operation might be tedious, time-consuming and error-prone. Research over the past decades has demonstrated the feasibility of extracting topographical information directly from raster-based digital elevation models (DEMs) (Doyle, 1978; O'Callaghan & Mark, 1984; Jenson & Dominique, 1988). In the field of hydrology and water resources, automatic evaluation of DEMs has mainly focused on watershed segmentation, definition of drainage divides and identification of river networks. This automatic extraction of network and subwatershed properties from the DEM represents a convenient and rapid way to parameterize a watershed (Garbrecht & Campbell, 1997).

The development of computer science and digital techniques such as DEMs provide a solid technical foundation for the revolution of hydrological modelling. The spatial distribution of land surface characteristics, such as topography, land-cover, soil, watershed divide, drainage network, catchment area, could be expressed digitally, so as to avoid using the conventional manual method. Thus the digital hydrological model emerges. This digital hydrological model is a modern, physically-based modelling technique containing a large amount of information. It is a DEM-based distributed model, and the combination of grid-based runoff generation model with streamflow routing model. In this study, the Xinanjiang model (Zhao, 1992) is applied for the development of digital rainfall–runoff model in each grid element that is generated and numbered automatically by the Digital Elevation Drainage Network Model (DEDNM; Martz & Garbrecht, 1992).

Techniques for solving hydrological problems are borrowed from several disciplines such as mathematics, statistics, operation research, computer science and information theory. Based on solution techniques, hydrology can be subdivided into different branches. Thus digital hydrology results from the intersection of computer science and hydrology. Intrinsically, being digital connotes being discrete in space or in time. In such a sense, it seems to us that digital hydrology focuses our attention on the spatial separation of a catchment as gridded cells or subcatchments, and the temporal discretization of hydrological processes within each cell or each subcatchment. Therefore, state values and instantaneous values of hydrological elements could be modelled at any site and at any time step, so as to show the whole hydrological process. In order to attain this goal, rainfall, as the input to hydrological model, is of great importance, especially for short-duration flood hydrograph modelling according to the past experience of rainfall–runoff modelling. The spatial variability of precipitation is very significant for runoff computation. Radar-based rain data have an advantage over highly spatial and temporal resolutions. Such an advantage may be fully and completely utilized in the grid-based hydrological model. In addition, the spatial and temporal distributions of state variables, such as soil tension water at any cell within the catchment at any time step, could be obtained by the model.

A catchment may be partitioned spatially as grid elements by the DEDNM. Thus a grid element on the ground within the catchment matches the atmospheric input, such as precipitation and evapotranspiration. It is of theoretical significance and of practical value to apply radar-measured rain data to digital hydrological modelling.

STUDY AREA AND DATA

The Shihe catchment within 31.2–32.3°N latitude and 115.25–115.75°E longitude is selected for the case study. The Huangnizhuang station is the outlet of that catchment with an area of 805 km^2. The Shihe River is the first-order southern tributary of the Huaihe River basin. There exists a highly mountainous area where the streamflow goes very fast. The Shihe catchment is selected as the intensified observation field for the HUBEX/GAME project supported financially from the National Natural Science Foundation of China during the ninth Five Year Plan.

DEM data were obtained from National Geophysical Data Center, USA, viz. Global Land One-kilometre Base Elevation (GLOBE) data. On the basis of DEM pre-processing, the steepest down slope path method is applied to the determination of flow vectors on each grid. Figure 1 shows the visual flow vectors over the Shihe catchment.

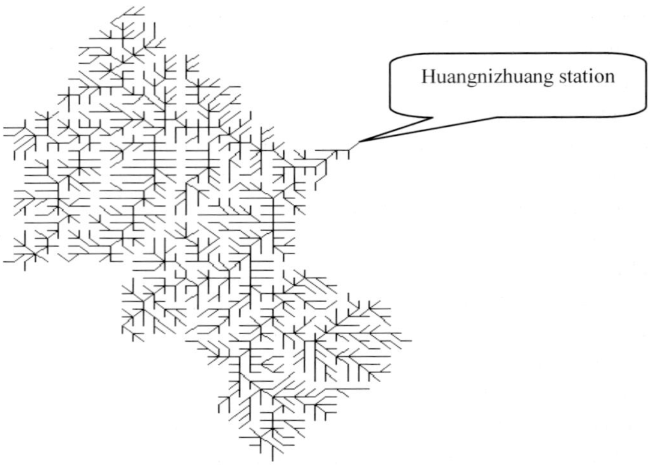

Fig. 1 Raster-based flow vectors in the upper area of Huangnizhuang station in the Shihe catchment, produced by the steepest descent method.

Radar reflectivity data were obtained from a C-band weather radar device with 5 cm of wavelength. It lies at 32.93°N latitude, 115.83°E longitude in Fuyang, Anhui Province. Fuyang radar scans a radius of about 250 km, within which the Shihe catchment lies. Radar reflectivity data are continuously observed by the volume scanning method. The radar scans one volume per 10 min with 1 km of spatial resolution. Every volume data set has 14–16 PPI. All PPI data were filtered and averaged within 1 h, after eliminating noise and correcting obstruction from cluster (Liu *et al.*, 2000).

METHODS

Determination of raster flow vectors

The Digital Elevation Drainage Network Model (DEDNM) developed by Martz & Garbrecht (1992), or so-called TOpographical PArameteriZation (TOPAZ), is applied in this presentation for assigning flow-vector directions over flat and depressional areas of the DEM, and for identifying stream network and watershed divides. The algorithm of the DEDNM is introduced very briefly as follows.

Depressions are sinks at the bottom of which the drainage network terminates. They are probably real storage volumes such as reservoirs, ponds, or artefacts of the horizontal and vertical DEM resolution, DEM generation method, and elevation data truncation. The program is able to distinguish between two types of depressions: sink-depressions and impoundment-depressions. Sink-depressions are caused by a group of raster cells at lower elevation than the surrounding landscape. Impoundment-depressions are caused by a narrow band of raster cells of higher elevation across drainage paths, similar to an obstruction or dam across a stream. In the latter situation, the program can breach the obstruction in the downstream direction of the drainage path, so as to let water pass it. Any remaining depression after breaching is regarded as a sink-depression, which is treated in the conventional manner by raising the elevations within the depression to the elevation of its lowest (breached) overflow.

The flat area refers to a group of raster cells with equal values of elevation and indeterminate drainage direction. The flat surface is inherent to the DEM, or comes from the above depression filling. A relief imposition algorithm proposed by Martz & Garbrecht (1992) is adopted for flow-vector designation over the flat area, which takes topographical characteristics around the flat area into account.

The program scans each cell of rectified DEM, and determines the direction of the steepest downward slope to an adjacent cell. This method is termed the D8 (deterministic eight-neighbours) method by Fairfield & Leymarie (1991). The final result is a new raster array containing flow vectors for all grid cells.

The row and column coordinates of a grid cell close to the watershed outlet must be given. With this information, the program displays a portion of catchment-area values around the initial outlet point to the monitor. On this display large upslope catchment-area values identify the channel network, and the user can then modify the watershed outlet coordinates. This updated value is used as the starting point to define the entire upstream drainage area by concurrently scanning flow vectors and watershed grids. The final result of this stage is a raster array defining cells inside and outside the watershed.

Development of hydrological model

The Xinanjiang model (Zhao, 1992) is characterized by the concept of runoff formation on repletion of storage, i.e. runoff is not produced until the soil moisture content of the aeration zone reaches field capacity, and thereafter runoff equals the rainfall excess without further loss. That concept could be understandable when applied to a point. The tension water storage capacity curve was introduced by Zhao in the 1960s while applied to a catchment. In this research, the parabolic curve is applied to each grid element.

Three kinds of spatially heterogeneous distributions are taken into consideration in the Xinanjiang model: (a) uneven distribution of tension water storage capacity throughout the subcatchment is expressed by a parabolic curve for partial-area runoff generation; (b) non-uniform distribution of free water storage capacity over partial area where runoff has been produced, is expressed also in terms of a parabolic curve for separation of runoff into surface flow, interflow and groundwater flow; and (c) the amount of free water storage is represented by the structure of linear reservoir for the sake of different velocities of different runoff components.

The Xinanjiang model (Zhao, 1992) is applied for runoff production in each grid element where rain data measured by radar at Fuyang station (Liu *et al.*, 2000) are used as the input of the hydrological model. The time step is taken as 1 h for radar-measured rain data. Hence, the time step is 1 h for the computation of the Xinanjiang model. Those elements are connected by flow vectors (as shown in Fig. 1) with the outlet of drainage catchment where runoff is routed by the successive sub-reaches of the

Table 1 Number of sub-reaches and channel length between the outlet (Huangnizhuang station) and each grid element within the Shihe catchment.

Grid ID	Length (m)	No. of sub-reaches	Grid ID	Length (m)	No. of sub-reaches
1	35222.4	10	11	32785	9
2	36084.3	10	...	...	...
3	34003.7	9	...	...	...
4	34865.6	10	...	...	...
5	35727.5	10	1078	42979.5	12
6	32785	9	1079	43841.4	12
7	32428.1	9	1080	43841.4	12
8	32785	9	1081	44198.2	12
9	33646.9	9	1082	44555.1	12
10	32785	9	1083	45060.1	13

Muskingum method from each grid element to the outlet, according to the channel length (see Table 1) between each grid and the outlet. The length is obtained by summing up the length of gridded cells along the raster-based flow vectors as shown in Fig. 1. Assuming that runoff flows at the speed of 1 m s^{-1}, the number of successive sub-reaches in the Muskingum method is obtained by the length divided by 3600 (see Table 1), since the time step is taken as 1 h.

RESULT ANALYSIS

As shown in Figs 2 and 3, the Nash-Sutcliffe model efficiency coefficient (Nash & Sutcliffe, 1970) is up to 92.41% from 31 May to 3 August 1998, and 85.64, 86.62, 92.57 and 83.91%, respectively, for the first, second, third and fourth flood event during the whole computational period.

The comparison of model efficiency coefficients was made as radar-measured and raingauge data were used in simulating hourly hydrograph at Huangnizhuang

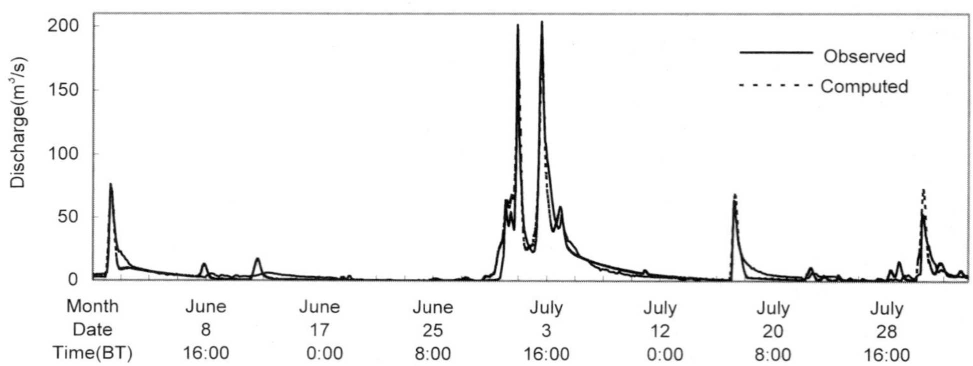

Fig. 2 Comparison between observed and computed discharges at Huangnizhuang station from 31 May until 3 August 1998, with 92.41% of model efficiency coefficient.

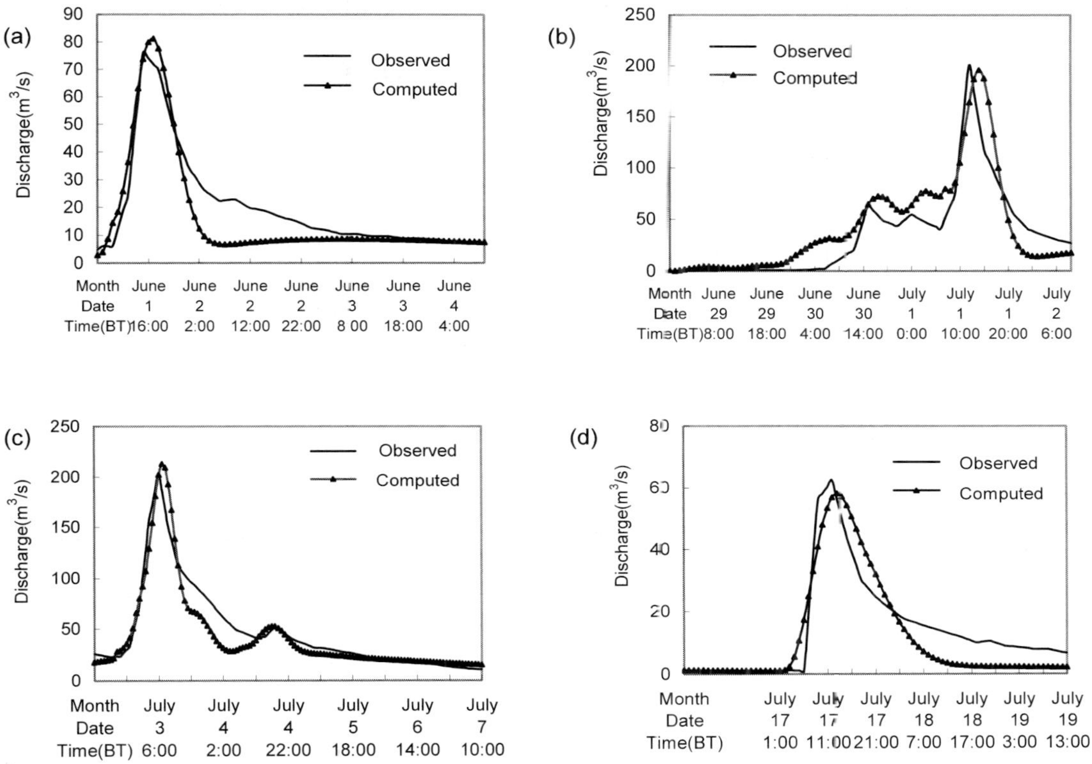

Fig. 3 Comparison between observed and computed discharges at Huangnizhuang station from June to July in 1998, with 85.64, 86.62, 92.57 and 83.91% of model efficiency coefficient, respectively, for the first, second, third and fourth flood event as shown in (a), (b), (c), (d).

Table 2 Comparison of model efficiency coefficients as radar-measured and raingauge data were used in simulating hourly discharge hydrograph at Huangnizhuang station in the Shihe catchment, respectively.

Computational period	Model efficiency coefficient R (%): Raingauge data used (R_1)	Radar data used (R_2)	Improvement index (%) $r = (R_2 - R_1)/(1 - R_1)$
First flood event	77.81	85.64	35.29
Second flood event	81.53	86.62	27.56
Third flood event	75.73	92.57	69.39
Fourth flood event	73.51	83.91	39.26
During the whole period from 31 May to 3 Aug.	84.44	92.41	51.22

station in the Shihe catchment, respectively. It is seen from Table 2 that the values of model efficiency coefficient R are higher as radar-measured data were used both for the whole computational period from 31 May to 3 August and for four flood events.

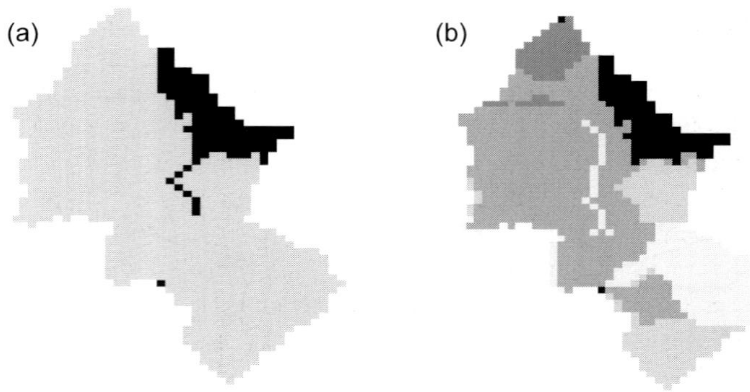

Fig. 4 Spatial distributions of (a) runoff depth and (b) soil tension water storage in each grid cell at 17:00 h (BT) on 19 June, 1998 (the darker indicates the larger value of runoff or soil water).

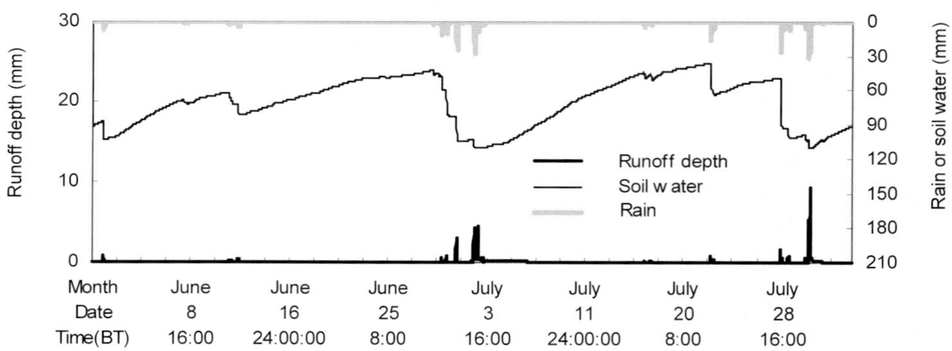

Fig. 5 Time series of rain, runoff depth, and soil tension water storage in the tenth grid cell within the Shihe catchment.

Therefore, the index of model efficiency improvement of Model 2 (radar-measured rain data being used) over Model 1 (raingauge data being used), expressed by r, arrives at 51.22% during the whole computational period, and is positive from 27.56% to 69.39% for four flood events. This shows that radar-measured data are superior to raingauge data as the input of hydrological modelling.

In addition, spatial distribution and time series of state variables, such as rain, runoff depth, soil tension water storage on each grid, could also be obtained from the Xinanjiang model. It can be seen from Fig. 4 that more runoff is generated where there is more soil tension water storage. In Fig. 5, runoff depth is temporally in agreement with rain and soil tension water storage.

CONCLUSIONS

The name of digital hydrology matches the reality, only as a catchment is spatially divided into grid elements and the whole hydrological process is recognized at any space/time scale. It is well known that rainfall plays an important part in hydrological process modelling. The spatial variability of precipitation is very significant for runoff computation. Radar can meet the requirement of hydrological modelling, which offers rain data with high spatial and temporal resolutions. Such an advantage may be fully and completely utilized in the grid-based hydrological model.

After a catchment is segmented spatially as grid elements by the DEDNM, a grid element on the ground within the catchment matches the atmospheric input, such as precipitation and evapotranspiration. The grid-based hydrological model provides a good platform for runoff computation, in which radar-measured rain data are taken as the input of the hydrological model. The model developed above performs better when radar-measured data, instead of raingauge data, were used in simulating hourly hydrograph at Huangnizhuang station in the Shihe catchment. In addition, the spatial and temporal distributions of state variables, such as soil tension water at any cell within the catchment at any time step, may be obtained by the model.

It should be pointed out that radar-measured data used in this study have been modified according to topographical characteristics and data observed by raingauge stations (Liu *et al.*, 2000). It is true that radar-measured rain data have an advantage of highly spatial and temporal resolutions. However, the greater the distance from the radar station, the lower the accuracy of rain data. It is worth being studied further. An approach to this problem is to build more radar stations, so that the stations can cover the whole study area. When an attempt is made for radar data to be used for the mesoscale or large-scale area, more investment is necessary. In such cases, the equilibrium point between investment and economic loss/risk should be found.

Acknowledgements This research was jointly supported financially by the National Natural Science Foundation of China under Grant no. 40171016 and 49794030. The author would like to give thanks to Dr Liu Xiaoyang at Beijing University for providing radar data.

REFERENCES

Beven, K. (1989) Changing ideas in hydrology—the case of physically-based models. *J. Hydrol.* **105**, 157–172.

Doyle, F. J. (1978) Digital terrain models: an overview. *Photogram. Engng & Remote Sens.* **44**(12), 1887–1896.

Fairfield, J. & Leymarie, P. (1991) Drainage networks from grid digital elevation models. *Water Resour. Res.* **27**(4), 29–61.

Garbrecht, J. & Campbell, J. (1997) TOPAZ: an automated digital landscape analysis tool for topographic evaluation, drainage identification, watershed segmentation and subcatchment parameterization. *TOPAZ User Manual*, US Dept of Agric.-ARS, Oklahoma, USA.

Jenson, S. K. & Dominique, J. O. (1988) Extracting topographic structure from digital elevation data for geographical information system analysis. *Photogram. Engng & Remote Sens.* **54**(11), 1593–1600.

Liu, X. Y., Mao, J. T. & Li, J. R. (2000) Comparison of runoff simulations from radar-estimated rainfall and raingauge measured rainfall in rainfall–runoff model. (Proc. Int. GAME/HUBEX Workshop, GAME/HUBEX Project Office), 147–155. Institute of Low Temperature Science, Hokkaido University, Sapporo, Japan.

Martz, W. & Garbrecht, J. (1992) Numerical definition of drainage network and subcatchment areas from digital elevation models. *Comp. Geosci.* **18**(6), 747–761.

Moore, I. D., Grayson, R. B., & Ladson, A. R. (1991) Digital terrain modelling: a review of hydrological, geomorphological and biological applications. *Hydrol. Processes* **5**, 3–30.

Nash, J. E. & Sutcliffe, J. V. (1970) River flow forecasting through conceptual models, Part I—a discussion of principles. *J. Hydrol.* **10**, 282–290.

O'Callaghan, F. & Mark, D. M. (1984) The extraction of drainage networks from digital elevation data. *Computer Vision, Graphics & Image Processing* **28**, 323–344.

Zhao, R. J. (1992) The Xinanjiang model applied in China. *J. Hydrol.* **135**, 371–381.

Weather Radar Information and Distributed Hydrological Modelling (Proceedings of symposium HS03 held during IUGG2003 at Sapporo, July 2003). IAHS Publ. no. 282, 2003.

Application of models with different types of modelling methodologies for river flow forecasting

H. A. PRASANTHA HAPUARACHCHI, LI ZHIJIA
College of Water Resources and Environment, Hohai University, Nanjing 210098, China
prasanth@jlonline.com

ALBERT WOLFGANG FLÜGEL
International Water Management Institute, No. 127, Battaramulla, Sri Lanka

Abstract In the present study, a conceptual watershed model, a distributed watershed model, and an artificial neural network (ANN) have been applied to river flow forecasting in the Kalu River upper catchment in Sri Lanka. The Xinanjiang watershed model has been used as a conceptual watershed model and the SWAT model (Neitsch, 2000) has been used with spatial data as a distributed model. Two types of ANN architectures, namely multi-layer perceptron network (MLP) and radial basis function network (RBF) have been implemented as "black box" type modelling methodology. Based on the application results, it seems that the conceptual watershed model could perform slightly better than the distributed model and the ANN for this watershed. It was clearly noted that the performance of distributed models strictly depends on the quality of input data (Arnold *et al.*, 1998) whereas the performance of conceptual models depends on the calibration (Duan *et al.*, 1992, 1993).

Key words artificial neural networks; conceptual watershed models; distributed watershed models; multi-layer perceptron network; radial basis function; Xinanjiang model

INTRODUCTION

At present, a number of modelling methodologies are available for watershed modelling and river flow forecasting. Basically they can be divided into three categories as conceptual, distributed, and "black box" type modelling. The application and performance of these types of models depend on the data existence, data quality, hydrological characteristics of the catchment, etc. Therefore the selection of an appropriate model should be done carefully, analysing these dependents. The main difficulty of using conceptual watershed models is the calibration. In general, evidence from previous conceptual rainfall–runoff (CRR) model calibration studies has indicated that the calibration problem should be solved by global optimization techniques, since inferior optimal solutions have been obtained by various local optimization techniques. On the other hand, distributed watershed models require more input data, which may not be readily available for every catchment. Generally, the performance of distributed models is strictly dependent on the quality of data (Arnold *et al.*, 1998). While conceptual or physically-based models are of importance in the understanding of hydrological processes, there are many practical situations where the main concern is with making accurate predictions at specific locations with less available data. In such situations it is preferred to implement a simple "black box" (data driven, or machine learning) model to identify a direct mapping between the input and output without detailed consideration of the internal structure of the physical process. The artificial neural network (ANN) is probably the most successful machine learning technique with a flexible mathematical structure which is capable of identifying complex nonlinear relationships between input and output data, without attempting to reach an understanding of the nature of the phenomena. Neural networks are particularly useful in situations where the underlying physical process relationships are not fully understood, or where the nature of the event being modelled (i.e. a flood) may display chaotic properties (Dibike *et al.*, 1999).

In the present study, the Xinanjiang model (Zhao *et al.*, 1980, 1995) as a conceptual watershed model, the SWAT model (Soil Water Assessment Tool, Arnold *et al.*, 1998) as a distributed watershed model and two types of ANN architecture named multi layer perceptron (MLP) network and radial basis function (RBF) network have been applied for predicting daily flow of the Kalu River upper catchment in Sri Lanka for comparing the behaviours of different types of models.

CATCHMENT CHARACTERISTICS

The catchment is located to the southwest and south of the central highlands of Sri Lanka, and lies between 80.40–80.60°N latitude and 6.53–6.80°E longitude. Generally elevation varies from 20 to 2225 m a.m.s.l. and the total area is about 603 km^2. Mountain ranges, high peaks, dissected plateaus, escarpments, cover a greater part of the area. On average (depending on their lithology and structure) slopes vary from 10° to 35° in the upland ridges. The slopes and the drainage system have largely been governed by its geological structure. Annual precipitation is above 4000 mm and there are eight precipitation-gauging stations in and around the catchment. The average annual temperature of the area ranges from 26.9°C to 27.8°C. Dominant soil types visible in this area are red brown earths and low humic grey soils, reddish brown earths and immature brown loams, red yellow podzolic soils, bog and half-bog soils, and alluvial soils. But nearly 86% of the area is covered with Red Yellow Podzolic soils (Karunanayake *et al.*, 1990). Two main vegetation types, tropical rain forests and mountain forests, can be seen in this area. Basically there are 13 types of land-use patterns. About 30.2% of the land has been used for Chena cultivation (Department of Survey, Sri Lanka, land-use map 1992, 1:50 000), which is a causative factor for soil erosion and deforestation. Tea and rubber are other major cultivations.

MODELS

Xinanjiang watershed model

The Xinanjiang rainfall–runoff model is a conceptual watershed model developed at Hohai University, China in the 1970s. It provides an integral structure, statistically describing the non-uniform distribution of runoff producing areas, which features it as one of the conceptual, semi-distributed hydrological models. By comparing with the

Table 1 Parameters of the Xinanjiang model.

Parameter	Abbreviation	
Evapotranspiration	*K*	Ratio of potential evapotranspiration to the pan evaporation
	WUM	Tension water capacity of upper layer
	WLM	Tension water capacity of lower layer
	C	Evapotranspiration coefficient of deeper layer
Runoff generation	*WM*	Areal mean tension water capacity
	B	Exponential of the distribution of tension water capacity
	IM	Ratio of impervious area to the total area of the basin
Runoff separation	*SM*	Free water storage capacity
	EX	Exponential of distribution water capacity
	KG	Out flow coefficient of free water storage to the ground water flow
	KI	Out flow coefficient of free water storage to the inter flow
Runoff concentration	*CG*	The recession constant of ground water storage
	CI	The recession constant of lower interflow storage
	CS	The recession constant of channel network storage
Flow routing	*XE*	Muskingum coefficient

Pitman model of South Africa, the Sacramento model of USA, the NAM model of Europe and the SMAR model of Ireland, Gan *et al.* (1996) proved that the Xinanjiang model did consistently better, even in dry catchments. The model consists of 15 parameters (Table 1) and the application results are best for the humid and semi-humid catchments.

All the parameters of Xinanjiang model should be calibrated before application. The hydrological data inputs of the model are areal mean rainfall *P*, and measured pan evaporation *EM*. Besides these, subcatchment areas, and initial state of the catchment are necessary for the calculations. The outputs are the discharge from the whole basin *TQ*. State variables are the areal mean free water storage *S* and the areal mean tension water storage *W*. Basically *W* is divided into three layers as upper, lower, and deeper having the tension water storage components *WU*, *WL*, *WD* and evaporation components *EU*, *EL*, *ED*, respectively (see Fig. 1). *RD* is the direct runoff from the impervious area and *FR* is the runoff contributing area factor, which is related to *W*. The runoff (*R*) produced from pervious area is divided into three components *RS*, *RI* and *RG* referred to as surface runoff, interflow and groundwater flow, respectively. The three components are further transferred into *QS*, *QI* and *QG* and together form the total inflow to the channel network of the basin. The outflow from each sub-basin is first simulated and then routed down through channels to the main basin outlet.

Artificial neural networks (ANN)

Neural nets are now widely used in many different areas of science to model complex problems as substitutes for, and in association with, conventional mathematical and statistical models. Normally, conventional models require a great deal of detailed data, for example, topographical maps, river networks and characteristics, soil rainfall and

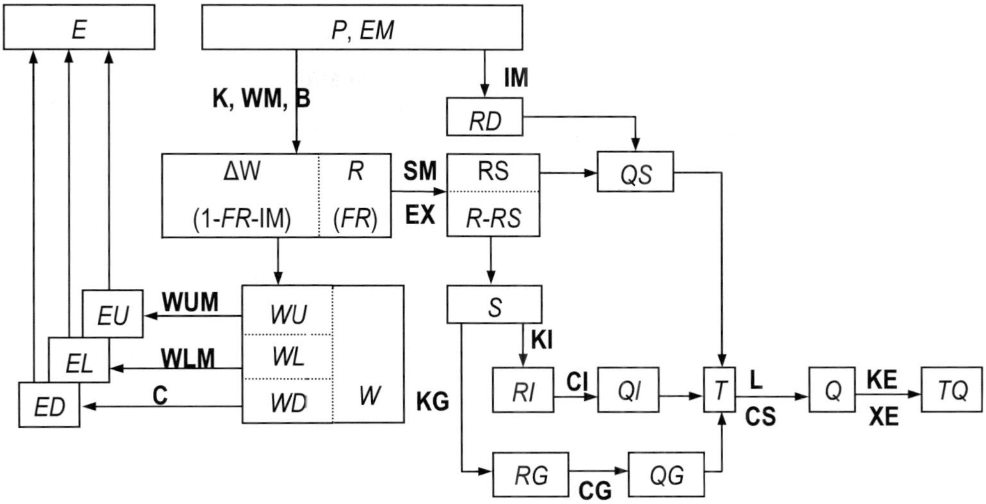

Fig. 1 Flowchart of the Xinanjiang model (All the symbols outside the blocks are parameters).

runoff data. Often, these data may not available and pose a great difficulty for model calibration. Neural networks, on the other hand, offer real prospects for a cheaper, more flexible, less assumption-dependent and adaptive methodology well suited to model flood processes, which by their nature are inherently complex, nonlinear, sometimes hard to predict because of nonlinearity and chaotic effects and sometimes life critical. In this study, two types of feedforward ANN architectures, multi-layer perceptron network (MLP) and radial basis function network (RBF) were implemented for downstreamflow forecasting. Details of these two methods can be found in Dibike *et al.* (1999) and Mason *et al.* (1996).

SWAT model

The Soil and Water Assessment Tool (SWAT) (Neitsch *et al.*, 2000) was developed by the US Department of Agriculture (USDA), Agricultural Research Service (ARS). It was developed to predict the impact of land management practices on water, sediment and agricultural chemical yields in large complex watersheds with varying soils, land-use and management conditions over long periods of time. Application of the model requires specific information about weather, soil properties, topography, vegetation, and land management practices in a watershed. SWAT is capable of handling with water movement, sediment movement, crop growth, and nutrient cycling with the above information. Therefore, watersheds with no monitoring data can be modelled, and the effect of changes in input data can be quantified. In the present study, only the hydrological processes embedded in SWAT are considered. A detailed description of the SWAT model and user manual can be found in Neitsch *et al.* (2000) or at the web site http://www.brc.tamus.edu/swat/.

APPLICATION AND RESULTS

Nine years of daily hydrological data (1987–1995) have been used for the calculations such that seven years of data (1987–1993) are used for calibrating or training and two years of data (1994–1995) for validating the models. The watershed was delineated into seven subcatchments and flow computations were done for each subcatchment.

Statistical indices

Two statistical indices have been selected to compare the results obtained from calibration and verification. They are the Nash-Sutcliffe coefficient D_y and the percentage of total error *%Err* in each year as:

$$D_y^2 = 1 - \frac{\sum_{i=1}^{m} w_i^2 \left[Q_{Obs,i} - Q_{Cal,i}\right]^2}{\sum_{i=1}^{m} w_i^2 \left[Q_{Obs,i} - \overline{Q}_{obs}\right]^2} \tag{1}$$

$$\%Err = \left[\frac{\sum_{i=1}^{m} Q_{Obs,i} - \sum_{i=1}^{m} Q_{Cal,i}}{\sum_{i=1}^{m} Q_{Obs,i}}\right] \times 100 \tag{2}$$

Where $\overline{Q}_{obs}$ is daily mean observed discharge, Q_{obs} is the observed discharge, Q_{cal} is the calculated discharge, m is the number of time steps in each year and w_i is a weighting factor.

Application of ANN

The precipitation data set was prepared by averaging the total daily precipitation of seven gauging stations in and around the basin. Evaporation was assumed to be the same in the whole catchment. Firstly, the networks were trained with concurrent rainfall and evaporation data as input and discharge as output. The efficiency was checked by trial-and-error process, with various combinations of precipitation, evaporation and discharge data. After a series of experiments, the best performance was obtained using two antecedent runoff values (observed), in addition to three rainfall and one evaporation value, as input to the network (some of the validation results are shown in Table 2). The number of nodes in the hidden layer was varied between two and seven and the one with three gave the highest efficiency in the verification stage. However, when additional antecedent runoff and rainfall data were added to the input pattern, it was noted that the performance of the network for the training period was improved, but its performance for validation period was deteriorated.

Table 2 Validation results of ANNs.

Network combination	Year	Q_{obs} (mm)	Multi-layer perceptron (MLP): Q_{cal} (mm)	*%Err*	D_y	Radial basis function (RBF): Q_{cal} (mm)	*%Err*	D_y
R10	1994	2105.63	2101.650	0.189	0.861	2210.279	–4.970	0.801
	1995	3039.80	2870.665	5.564	0.842	2941.006	3.250	0.836
P2, R1, E1	1994	2105.63	2510.121	–19.213	0.849	2224.598	–5.650	0.819
	1995	3039.80	3286.601	–8.119	0.873	2986.968	1.738	0.831
P3*, R2†, E1‡	1994	2105.63	2167.240	–2.926	0.876	2099.813	0.276	0.847
	1995	3039.80	2979.118	1.996	0.875	2967.643	2.374	0.816
P5, R2, E1	1994	2105.63	2310.697	–9.739	0.835	2215.038	–5.196	0.823
	1995	3039.80	3022.590	0.566	0.896	2990.585	1.619	0.844
P5, R3, E1	1994	2105.63	2087.489	0.862	0.854	2077.381	1.342	0.832
	1995	3039.80	2927.813	3.684	0.860	2987.477	1.721	0.822
P7, R5, E3	1994	2105.63	1918.734	8.876	0.853	2196.530	–4.317	0.814
	1995	3039.80	2688.946	11.542	0.872	2948.697	2.997	0.824
P9, R9, E9	1994	2105.63	2125.022	–0.921	0.595	2216.280	–5.255	0.457
	1995	3039.80	2823.184	7.126	0.556	2917.782	4.014	0.570

*Three antecedent precipitation data sets; †Two antecedent runoff data sets; ‡One evaporation data set.

Table 3 Calibration and validation results of the Xinanjiang model.

From	To	Q_{obs} (mm)	Q_{cal} (mm)	*%Err*	D_y
1987.01.01	1987.12.31	1820.686	1865.743	–2.475	0.853
1988.01.01	1988.12.31	3230.345	3130.724	3.084	0.848
1989.01.01	1989.12.31	3050.908	2951.325	3.264	0.946
1990.01.01	1990.12.31	2260.360	2215.434	1.988	0.901
1991.01.01	1991.12.31	2095.514	2082.390	0.626	0.862
1992.01.01	1992.12.31	2451.887	2389.625	2.539	0.919
1993.01.01	1993.12.31	3404.041	3340.351	1.871	0.908
1994.01.01	1994.12.31	2105.631	2070.130	1.686	0.894
1995.01.01	1995.12.31	3039.802	2980.708	1.944	0.885

Application of the Xinanjiang model

Hapuarachchi *et al.* (2001) calibrated all the parameters of the Xinanjiang model using the SCE-UA method (Duan *et al.*, 1992, 1993, 1994). The calibration and verification results of the Xinanjiang model are shown in Table 3. According to the results, *%Err* is considerably small ($-3 < \%Err < 4$) and the Nash coefficient D_y is grater than 84% for each year, in both the calibration and verification stages.

Application of the SWAT model

Firstly, the watershed was delineated into 15 subcatchments using a DEM (200 × 200 m^2 cell size) and a digitized river network. The number of subcatchments

Table 4 Calibration and validation results of the SWAT model.

From	To	Q_{obs} (mm)	Q_{cal} (mm)	*%Err*	D_y
1987.01.01	1987.12.31	1820.683	1483.143	18.539	0.720
1988.01.01	1988.12.31	3230.340	2552.167	20.994	0.703
1989.01.01	1989.12.31	3050.903	2748.810	9.902	0.762
1990.01.01	1990.12.31	2260.356	1882.359	16.723	0.774
1991.01.01	1991.12.31	2095.511	1774.940	15.298	0.759
1992.01.01	1992.12.31	2451.883	2197.655	10.369	0.782
1993.01.01	1993.12.31	3404.035	2409.872	29.205	0.763
1994.01.01	1994.12.31	2105.631	1470.104	30.182	0.715
1995.01.01	1995.12.31	3039.802	2374.725	21.879	0.705

generated depends on the threshold limit of flow accumulation. The SWAT was applied for the period 1987–1995 for daily streamflow generation. However, we are only interested in comparing results of two years, 1994 and 1995, that are common to the validation periods of ANNs and the Xinanjiang model. The application results are shown in Table 4. The *%Err* (*%Err* > 9) is considerably high for each year indicating the water imbalance of the catchment. Furthermore, the highest Nash coefficient that could obtain was 78% for the data year of 1992.

DISCUSSION AND CONCLUSIONS

The compiled results of performance of the three models in the validation stage, are shown in Table 5. A careful inspection revels that the results of ANNs and the Xinanjiang model are acceptable. Based on the results (Table 5), the performance of Xinanjiang model is better than ANN and the SWAT model for daily river flow predictions of the Kalu River upper catchment. Thereby it is possible to conclude that the Xinanjiang model can be successfully applied in humid or semi humid catchments in Sri Lanka for computing river flow. The input data requirement for the Xinanjiang model are only precipitation, pan evaporation, and observed discharge. Therefore it is suitable for modelling many catchments where sophisticated data are not readily

Table 5 Compiled results of different models.

Model		Xinanjiang	ANN		SWAT
Criteria	Year	Model	MLP	RBF	Model
Q_{obs}	1994	2105.631	2105.63	2105.63	2105.631
(mm)	1995	3039.802	3039.80	3039.80	3039.802
Q_{cal}	1994	2070.130	2167.240	2099.813	1470.104
(mm)	1995	2980.708	2979.118	2967.643	2374.725
%Err	1994	1.686	–2.926	0.276	30.182
	1995	1.944	1.996	2.374	21.879
D_y	1994	0.894	0.876	0.847	0.715
	1995	0.885	0.875	0.816	0.705

available. The main difficulty of using the Xinanjiang model is its calibration. The manual calibration requires detailed understanding of the model's behaviour, which can only be obtained through long-term calibration experience and data. Autocalibration is efficient, but it is difficult to incorporate the expert knowledge in the calibration process.

When considering the two ANNs, MLP and RBF, the performance of the MLP network is slightly better than the RBF network (Table 5). The *%Err* of the Xinanjiang model, MLP and RBF networks are acceptable (<5%) proving that the water utilization and losses balance in the catchment. However, the main disadvantage of using ANNs is the absence of criteria that can optimize the network parameters, dominant model inputs and, input data combinations for making the model output optimum. In this case study, we have carried out many trials (Table 2) for finding the optimum combination. It was observed that the performance of the networks in the validation period was reduced when only runoff data (R10) or more additional runoff and rainfall data are added (P9, R9, E9) to the input pattern (Table 2). However, implementing the ANN is easier and the input data requirement is small.

In this case study the performance of the SWAT model is poor compared to the other two models. Although the efficiency D_y is acceptable (Table 5), the *%Err* is high indicating that the model is unable to satisfy the water balance in the catchment. It may be partly attributable due to the poor quality of the data since the output of distributed models strictly depends on the quality of data. The main disadvantage of using such distributed models for river flow predictions is the higher requirement of data. In addition, many people in Sri Lanka use well water for their domestic use. When considering a catchment as a whole, it is normally a very large area, and therefore it is not possible to record or count all the individual minor scale water utilizations in detail such as small irrigation, animal husbandry in minor scale, industrial water utilizations in minor scale, etc. The cumulative value of such water utilizations might be large. The absence of these data may especially affect the distributed models in water balancing. But the conceptual watershed models and ANNs are capable of adjusting their parameters while calibrating or training, according to the situation, since most of their parameters have no physical background. As a result, a conceptual watershed model and ANNs show better performance than a distributed model where the catchment characteristics and model inputs are limited or incomplete. However, distributed watershed models are excellent for understanding the physical processes of the catchment.

Acknowledgements This project was funded by the International Water Management Institute (IWMI). The support received from the IWMI in data collection and preparation is gratefully acknowledged.

REFERENCES

Arnold, J. G., Srinivasan, R., Muttiah, R. S. & Williams, J. R. (1998) Large area hydrologic modelling and assessment part I: model development. *J. Am. Water Res. Assoc.* **34**(1), 73–89.

Dibike, Y. B., Solomatine, D. P. & Abbott, M. B. (1999) On the encapsulation of numerical-hydraulic models in artificial neural network. *J. Hydraul. Res.* **37**(2), 147–161

Duan, Q. Y., Gupta, V. K. & Sorooshian, S. (1993) Shuffle complex evolution approach for effective and efficient global minimization. *J. Opti. Theory Appl.* **76**(3), 501–521.

Duan, Q., Sorooshian, S. & Gupta, V. K. (1992) Effective and efficient global optimization for conceptual rainfall–runoff models. *Water Resour. Res.* **28**(4), 1015–1031.

Duan, Q., Sorooshian, S. & Gupta, V. K. (1994) Optimal use of SCE-UA global optimization method for calibrating watershed models. *J. Hydro.* **158**, 265–284.

Gan, T. Y. & Biftu, G. F. (1996) Automatic calibration of conceptual rainfall–runoff models: optimization algorithms, catchment conditions and model structure. *Water Resour. Res.* **32**, 3513–3524.

Hapuarachchi, H. A. P., Li, Z. J, Ranjit, M. & Wang, Q. J. (2001) Application of global optimization technique for calibrating the Xinanjiang watershed model. *Lowland Tech. Int.* **3**, 43–57

Karunanayake, M. M. & Katupotha, J. (1990) An environmental profile of the Ratnapura district. Tech. report, Central Environmental Authority & NORAD, Sri Lanka.

Mason, J. C., Price, R. K. & Tem'me, A. (1996) A neural network model of rainfall–runoff using radial basis functions. *J. Hydraul. Res.* **34**, 537–548

Neitsch, S. L, Arnold, J. G., Kiniry, J. R., Srinivasan, R. & Williams, J. R. (2000) *SWAT User Manual 2000*. Grassland, Soil and Water Research Laboratory, Agricultural Research Service, USA.

Zhao, R. J., Zhuang, Y. L., Fang, L. R., Liu, X. R. & Zhang, Q. S. (1980) The Xinanjiang model. In: *Hydrological Forecasting.* (Proc. Oxford Symp., April, 1980), 351–356. IAHS Publ. no. 129.

Zhao, R. J. & Liu, X. R. (1995) *Computer Models of Watershed Hydrology: The Xinanjiang Model* (ed. by V. P. Singh), 215–232. Water Resources Publications, Highlands Ranch, Colorado, USA.

Effects of sub-basin scale on runoff simulation in distributed hydrological model: BTOPMC

TIANQI AO[1], JUNICH YOSHITANI[1], KUNIYOSHI TAKEUCHI[2], KAZUHIKO FUKAMI[1], TADASHI MUTSUURA[1] & HIROSHI ISHIDAIRA[2]

1 *Hydrological Engineering Team, Hydraulic Engineering Research Group of Public Works Research Institute, 1-6 Minamihara, Tsukuba, Ibaraki 305-8516, Japan*
t-ao55@pwri.go.jp

2 *Institute of Material and Environment Technology, Graduate School, Yamanashi University, Takeda 4-3-11, Kofu, Yamanashi 400-8511, Japan*

Abstract Scale problems result in uncertainty in hydrological modelling and are far from being solved. A distributed hydrological model BTOPMC (block-wise use of TOPMODEL with Muskingum-Cunge method) analyses the effects of sub-basin scale on runoff in the Fuji-kawa and the Nakagawa basins in Japan. Study basins were subdivided into natural sub-basins using the Pfafstetter method. Results indicate that smaller average sub-basin size produces higher peak discharge, larger baseflow and total runoff for floods, while similar effects on annual runoff do not show any determined tendency and are not obvious. This shows the effect of averaging scale of the topographic index. Furthermore, threshold values of average sub-basin areas were found for study basins, around which simulation results seem stabilized, indicating that there may exist a proper sub-basin scale for BTOPMC, at which uncertainty due to subdivision level can be reduced to quite a low level.

Key words basin subdivision; Fuji-kawa basin; Japan; Nakagawa basin; Pfafstetter numbering; physically based distributed hydrological model BTOPMC; scale issue

INTRODUCTION

It has been widely recognized that spatial and temporal scales generally lead to predictive uncertainty in distributed hydrological modelling (e.g. Sivapalan & Kalma, 1995; Blöschl & Sivapalan, 1995). Model parameter dependence on sub-basin size and/or grid-cell size is an example. Although such difficult and serious scale problems are yet far from any form of solution, many valuable ideas have been proposed to attempt to solve them and hence improve model reliability. Examples are the simple scaling and multiscaling frame (e.g. Gupta *et al.*, 1994); the REA (Representing Element Area) concept (e.g. Wood *et al.*, 1988); the GLUE (Generalized Likelihood Uncertainty Estimation) framework (e.g. Beven & Binley, 1992); the HRU (Hydrological Response Units) concept of Flügel (1995); as well as the basin-scale model equations (e.g. Kavvas *et al.*, 1998). In addition, many efforts have been focused on the effects of grid size on model parameters and performances (e.g. Quinn & Beven, 1991; Franchini *et al.*, 1996; Saulnier *et al.*, 1997; Sunada *et al.*, 2001; Horritt & Bates, 2001).

In the case of a physically-based distributed hydrological model, BTOPMC (Block-wise use of TOPMODEL with Muskingum-Cunge method), the topographic-spatial scale of the sub-basin and the grid-cell size are closely related to runoff to be generated. In this study, the effects of subdivision level (sub-basin scale) on runoff simulation were investigated, attempting to find a proper sub-basin scale for applying BTOPMC to large watersheds, especially to ungauged/data-poor basins.

THE SIGNIFICANCE OF SUB-BASIN SCALE IN BTOPMC MODEL

The BTOPMC model is developed for hydrological simulation for large river basins (Takeuchi *et al.*, 1999; Ao *et al.*, 1999). In this model, runoff generation is based on TOPMODEL (Beven & Kirkby, 1979; Quinn *et al.*, 1995) and flow routing is carried out by the Muskingum-Cunge method (Ao *et al.*, 2000). Drainage networks are generated using the automated pit-removal method by Ao *et al.* (2001). Study basins are automatically subdivided into natural sub-basins by the Pfafstetter numbering system (Verdin & Verdin, 1999) or rectangular blocks, over which groundwater balance is taken. Model parameters are automatically calibrated using the SCE-UA optimization algorithm (Duan *et al.*, 1992).

BTOPMC is a grid- and topography-based model, in which quantitative effects of topography on runoff generation are indirectly represented by the following equation:

$$S(k,i) = \{S_{av}(k) + m(k)[\lambda_{av}(k) - \lambda(k,i)]\}^{+} \tag{1}$$

where k is sub-basin number; i is grid number; $\{\cdot\}^{+}$ indicates that the value of $\{\cdot\}$ is non-negative and zero if negative; $S(k,i)$ is saturation deficit of grid i; $S_{av}(k)$ is average saturation deficit of sub-basin k; $m(k)$ is decay factor of saturated soil transmissivity of sub-basin k; and $\lambda_{av}(k)$ is average topographic index of sub-basin k, which is defined as an arithmetic mean of the local topographic index $\lambda(k,i)$ given by:

$$\lambda(k,i) = \ln[a(k,i)/\tan\beta_i] \tag{2}$$

where a is the area of the hillslope per unit contour length that drains through grid i; $\tan\beta_i$ is the local topographic slope.

In this model, because the local saturation deficit $S(k,i)$ in equation (1) is directly used for overland flow and baseflow calculations, the difference between average and local topographic index, $\lambda_{av}(k) - \lambda(k,i)$, will probably have significant impacts on runoff generation. Here, we define:

$$\Delta\lambda(i) = \lambda_{av}(k) - \lambda(k,i) \tag{3}$$

as effective topographic index. In this equation, as local value $\lambda(k,i)$ is a constant for a given grid size, thereby the spatial distribution of effective topographic index $\Delta\lambda$ and hence runoff generation could be significantly affected by the mean value $\lambda_{av}(k)$, i.e. by the shapes and sizes of sub-basins. Therefore, in BTOPMC (and also in TOPMODEL), in addition to grid size, the subdivision method and subdivision size are also very important topographic scale problems to be studied.

STRATEGY INVESTIGATING SUB-BASIN SCALE PROBLEM AND CASE STUDY

In order to focus on the effects of sub-basin scale, the following three aspects were taken into account. Firstly, as we have seen from equations (2) and (3), both sub-basin scales and grid sizes could have great impact on the spatial distribution pattern of the effective topographic index $\Delta\lambda(i)$. Therefore, grid size needs to be fixed in exploring the effects of sub-basin scales. Considering the worldwide comprehensive availability of DEMs (digital elevation models) and for reducing data requirements for large river basins, GTOPO30 DEMs were selected here. Secondly, the Pfafstetter method was used to subdivide study areas into natural sub-basins to reflect the actual drainage boundary. Thirdly, assuming soil and vegetation types to be homogeneous, parameter optimizations for study basins were carried out only for the case of no subdivision, and all parameters were kept constant for runoff simulations of all subdivision levels. This implies the simulated hydrographs of other subdivision levels will likely have worse fitting to the observed one than that of no subdivision. Therefore, the emphasis was placed on the comparison of simulation results of different subdivision levels, rather than the comparison with observed hydrographs.

Based on the above conditions and considerations, BTOPMC was applied to the Fuji-kawa basin (3500 km^2) and the Nakagawa basin (3270 km^2) in Japan. Figure 1 shows their drainage networks and the study sites. The upstream of Shimizubata and Noguchi were subdivided into 1–249 and 1–153 sub-basins (units), respectively. Figure 2 shows the density distribution curves of the effective topographic index $\Delta\lambda(i)$ corresponding to each subdivision level for Shimizubata. In Fig. 2, the case of one unit means no subdivision; and the larger the number of total sub-basins (units), the smaller the scale (average area) of sub-basins. From this figure it can be seen that subdivision level (sub-basin scale) have significant impact on the spatial distribution of $\Delta\lambda(i)$ used in runoff calculations.

Corresponding to these subdivision levels, a flood simulation in September 1993 and an annual runoff simulation from 1990 to 1991 at the Fuji-kawa basin, and a flood simulation in 1989 at the Nakagawa basin were conducted by using spatially-distributed hourly rainfall datasets. In all simulations, rainfall input for a grid cell was always given the record of its nearest raingauge and not affected by subdivision level. The only difference of computational conditions for each simulation is the subdivision level.

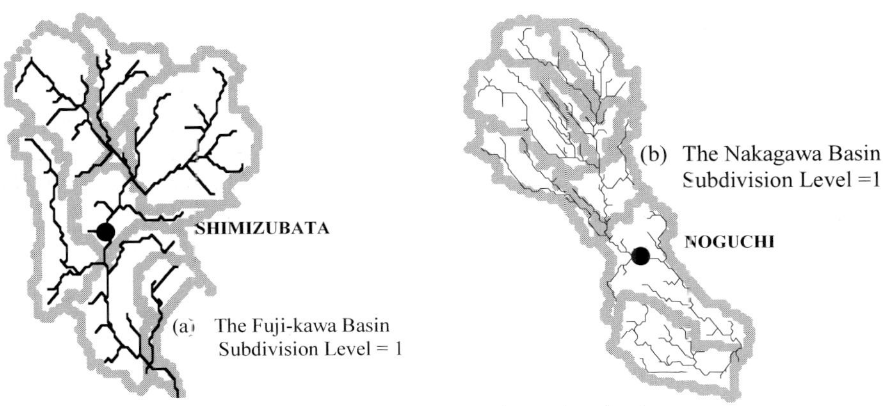

Fig. 1 Stream networks of the study basins and study sites.

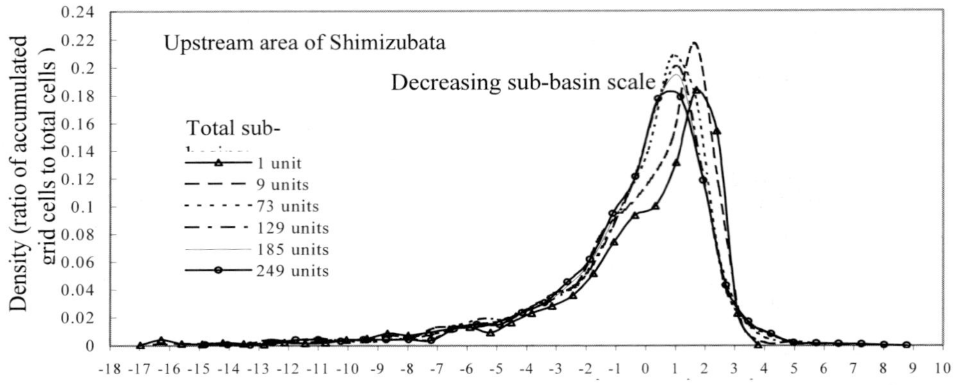

Fig. 2 Effects of subdivision level on the density distribution of the effective topographic index over the upstream area of Shimizubata.

RESULTS AND CONCLUSIONS

Through analysing all hydrographs and model performance indices such as the volume ratio (the proportion of simulated to observed total runoff amount) and the percentage of baseflow, the effects of sub-basin scale on runoff simulations are summarized as follows:

Effects of sub-basin scale on peak time The simulation results indicated that even if subdivision level increased from no subdivision to the highest level, the maximum difference of peak time was only one time step earlier in the case of flood simulations, and no difference of peak time appeared for the two-year runoff simulation. Moreover, from the hydrographs shown in Figs 3 and 4(a), almost no obvious difference of peak time is observed in spite of the magnitude of flood peaks.

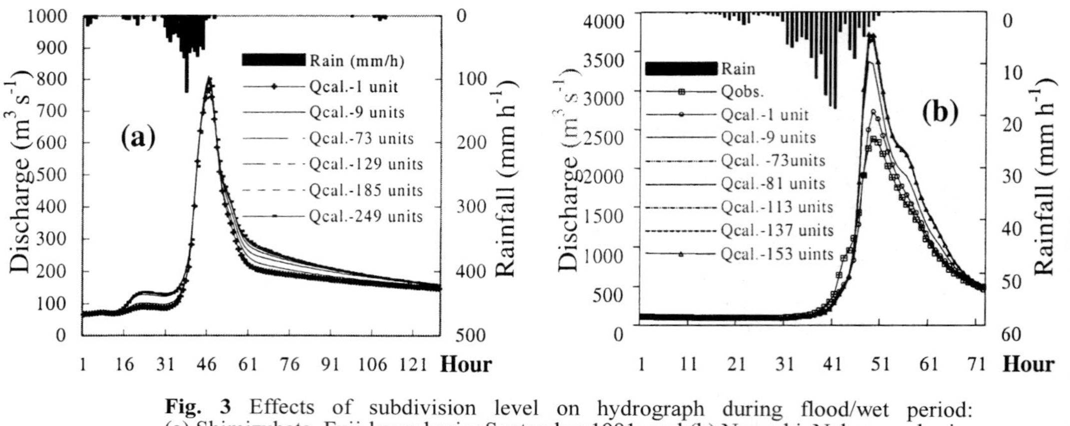

Fig. 3 Effects of subdivision level on hydrograph during flood/wet period: (a) Shimizubata, Fuji-kawa basin, September 1991; and (b) Noguchi, Nakagawa basin, August 1989.

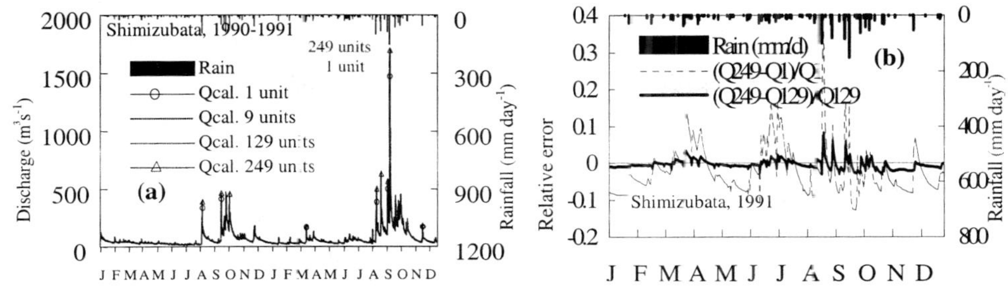

Fig. 4 Effects of subdivision level on annual runoff simulations for Shimizubata, Fuji-kawa basin, 1990–1991: (a) simulated hydrographs; (b) relative error among different sub-basin scales.

Effects of sub-basin scale on total runoff and hydrograph As shown in Figs 3 and 4, as the sub-basin area decreases, discharges are generally increased during flood/wet period while decreased during dry season. Consequently, due to such temporal distribution characteristics, higher subdivision levels produce larger total runoff for floods (Fig. 5(a), (b)), whereas differences between total runoff are not obvious and almost negligible in the case of annual runoff simulations (Fig. 5(c)).

Effects of sub-basin scale on flow components and response processes. Figure 5(a), (b) and (d) indicates that, during flood/wet periods, the models with smaller sub-basin scales generate higher baseflow (e.g. 3–7% higher) and lower saturation ratio.

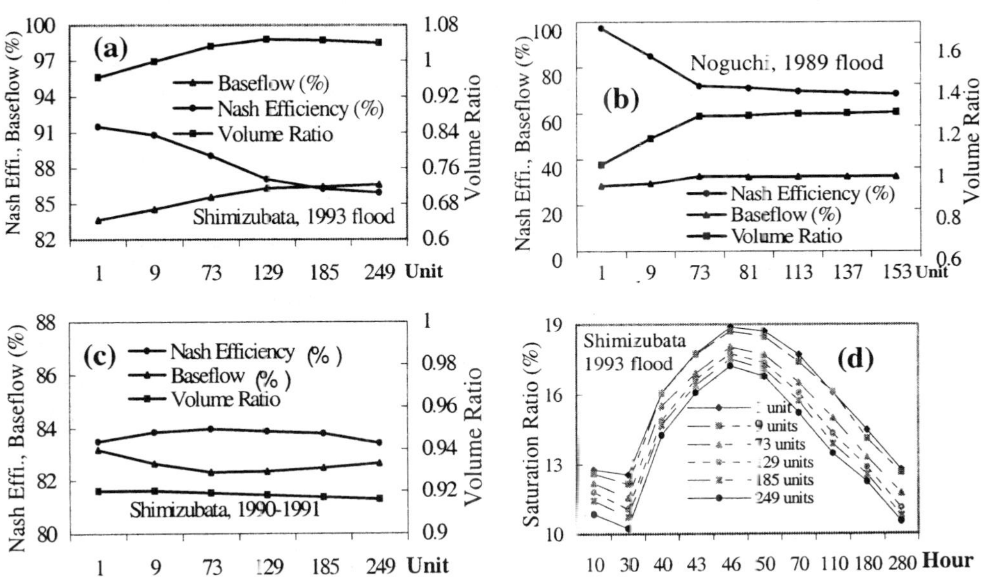

Fig. 5 Effects of sub-basin scale on model evaluation indices and basin response processes: (a) flood simulation at Shimizubata in 1993; (b) flood simulation at Noguchi in 1989; (c) two-year runoff simulation at Shimizubata during 1990–1991; and (d) temporal change of saturation ratio in flood simulation at Shimizubata in 1993.

On the other hand, during the dry season, as shown in Fig. 4(b), the models with smaller sub-basin scales produce lower baseflow since almost no overland runoff is contained in discharges. Because of such tendency, the flow components of annual runoff behave insensitive to sub-basin scales as shown in Fig. 5(c).

Stabilized subdivision level By further analysing Figs 3–5, it can be observed that, although sub-basin scale does have varying effects on runoff simulation results as stated above, the differences among hydrographs and among corresponding evaluation indices of model performances become smaller and smaller as subdivision levels increase (average sub-basin size decreases); and when even further subdivisions are made, simulation results appear stabilized. These behaviours suggest that there may exist an average sub-basin scale at which uncertainty in modelling due to subdivisions can be reduced to quite a low level. Here we define such average sub-basin size as the threshold of subdivision. As shown in Figs 5(a) and (b), the subdivision threshold for Shimizubata may be estimated as about 1/130 of its drainage area and for the Nakagawa basin 1/75. Here the threshold appears to depend on the basin and the structure of stream networks. Therefore, more case studies are needed to explore a common threshold for different basins. As a conservative estimate for the current two study basins, their common threshold may be set as 1/130–1/150 of their drainage areas, by which simulation stabilities and avoiding too many sub-basins can be achieved.

From the preliminary results described above, it is concluded that the main effects of sub-basin scale appear in flood simulations. This is because different subdivision levels generate different spatial distribution pattern of the effective topographic index. The distribution pattern tends to be almost the same when the mean area of sub-basins is smaller than a certain threshold. This phenomenon suggests that there may exist a proper subdivision level for BTOPMC that may be used both in modelling large river basins and in establishing the relationship between model parameter values and basin features. However, further study is required to explore the basin dependence of the subdivision threshold.

Acknowledgements The authors thank the Japan Society for the Promotion of Science for providing them with financial support for this research project.

REFERENCES

Ao, T. Q., Ishidaira, H. & Takeuchi, K. (1999) Study of distributed runoff simulation model based on block type TOPMODEL and Muskingum-Cunge method (in Japanese). *Ann. J. Hydraul. Engng JSCE* **43**, 7–12.

Ao, T. Q., Takeuchi. K. & Ishidaira, H. (2000) On problems and solutions of the Muskingum-Cunge routing method applied to a distributed rainfall runoff model (in Japanese). *Ann. J. Hydraul. Engng JSCE* **44**, 139–144.

Ao, T. Q., Takeuchi. K. & Ishidaira, H. (2001) On the method of generating artificial stream networks of large river basins and its effect on runoff simulation (in Japanese). *Ann. J. Hydraul. Engng JSCE* **45**, 139–144.

Beven, K. J. & Kirkby, M. J. (1979) A physically based, variable contributing area model of hydrology. *Hydrol. Sci. Bull.* **24**(1), 43–69.

Beven, K. J. & Binley, A. M. (1992) The future of distributed models: model calibration and predictive uncertainty. *Hydrol. Processes* **6**, 279–298.

Blöschl, G. & Sivapalan, M. (1995) Scale issues in hydrological modelling: a review. *Hydrol. Processes* **9**, 251–290.

Duan, Q., Sorooshian, S. & Gupta, V. K. (1992) Effective and efficient global optimization for conceptual rainfall–runoff models. *Water Resour. Res.* **28**(4), 1015–1031.

Flügel, W.-A. (1995) Delineating hydrological response units by geographical information system analyses for regional hydrological modelling using PRMS/MMS in the drainage basin of the river BRÖL, Germany. In: *Scale Issues in Hydrological Modelling* (ed. by J. D. Kalma & M. Sivapalan), 181–194. John Wiley & Sons Ltd, Chichester, West Sussex, UK.

Franchini, M., Wendling, J., Obled, Ch. & Todini, E. (1996) Physical interpretation and sensitivity analysis of the TOPMODEL. *J. Hydol.* **175**, 293–338.

Gupta, V. K. & Dawdy, D. R. (1994) Regional analysis of flood peaks: multi scaling theory and its physical basis. In: *Advances in Distributed Hydrology*. Water Resources Publications, Fort Collins, Colorado, USA.

Horritt, M. S. & Bates, P. D. (2001) Effects of spatial resolution on a raster based model of flood flow. *J. Hydrol.* **253**, 239–249.

Kavvas, M. L., Chen, Z.-Q., Tan, L., Soong, S.-T, Terakawa, A., Yoshitani, J. & Fukami, K. (1998) A regional-scale land surface parameterization based on areally-averaged hydrological conservation equations. *Hydrol. Sci. J.* **43**(4), 611–631.

Quinn, P. F. & Beven, K. J. (1991) The prediction of hillslope flow parths for distributed hydrological modelling using digital terrain models. *Hydrol. Processes* **5**, 59–79.

Quinn, P. F., Beven, K. J. & Lamb, R. (1995) The ln(a/tanβ) index: how to calculate it and how to use it in the TOPMODEL framework. *Hydrol. Processes* **9**, 161–182.

Saulnier, G.-M., Obled, C. & Beven, K. (1997) Analytical compensation between DTM grid resolution and effective values of saturated hydraulic conductivity within the TOPMODEL framework. *Hydrol. Processes* **11**(9), 1331–1346.

Sivapalan, M. & Kalma, J. D. (1995) Scale problems in hydrology: the contributions of the Robertson workshop. In: *Scale Issues in Hydrological Modelling* (ed. by J. D. Kalma & M. Sivapalan), 1–8. John Wiley & Sons Ltd, Chichester, West Sussex, UK.

Snada, K., Aoki, K. & Fujimura, T. (2001) Effects of scale of catchment element on runoff response of the distributed model (in Japanese). *Ann. J. Hydraul. Engng, JSCE* **45**, 145–150.

Takeuchi. K., Ao, T. Q. & Ishidaira, H. (1999) Introduction of block-wise use of TOPMODEL and Muskingum-Cunge method for the hydro-environmental simulation of a large ungauged basin. *Hydrol. Sci. J.* **44**(4), 633–646.

Verdin, K. L. & Verdin, J. P. (1999) A topographical system for delineation and codification of the Earth's river basins. *J. Hydrol.* **218**, 1–12.

Wood, E. F., Sivapalan, M., Beven, K. & Band, L. (1988) Effects of spatial variability and scale with implications to hydrologic modelling. *J. Hydrol.* **102**, 29–47.

Weather Radar Information and Distributed Hydrological Modelling (Proceedings of symposium HS03 held during IUGG2003 at Sapporo, July 2003). IAHS Publ. no. 282, 2003.

A grid-cell based distributed flood runoff model and its performance

TOSHIHARU KOJIMA & KAORU TAKARA
Disaster Prevention Research Institute, Kyoto University, Gokasho, Uji, Kyoto 611-0011, Japan
kojima@rdp.dpri.kyoto-u.ac.jp

Abstract Distributed parameter runoff models were recently developed and are expected to be used in practice. They have the ability to define model parameter values with spatially-distributed data such as topographic data, land-use data, radar rainfall and remote sensing images. This paper introduces a grid cell based kinematic wave runoff (GC-KWR) model that was developed by the authors in 1998, and discusses the effect of spatial resolution of the grid cell based distribution runoff model. The authors compare the 50-m resolution model with the 250-m model. The 250-m model tends to give earlier peak time and larger peak discharge. This paper also compares the distributed runoff model with two models conventionally used in Japan: the storage function model and the slope-channel system KWR (SC-KWR) model. The GC-KWR model and SC-KWR model can reproduce the observed hydrographs well, while the performance of the storage function model depends on flood events and study areas.

Key words DEM; flood precipitation; GIS; grid cell based distributed model; kinematic wave model; storage function model

INTRODUCTION

Distributed parameter runoff models have recently been developed and are expected to be used in practice. They have the ability to define model parameter values with spatially distributed data such as topographical data, land-use data, radar rainfall and remote sensing images. The authors developed a grid-cell based distributed kinematic wave runoff (GC-KWR) model in 1998. This distributed runoff model needs geographical data such as flow direction, slope and land cover on each grid cell for the hydrological modelling. The model parameters, equivalent roughness coefficients, are defined using land cover/use developed by the Geological Survey Institute Japan (GSI) or generated from remote sensing imageries. This paper introduces the development of the GC-KWR model and describes its applications to two basins. The performance of the GC-KWR model is compared with two conventionally used models in Japan: the storage function model and the slope-channel system KWR (SC-KWR) model.

GRID CELL BASED KINEMATIC WAVE RUNOFF (GC-KWR) MODEL

The grid cell based KWR (GC-KWR) model has the following features:

(a) A square area on a node point of a DEM is considered as a sub-basin, which is called a grid cell. A river basin is modelled as a network of grid cell.

(b) Grid cells are connected to each other with a drainage path defined by the steepest direction. Each grid cell receives flows from upper grid cells and rainfall on it.

Fig. 1 Drainage paths with 250-m resolution at the Yada River basin.

(c) Runoff from each grid cell is calculated by the kinematic wave model.

Necessary model parameters for each grid cell are equivalent roughness value and topographic properties. Equivalent roughness values are assigned according to land-use/cover. Topographic properties such as slope angle and length are defined with a DEM. The grid cell based model is also applied to sediment yield prediction in an Indonesian river basin covered with volcanic ash (Kojima *et al.*, 2002)

APPLICATION TO THE YADA RIVER BASIN

The GC-KWR model is applied to the Yada River basin (110 km^2), tributary of the Shonai River in Japan. Figure 1 shows the drainage paths with 250-m resolution. The effect of grid cell resolution is investigated. Figure 2(a) and (b) shows the hydrographs with the 50-m and 250-m resolution models, respectively. Both models use the same model parameters (equivalent roughness). The 250-m model has a tendency to give earlier peak time and larger peak discharge than the 50-m model. The reason is that the total length of the 250-m drainage paths is shorter than the 50-m model.

Using the 250-m model the effect of rainfall distribution on runoff at the outlet is examined. The spatial resolution of radar rainfall is about 2 km. Figure 2(b) and (c) shows the hydrographs with the spatially distributed radar rainfall and spatially-averaged (uniform) rainfall, respectively. The model with the spatially uniform rainfall estimates the larger peak discharge than that with spatially distributed radar rainfall. Figure 3 shows the movement of the rainfall field. In this case the heavy rainfall area moved from the downstream to the upstream. Thus, ignoring the movement of the rainfall area overestimates the peak discharge in this flood event.

APPLICATION TO THE SHONAI RIVER BASIN

The GC-KWR model is applied to the Shonai River (Shonai-gawa) basin (1010 km^2, see Fig. 4) and compared with a conventional lumped model and a traditional kinematic wave model. Simulated discharges are evaluated at the outlets of sub-basins Shidami, Tajimi, Toki and Mizunami for seven flood events between 1990 and 1995.

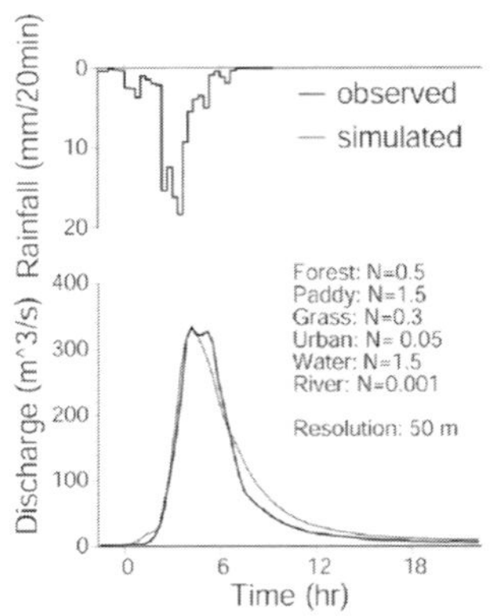

(a) 50-m resolution with spatial distributed rainfall.

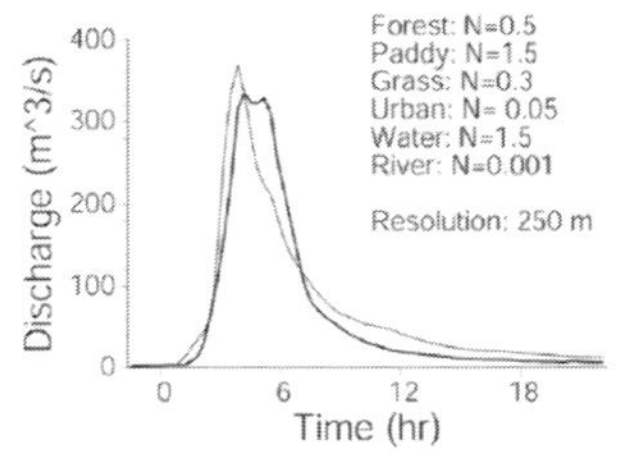

(b) 250-m resolution with spatial distributed rainfall.

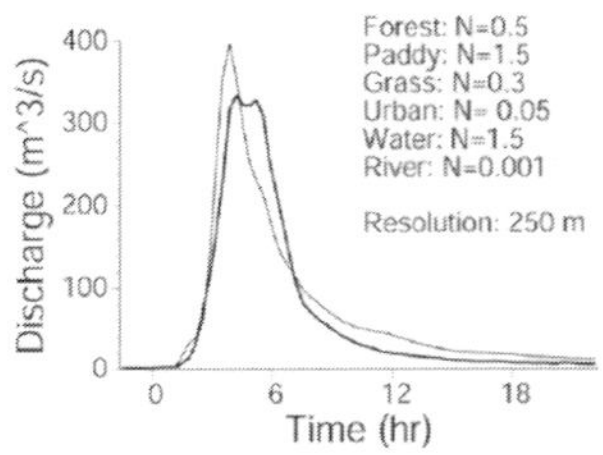

(c) 250-m resolution with spatial averaged rainfall.

Fig. 2 Comparison of simulated and observed hydrographs.

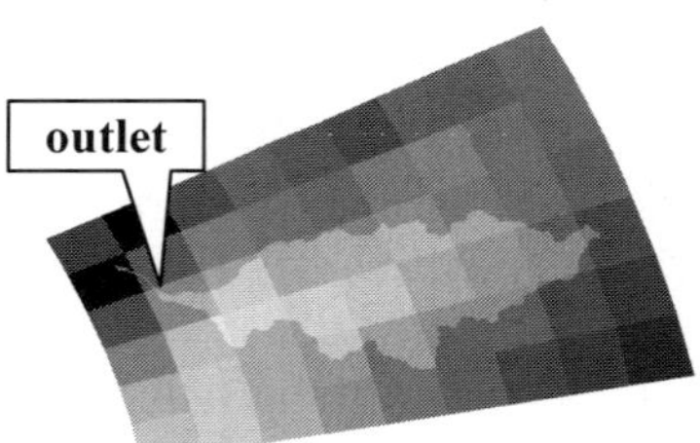

(a) At 3:00

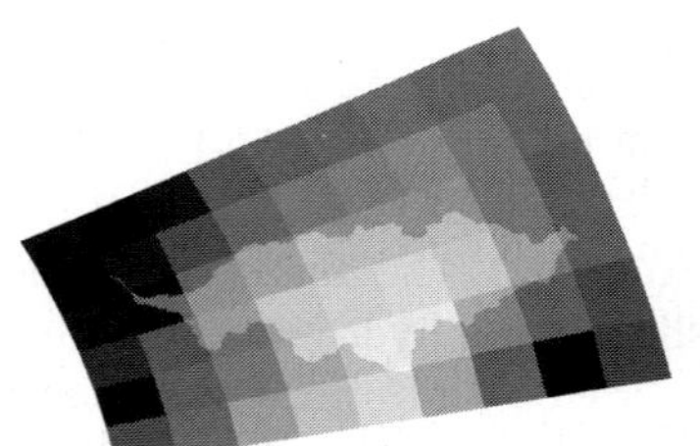

(b) At 3:20

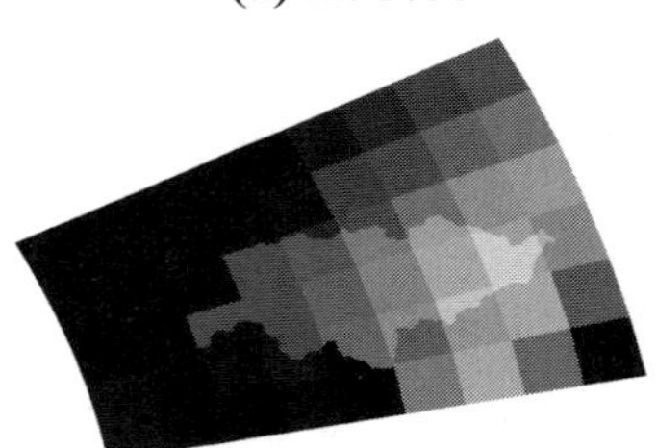

(c) At 3:40

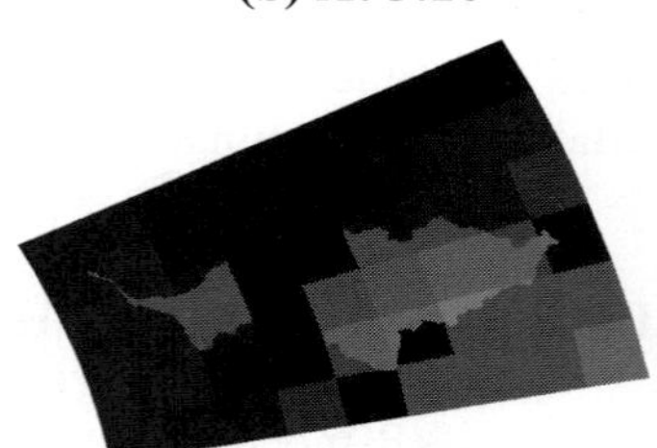

(d) At 4:00

Fig. 3 Movement of rainfall fields.

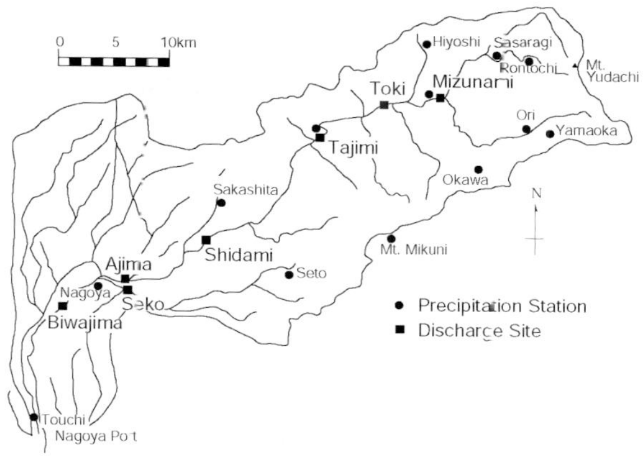

Fig. 4 The Shonai River basin.

Storage function model

In the storage function runoff model, the storage-discharge relation is given as follows:

$$S = Kq^{P} \tag{1}$$

$$\frac{dS}{dt} = r_e(t - T) - q \tag{2}$$

where S is volume of storage, q is discharge, K and P are model parameters, $r_e(t)$ is effective rainfall, t is time and T is lag time.

The model has been applied and used for flood prediction in the Shonai River by the Shonai River Work Office, Ministry of Land, Infrastructure and Transport (MLIT), Japan. Based on the model parameter values K and P calibrated by the office, the authors simulated the hydrographs with the fixed values of P and K multiplied by the magnification factor in the study area. Figure 5 shows the variations of the three evaluation indices such as correlation coefficient between observed and simulated hydrographs during the runoff event (COR), mean relative error (MRE) and relative root mean square error (R.RMSE). Figure 5 shows that if we use $0.2 \times K$ as the value of K, we may minimize the error between observed and simulated hydrographs for Flood 1.

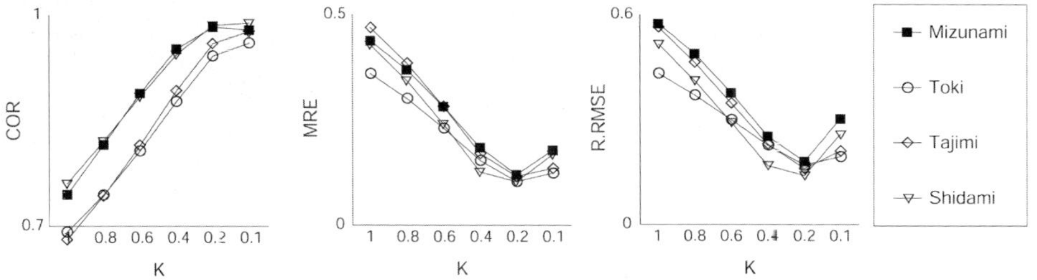

Fig. 5 Variations of the evaluation indices of the storage function model applied for Flood 1.

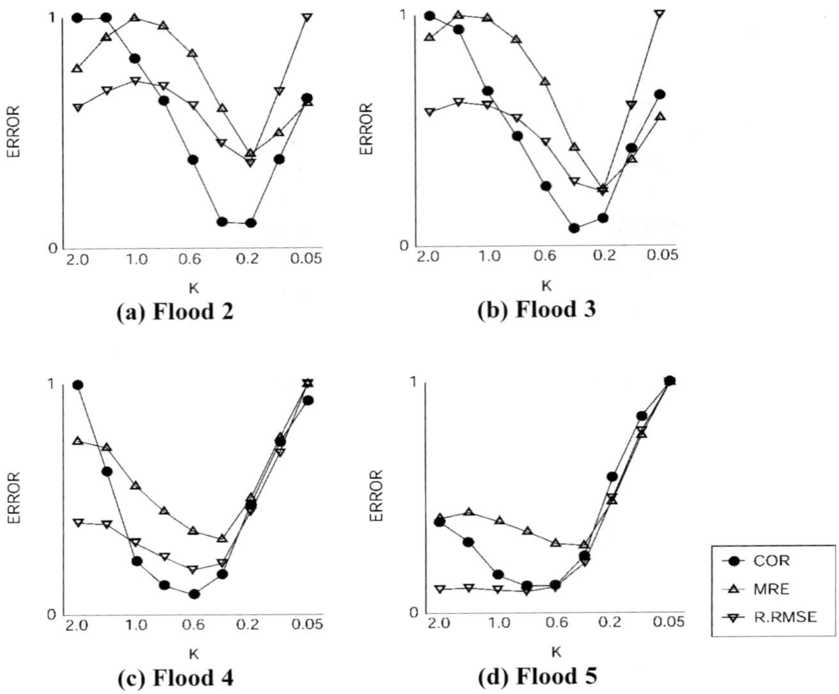

Fig. 6 Variations of the evaluation indices of the storage function model for the several flood events at Shidami.

Figure 6 shows the variations of evaluation indices for several flood events at Shidami. Figure 6 shows that the optimum magnification factors of K are 0.2 for Flood 2, 0.2–0.4 for Flood 3, 0.4–0.6 for Flood 4 and 0.4–0.6 for Flood 5. The optimum parameters of the storage function model change much depending on the flood events.

Slope-channel system KWR model

The Slope-channel system KWR (SC-KWR) model consists of sub-basins with a river channel section and rectangular slopes connected along the river channel section, on which the kinematic wave model is applied to simulate overland flow. This runoff model requires the topographic properties such as slope angle, slope length and river length. These topographic properties are automatically calculated by tracing the drainage path based on 50-m DEM. The equivalent roughness coefficients are defined by land-use class of each slope. The authors verified the effect of the urban area roughness value. The fixed roughness values ($m^{-1/3}$ s) of each land-use class are 1.0, 0.05, 0.3, 1.5 and 0.025 for forest, urban area, grass land, paddy field and river, respectively. These values are conventionally used in Japan (River Bureau, Ministry of Construction, 1998).

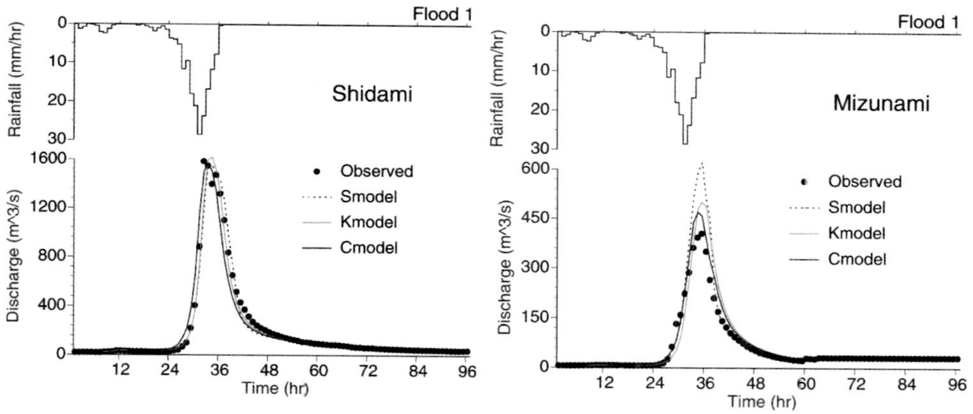

(a) Shidami (b) Mizunami

Fig. 7 Comparison of the simulated and the observed hydrographs for Flood 1.

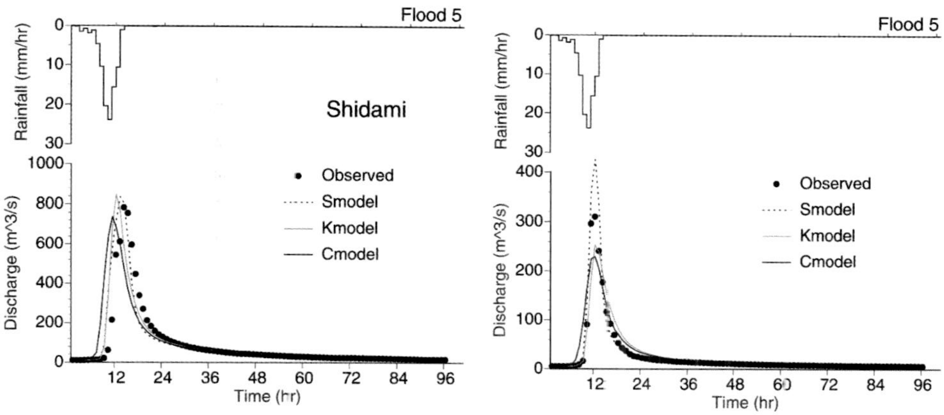

(a) Shidami (b) Mizunami

Fig. 8 Comparison of the simulated and the observed hydrographs for Flood 5.

MODEL PERFORMANCE COMPARISON

Figures 7 and 8 are simulated hydrographs at Shidami and Mizunami with three runoff models for Flood 1 and Flood 5. For Flood 1, any runoff models can simulate hydrographs well at the outlet of the study area, Shidami, while the storage function model (S model) overestimates the peak discharge at the upper sites such as Mizunami and Toki. The GC-KWR model (C model) simulates the hydrographs well with slightly early peak time. The SC-KWR model (K model) can reproduce the observed hydrographs well with conventionally used roughness coefficients.

CONCLUSIONS

In this paper, we obtained the following findings:

(a) The 250-m resolution GC-KWR model tends to give earlier peak time and larger peak discharge than the 50-m model.

(b) The GC-KWR model represents the effect of the storm movement. Time and space rainfall distribution is effectively incorporated into the runoff model.

(c) The optimum model parameters depend on flood events in the storage function model.

(d) The SC-KWR model can reproduce the observed hydrographs well.

(e) The GC-KWR model can also reproduce the observed hydrographs well with a slightly earlier peak than the observed one. Further investigation is necessary to improve this tendency.

REFERENCES

Kojima, T., Takara, K., Oka, T. & Chitose, T. (1998) Effect of resolution of raster spatial data on flood runoff simulation. *Ann. J. Hydraul. Engng JSCE* **42**, 157–162 (in Japanese).

River Bureau, Ministry of Construction (1998) River/Sabo Technical Standard (Proposal) Investigation Section, Sankai-do (in Japanese).

Kojima, T., Takara, K. & Tachikawa, Y. (2002) A distributed runoff model for flood prediction in ungauged basin. In: *The IAHS Decade on Prediction in Ungauged Basins* (Proc. Kick-off Workshop). Brasilia, Brazil.

Weather Radar Information and Distributed Hydrological Modelling (Proceedings of symposium HS03 held during IUGG2003 at Sapporo, July 2003). IAHS Publ. no. 282, 2003.

Evaluating land-use change effects on flood peaks using a distributed rainfall–runoff model in Yasu River, Japan

TUMAINI A. KIMARO

Disaster Prevention Research Institute, Kyoto University, Uji, Kyoto, 611-0011 Japan
kimaro@rdp.dpri.kyoto-u.ac.jp

YASUTO TACHIKAWA & KAORU TAKARA

Disaster Prevention Research Institute, Kyoto University & CREST, JST, Uji, Kyoto, 611-0011 Japan

Abstract The effects of land-use changes on flood peaks in the Yasu River basin were investigated using a distributed rainfall–runoff model. The land-use changes over a 20-year period were determined using 100-m spatial resolution data from 1976 and 1997. The main land-use classes in the basin were forest, paddy fields, urban and water bodies. There was a decrease in areal coverage by forests of 2.5% and paddy fields by 1.4% compared to 1976. The coverage of urban area increased by 2% and that of agricultural land by 0.2%. A new land-use category "golf courses" emerged in 1997, covering 2.3% of the basin in areas covered by forest in 1976. Simulation results show that 1997 land-use data produce 6% higher flood peaks and 16-min faster travel times compared to 1976 land-use data. The changes detected indicate the gradual effects of urbanization in the study basin.

Key words distributed rainfall–runoff model; effect of land-use change; Yasu River basin; Japan

INTRODUCTION

The effects of land-use changes on river flow have been a major concern for hydrologists for a long time. Recently, it has become even more important to study these effects in anticipation of global climate changes which are expected to result in increased frequency of extreme events such as floods and droughts. It follows naturally that planners and managers need to be well informed in advance of the possible impacts so that they can prepare better disaster management plans.

The documentation of effects of land-use changes on rainfall–runoff relationships mainly involve those involving studies on deforestation and urbanization. These are considered as the most important manmade impacts to the hydrological cycle. Douglas (1987) showed that clearing 19% of the forest in a basin can increase the peak flows by 13.5%. Several other researchers at different times and places have arrived at the same conclusion including Hewlet & Helvey (1970) in the USA, 11%, Swanson & Hillman (1977) in Canada, 200%, and Nagano (1967) in Japan, 114%. These studies show that changes in the streamflow are expected when modifications in the land-use occur. In some cases such modification can occur over a short period such as harvesting a forest in a drainage basin, but they can also occur gradually as the land-use changes with the socio-economic developments in the basin. Our primary concern is the changes in flood peaks introduced by changes in the land-use over a long time. The investigation seeks to quantify the land-use changes in the Yasu River basin and evaluate their effects on flood peaks and travel time.

STUDY AREA

The Yasu River basin (Fig. 1) is located between 34°50′ to 35°10′N and 136°0′ to 136°30′E, near Lake Biwa in Shiga Prefecture, Honshu Island, Japan. The catchment area upstream of the Yasu discharge station is 377.42 km^2. The basin receives rainfall in summer from May to July and during the typhoon season from the end of August to early October. There is a clear trend of high precipitation in the high elevation areas which requires careful consideration in generating spatial rainfall input for a hydrological model. Typical storm hyetographs show that during storms, the mountainous stations can receive up to twice the rainfall recorded in the lower stations. Most floods occur during the typhoon season which has high intensity rainfall over a short period, normally less than three days. Floods occur immediately after the rain storms, within 2–3 h of the peak rainfall. The major land-use class in the basin is forest (60.9%), followed by paddy fields (17.3%), according to 1997 land-use data. There are five gauging points where the river stage is measured and 10 rainfall stations, which are well distributed within the basin, as seen in Fig. 1. In addition to hourly river water level measurements, river discharge is measured during floods every year at the river gauging stations.

RAINFALL–RUNOFF MODEL

One of the most important uses of a distributed rainfall–runoff model is to predict the effects of future land-use and climate changes on basin hydrology, particularly the extreme events and water yield for water resources evaluation (Beven, 2000). Land-use changes have different effects on the local hydrological cycle depending on the nature

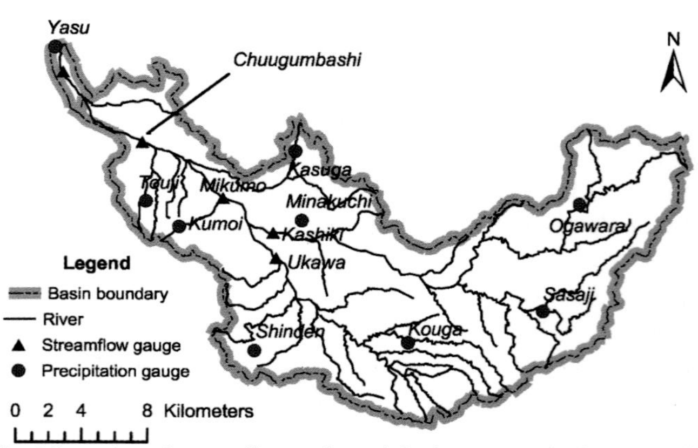

Fig. 1 Location of streamflow and precipitation gauges in the Yasu River basin.

of the land-use cover that existed and that one which results after the changes. To assess the hydrological impacts of land-use changes, temporal and spatially highly detailed land-use data are required. In addition to this, the assessment of the hydrological impacts must be done with a distributed hydrological model which represents the spatial variability in the land cover with appropriate parameters such that the spatially variable response is taken into account in total basin simulation.

In this study a distributed rainfall–runoff model developed by the authors has been used. The model utilizes the kinematic wave equation in routing the overland flow from each cell. The output of the upstream cell is given to the downstream cell and routed together with rainfall over that cell to its outlet. The process continues on with each upstream cell sending output to its downstream neighbour up to the most downstream cell, which represents the basin outflow point. The model is built using the object oriented approach which defines each cell as an element model containing data and functions for converting rainfall to runoff. The connection between grids is defined by a flow direction map derived from a digital elevation model (DEM) of the area. This map is stored in UTM coordinates and is used as a reference point for all point and spatial data. The flow map is based on the selected DEM resolution and it is extracted automatically from DEM data at any feasible resolution. In the current study a 500-m grid flow map was chosen in order to reduce the computational work load. In this case all spatial inputs are rescaled to this grid size.

The model has utility programs for preparation of spatial land-use and rainfall data. Raster files of land-use data and rainfall are created as input to the model and for display in the GIS environment. The spatial rainfall can be allocated by ordinary kriging, Thiessen polygons or inverse distance interpolation (Guillermo & Salas, 1985). The rainfall processing utilities allow the production of the storm hyetograph and cumulative rainfall raster maps to show the evolution of the storm during the simulation. This allows any errors in the data to be identified and the results of the simulated hydrograph to be interpreted properly. The results of the model can be stored in time series of river flows at any desired location on the basin and raster maps of runoff depth distribution over the catchment. The architecture of the model is flexible and permits different process models to be applied for each grid. In order to investigate the effects of land-use changes effectively, the model is also equipped with utilities to extract land-use data based on the DEM-derived flow map and to analyse the land-use changes in pattern and magnitude over the years.

LAND-USE AND HYDROLOGICAL DATABASE FOR CHANGE DETECTION

The successful detection of land-use changes and their hydrological impacts requires the maintenance of a long-term accurate and easily accessible database of hydrological variables and land-use data. It should be possible to easily access this database and extract summary information corresponding to observed floods. This information may include hyetographs of the storm which caused the flood, raster maps of the cumulative rainfall depth before the flood, and river discharges. Such information is useful for documentation and it gives an indication of the variable basin response to different storms over the years. In this study we compiled a database which includes the long-term records (30 years) of rainfall, river flows, observed floods and land-use data. Rainfall and river flow data used were obtained from the local office of the Ministry of Land, Infrastructure and Transportation (MLIT). The river flow data at Yasu, the outlet of the study basin, consist of water level from January 1975 to the present, and rating curves for the same period.

The land-use data was obtained from the Japan Geological Survey Institute for the years 1976, 1987, 1991 and 1997. These data are classified in 15 land-use classes for 1976, 12 for 1987 and 11 for 1991 and 1997, and spatial resolution is 100 m. For our analysis we regrouped those land-use classes into five groups representing land covers with different Manning coefficients. The groups and their Manning's roughness coefficient values ($m^{-1/3} s^{-1}$) in brackets were paddy field (2.5), forest (0.5), water body (0.01), urban area (0.02) and agricultural land (0.3). The values of coefficients for different classes were based on published values (Ministry of Construction, 1997) currently (MLIT). In the preliminary investigations of these data it was found that the rate of land-use changes is small and can only be effectively described with data over a long period of time. In regard to this, all subsequent analyses are based on data for 1976 and 1997.

The primary use of the land-use data was for allocating a spatially variable Manning's roughness coefficient for the kinematic wave model used to route overland flow from each grid. It may be argued that there is uncertainty involved in assigning this coefficient to a specific grid with a certain land-use because the exact value of the coefficient for each landcover is not exactly known. However, it is still true that most of the difficulty faced is to identify the land cover at a certain location rather than the value of the coefficient to be assigned. Once the land-use type of a grid is known the error reduces greatly because the coefficient is selected within a small range of feasible values. Our aim is to study the effect of land-use change by comparing hydrographs simulated with two different land-use data sets using consistent roughness values assigned according to land-use class.

ANALYSIS AND RESULTS

Land-use changes

The changes in land-use were determined by comparing the land-use data in 1976 and 1997. The results are shown in Fig. 2, as well as in Tables 1 and 2. The most visible change between the two periods is the conversion of forest areas into golf courses, as seen on the 1997 land-use map. This shift of land-use has contributed mostly to the decrease of the percentage of the basin covered by forests, which amounts to 2.5%. The golf courses covered 2.3% of the basin area in 1997. The area covered by paddy fields decreased by 1.4%, while the built-up area coverage increased by 2%. A closer examination shows that the expansion of the urban area took place in the lower part of the basin downstream of Kashiki and Ukawa gauging stations. The 1997 land-use map shows an extension of a transportation line further to the southwestern part of the basin, close to the Kashiki River. This explains the growth of urbanized area in this

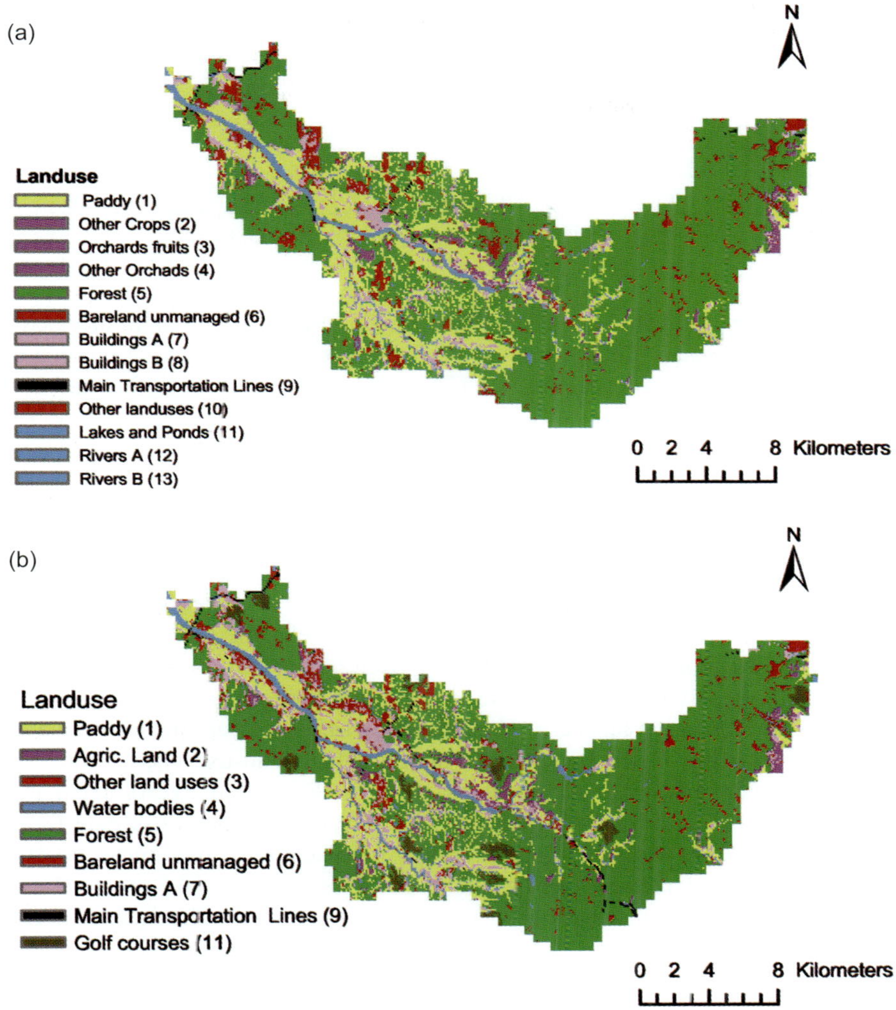

Fig. 2 Land-use map of the Yasu River basin (a) in 1976, and (b) in 1997.

Table 1 Percentage of area of the Yasu basin covered by different land-uses in 1976 and 1997.

Landuse	Coverage in 1976 (%)	Coverage in 1997 (%)
Forests	63.4	60.9
Paddy field	18.7	17.3
Buildings (built up area)	4.4	6.4
Water bodies	3.6	3.8
Other land-uses	4.2	3.4
Agric. land	2.6	2.3
Bare land (unmanaged)	2.9	2.8
Golf courses	0	2.3
Main transport lines	0.27	0.5

Table 2 Distribution of converted land-use for selected different classes from 1976 to 1997.

	Percentage of change contributed by conversion to given land-use in 1997:								
Land-use code in 1976	1	2	3	4	5	6	7	9	11
1		5.2	22.5	7.1	7.2	4.1	38.3	4.3	11.3
5	12.2	6.6	17.8	4.4		3.7	12.1	2.7	40.5
6 & 10	1.6	5.5	46.1	2.9	8.0		18.8	0.6	16.5

Land-use codes are shown in the land-use maps in Fig. 2.

part of the basin. Table 2 also reveals that land-use class that is affected most by urbanization is paddy field. There is only a slight effect of urbanization on the forested area where urbanization accounts for 12% of the change between 1976 and 1997.

Implications for flood runoff

The specific effect of the land-use changes on the hydrology of a given local area are complex and can only be evaluated with high spatial resolution land-use and hydro-meteorological data. The effect of land-use changes on hydrological response is not represented by the change the individual land-use classes, but by the combination of changes of the individual classes replacing each other, forming a new land-use pattern. The resulting pattern may increase or decrease the peak flood, depending on the replaced and replacing land-use. This means that it is not possible to make a prediction of runoff changes based on changes in individual land-use classes and a probable behaviour can only be ascertained by using a distributed hydrological model that specifically accounts for each land-use type. We used the land-use data for 1976 and 1997 to create two different raster maps of Manning's roughness coefficient for overland flow calculations. In assigning the roughness coefficient, land-use classes with similar hydrological responses were given the same roughness coefficient. Due to scale differences in the land-use raster resolution and the runoff model grid resolution, the roughness coefficient for each grid was determined by averaging roughness coefficients of all smaller grids (100 m) that fall inside the model grids (500 m).

The results of this simulation with different storms over the years showed a consistent change in the simulated hydrograph, developing higher and faster peaks using the 1997 land-use data as compared to floods simulated using the 1976 land-use data. A typical case of simulation result with different land-use data is shown in Figs 3 and 4. In these figures the rise in peak flow reflects the changes observed in land-use data between 1976 and 1997. In all cases the hydrographs exhibit higher peaks and shorter travel time. The attenuation of peak clearly indicates the potential effects of land-use changes in the study basin. Although the effects are not so severe at the moment it is still necessary to take precautions by conducting similar studies from time to time and evaluating the impacts.

Based on the land-use data, we note that major land-use classes forest and paddy fields did not experience much change during the study period. This may mean that the simulation results indicate the changes brought about by redistribution of land-use in the lower part of the basin, mainly from paddy fields to building plots. The supposition

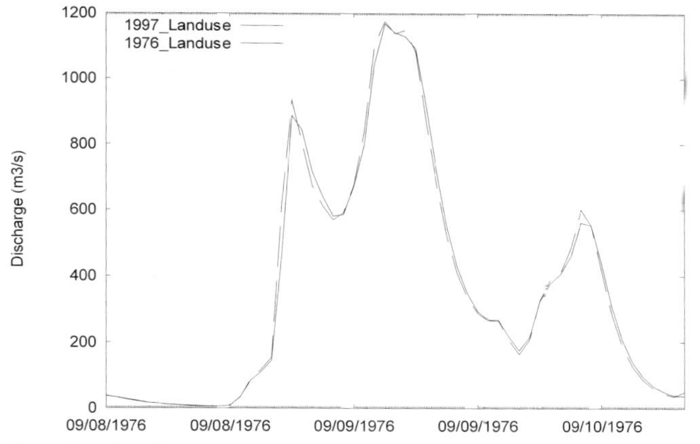

Fig. 3 Simulated hydrograph for a typical storm in 1976 using 1976 and 1997 land-use data.

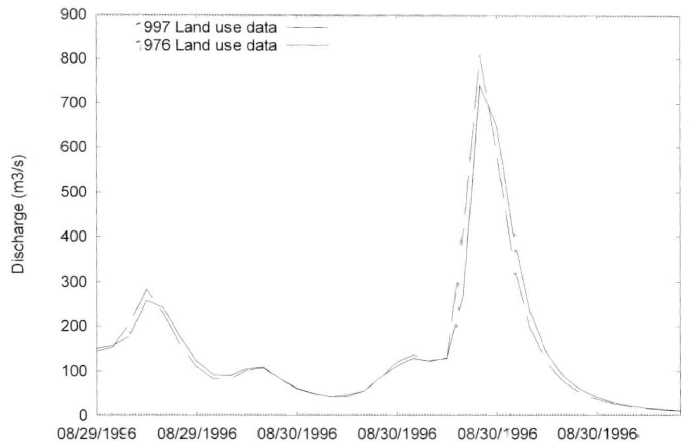

Fig. 4 Simulated hydrograph for a typical storm in 1996 using 1976 and 1997 land-use data.

is in agreement with land-use data which show the largest (38%) change of paddy fields between 1976 and 1997 is attributed to conversion to building plots. On the other hand the changes in forested land were mainly due to conversion to golf courses (40%). The roughness coefficient of golf course assumed in this study (0.3 $m^{-1/3}$ s^{-1}) is not very different to the one assumed for forest (0.5 $m^{-1/3}$ s^{-1}). It was also found that most of the area classified as other land-uses in 1976 was converted to built-up area in 1997. Coincidentally, this class was also assigned the same coefficient as the built up area, assuming that it is likely to be located in built-up areas. A summary of interchange of different land-use classes in 1976 to different classes in 1997 is presented to further highlight this point in Table 2. It is clear from these data and the simulated hydrographs that the effects of land-use changes on the flood peaks can be modelled and explained by using fine resolution land-use data and the distributed rainfall–runoff model.

CONCLUSION

Evaluation of the effects of land-use changes on flood peaks has been performed using land-use data and a distributed rainfall–runoff model. The changes over the 20 years indicate increase in flood peaks by 6% and decrease in travel time by 16 min. It is further shown from land-use data that these changes reflect mainly developments in the lower part of the basin and are due to urbanization. The results of this work show a high potential of distributed hydrological models for explaining changes in streamflows if high spatial land-use data are used. From the experience gained in this study we found out that the response of the catchment to rainfall for a certain land-use change is also dependent on the spatial distribution of the storm used during the study. To gain a full understanding of all possible response be necessary it may be subject to the changed land-use pattern to all possible storm patterns. In anticipation of global climate changes which may change the seasons and patters of storm distribution, such experiments are recommended to evaluate the worst possible response of a changed land-use to rainfall. The effect of land-use changes on flood peaks is a combination of individual changes in different land covers over the whole basin. Land-use data can be used to identify the most probable origin of the effects and proper measures can be taken to address the situation. Remotely sensed land-use data provide hydrologists with valuable data for hydrological models. However, there are few models which can utilize this data directly and estimated parameters still have large uncertainties and hinder more deep analyses into land-use change studies. It is proposed that future hydrological models should concentrate in utilizing high spatial land-use data such as provided by remote sensing images in order to model effectively the effects of changes in land-use.

Acknowledgements We wish to thank Biwako work office of the MLIT, Mr Kudo and Ms Morita at Nippon Construction Consultant for providing hydrological data used in the study. This research was supported by DPRI, Kyoto University, Nissan Science Foundation and CREST program of the Japan Science and Technology Corporation.

REFERENCES

Beven, K. J. (2000) *Rainfall–Runoff Modelling: The Primer*. John Wiley & Sons, Chichester, West Sussex, UK.

Douglas, L. (1987) Changes in streamflow peaks following timber harvest of a coastal British Columbia watershed. In: *Forest Hydrology and Watershed Management* (ed. by R. H. Swanson, P. Y. Bernier & P. D. Woodard) (Proc. Vancouver Symp., August 1987), 509–517. IAHS Publ. no. 167.

Guillermo, Q. T. & Salas, D. J. (1985) A comparative analysis of techniques for spatial interpolation of precipitation. *Water Resour. Bull.* **21**(3), 365–380.

Hewlett, J. D. & Helvey, J. D. (1970) Effects of forest clearing on the storm hydrograph. *Water Resour. Res.* **6**, 768–782.

Ministry of Construction (1997) *Manual for River Works in Japan-Survey*. Sankaido, Tokyo, Japan (in Japanese).

Nagano, H. (1967) Effects of changes of forest conditions on water yield, peak flow and direct runoff of a small watershed in Japan. In: *Proc. Int. Symp. Forest Hydrol.* (ed. By W. E. Sopper & H. W. Lull), 551–564.

Swanson, R. H & Hillman, G. R. (1977) Predicted increased water yield after clear-cutting verified in west-central Alberta. Report NOR-X-198. Fish & Environ., Canada Survey North Forest Res. Center, Canada.

4 Application of Distributed Hydrological Models in Watershed Management

Operational deployment of a physics-based distributed rainfall–runoff model for flood forecasting in Taiwan

BAXTER E. VIEUX, JEAN E. VIEUX
Vieux & Associates, Inc., 1215 Crossroads Blvd, Suite 118, Norman, Oklahoma 73072, USA
bv@vieuxinc.com

CHIARONG CHEN
Central Weather Bureau, Taipei, Taiwan

KENETH W. HOWARD
NOAA-National Severe Storms Laboratory, Norman, Oklahoma, USA

Abstract The Central Weather Bureau in cooperation with the Water Resources Agency of Taiwan, among others, seeks to improve technology and techniques employed to monitor and predict rainfall–runoff throughout Taiwan. The rainfall and runoff techniques and technologies are critical to the potential mitigation of specific hazards such as flash floods, river floods, debris flow, and landslides. A programme initiative for the research and development of a flood alert and water resources management system is to unify monitoring and prediction of floods within a single system. Enhancing the accuracy and efficiency of information disseminated from the central government to the public, and to regional and local water management and emergency response agencies is the major goal of this project.

Key words distributed modelling; floods; hydrology; hydrological modelling; radar rainfall; real-time flood monitoring operations

INTRODUCTION

Operational deployment of a physics-based fully distributed rainfall–runoff model, called V*flo*™, coupled in real time with quantitative precipitation estimates (QPE) from radar, satellite and raingauges, allows monitoring and prediction of river stage at any geographic location. QPE using multiple sensors provides hydrological model input for improving prediction of rainfall runoff and resulting flood hazards. V*flo*™ relies on digital elevations and other geographical information to predict discharge and stage at any selected location within a drainage network described by Vieux & Vieux (2002). This implementation of the model relies on the QPE derived from multiple sensors (raingauges, satellite, and radar); an approach developed by the US National Oceanic and Atmospheric Agency (NOAA)—National Severe Storms Laboratory (NSSL) called QPESUMS (Gourley, 1998; Howard *et al.*, 1997).

V*flo*™ and QPESUMS were installed in June 2002, and were operational within a few days before rain bands from Typhoon Rammasan struck the island the first week of July. This event caused localized flooding and some landslides in mountainous areas, while ending water usage restrictions by partially filling reservoirs that supply Taipei, the capital city of Taiwan. One of the most damaging of recent tropical cyclones was Typhoon Nari, which occurred in September 2001. Typhoon Nari resulted in 93 dead, 10 people missing and extensive damage to the northern part of the island and to the city of Taipei. This event is the subject of retrospective analysis and event reconstruction for calibration of the flood alert system, which is the subject of this paper. Analysis of land–sea interactions that may have augmented precipitation processes in Typhoon Nari causing severe flooding in Taipei is reported by Sui *et al.* (2002). Herein we present an overview of the flood warning system design and operational requirements, a sensitivity study, and simulation of river stage compared with observed streamflow using an operational distributed hydrological model with multiple radar input.

Monitoring and prediction technologies and techniques are critical, not only for hazards associated with floods, but also for debris flow, landslides and water resources management. The 35 980 km^2 Island of Taiwan is located north of the Philippines, off the southeastern coast of China; and surrounded by the East China Sea, Philippine Sea, South China Sea, and Taiwan Strait. The climate is typical of tropical maritime conditions with a rainy season during the southwest monsoon period from June to August. Typhoons average 3.6 per year often bringing damage from wind and flooding, but also filling reservoirs that supply water after the monsoon season ends. Average annual rainfall is in excess of 2500 mm. The terrain is mountainous in the interior with flat to gently rolling plains in the west along the coast. The highest elevation is 3997 m above mean sea level. The terrain is complex with extremely steep gradients in the mountainous areas. As the drainage network encounters coastal areas, progressively flatter gradients affect velocities. Channelization in urban areas restricts inundation until dikes or walls are overtopped. With few rivers exceeding 30 km, runoff generated in the mountains arrives at sea level in less than a few hours.

The Central Weather Bureau (CWB) is responsible for issuing forecasts of typhoon tracks and heavy rains which may result in floods throughout the island. Flooding is prevalent during the typhoon season because of intense and prolonged rainfall coupled with orographic influences creating torrential runoff. Once runoff reaches the flatter coastal plains, where most of the population and economic development are concentrated, flood disasters result. Recent advancement in radar-based QPE offered the opportunity to improve the monitoring of conditions that may result in flooding on very small time and space scales. The central government desired monitoring and prediction capability encompassing all ten river basins. Central planning and emergency relief efforts benefit from having timely information on flooding locations; and not just locations of heavy rainfall. A comprehensive system was required that integrated multiple sensor-based QPE with a distributed model for flood prediction. Because flooding is both a hydrometeorological and hydrological phenomenon, cross-governmental cooperation between CWB and WRA became an important first step to developing an integrated flood warning system. The flood prediction system now deployed in Taiwan is being upgraded and improved over a multi-year period. The architecture of the system provides monitoring and hydrological prediction in addition to analysis functions. There are several key components in the overall system, which is described below.

NSSL has developed QPESUMS, which is a suite of algorithms that utilizes radar and satellite data to provide precipitation estimates. Depending on the region and application, these sensors may include raingauge, lightning, surface observation data, and satellite imagery. The QPE is produced within a GIS framework consisting of three components: (a) data integration, radar quality control; (b) three-dimensional (3-D) reflectivity mosaics of storm structure; and (c) quantitative precipitation estimation. V*flo*™ accepts input from a variety of common radar and gauge formats and can operate independently using radar reflectivity processing algorithms and raingauge calibration for real-time operation or post-analysis. The deployment in Taiwan required integration of four radars of different manufacturer specifications and scanning strategies. Besides radar, an extensive raingauge network of over 200 sites provides information for real-time bias adjustment. The resulting QPE comprised of radar and raingauges are processed, providing input to the hydrological model.

V*flo*™ MODEL DESCRIPTION AND ATTRIBUTES

V*flo*™ is a fully distributed, physics-based hydrological model capable of utilizing geographic information and multi-sensor input to simulate rainfall runoff from major river basins to small catchments. V*flo*™ is a *hydraulic* approach to hydrological analysis and prediction. Overland flow and channels are simulated using the kinematic wave analogy. With the addition of Muskingum-Cunge, and Modified Puls, the model is extended to larger rivers and more complex hydraulics defined by rating curves or measured cross-sections. The theoretical basis for the distributed modelling that underlies V*flo*™ is given by Vieux (2001), Vieux & Moreda (2002), Vieux & Vieux (2002) and Vieux (2002).

Real-time flood forecasting requires ingestion of radar rainfall data directly into V*flo*™. The model is implemented in Java to take advantage of secure Servlet/Applet technology for multi-user access via the Internet or intranets. V*flo*™ utilizes GIS grids to represent the spatial variability of factors controlling runoff. Runoff production is from infiltration excess and is routed downstream using the kinematic wave analogy. Computational efficiency of the fully distributed physics-based model is achieved using finite elements in space and finite difference in time. Evolution of rain fields during a tropical storm is measured by radar and updated every 10 minutes.

The finite element approach is based on a drainage network, which allows flexibility in selecting the location for reporting streamflow information (see inset in Fig. 1). Configuring hydrological prediction on a drainage network basis makes the model scalable from small catchments to major river basins, eliminating the need for separate models for different scales of catchment. Terrestrial parameters are derived from commonly available sources of digital geospatial data. Parameters include topography and drainage networks derived from a digital elevation model (DEM), infiltration parameters derived from soils, and hydraulic roughness derived from land use/cover (e.g. Landsat). Parameterization proceeds by importing GIS grids to the model, then adjusting parameters with a multiplicative factor. Once the model is calibrated, the settings and parameters are saved in a basin-overland-properties (bop) file. Model application to a particular basin is achieved by comparing observed discharge and stage with simulated at select locations. A limited sensitivity and calibration study was undertaken and is described in the following sections.

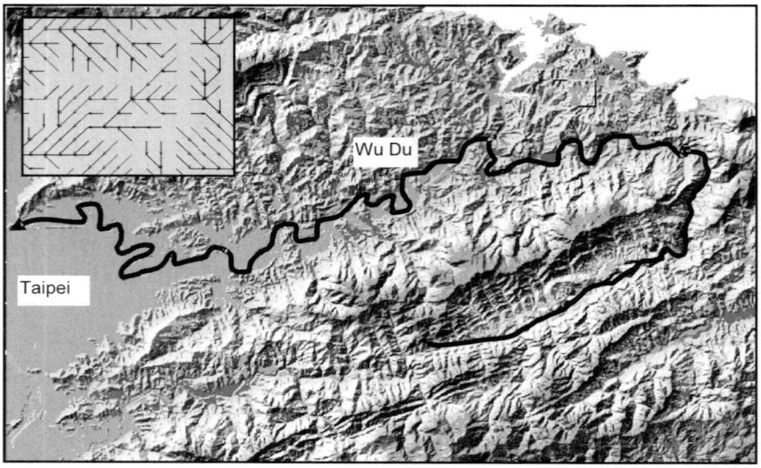

Fig. 1 Shaded relief map shown here at 40-m resolution, Keelung River (black line) and Wu Du bridge location. The inset shows a typical finite element representation of the drainage network defined by the DEM (upper left).

STUDY AREA

The initial study area selected is the Keelung River because of its tendency to flood in critically sensitive social and economic areas. The headwaters originate in the mountains to the southeast of Taipei, the runoff typically moves rapidly downstream to where it joins the Tanshui River in Taipei. No suitable sites for flood control reservoirs exist in the Keelung due to the topographical characteristics of the basin. Accurate monitoring is one of the key alternatives for mitigating flooding experienced by the city of Taipei from this river. The Keelung River is shown in Fig. 1. Beginning in the highlands (see black line), the river takes a northeasterly route with very steep gradients on the order of 10%, then takes a sharp turn to the west and then proceeds in a southwesterly route through flatter terrain as it approaches Taipei. The arrows shown in the inset indicate the finite elements that connect each digital elevation grid cell. The drainage network composed of overland and channel cells is parameterized according to expected hydraulic characteristics. These parameters are then adjusted on a global, basin, or an incremental basis to affect peak and timing. The ordered adjustment of the parameters controlling the runoff process follows the OPPA methodology described by Vieux & Moreda (2002).

MODEL SENSITIVITY

The sensitivity of simulated stage and discharge to inputs and parameters helps guide the development of the model. Knowing which parameter causes the most response in stage or discharge identifies where effort should be expended to improve parameter specification. A limited sensitivity study was conducted for the Keelung with input from Typhoon Nari. Multiple locations were used to test the sensitivity with results shown for four bridge locations along the Keelung. Two sensitivity tests are shown: sensitivity to infiltration, and to rainfall. Infiltration is treated as a constant loss rate equivalent to saturated hydraulic conductivity in the Green and Ampt equation. Given the short residence time of runoff on the land surface before it arrives in a channel, infiltration may not be as important as other factors. Clay soils and intense rainfall also limit the importance of infiltration rates on controlling runoff in many areas including Taiwan. The infiltration rate, I (mm h^{-1}) is adjusted by scalar multipliers in steps of 0.2 from 0.4 to 2.0. The multiplication of the scalar times on the parameter map is indicated in the legend of the respective figures, e.g. infiltration, I multiplied by 2.0 is indicated as: *I_2.0*. Selected infiltration trials are shown for scalars of 0.4, 1.0, 1.4, and 2.0 in Fig. 2. The narrow range of variation in simulated stage is indicative that runoff production is not dominated or significantly affected by infiltration rates at this location in the basin.

The radar rainfall is reproduced from reflectivity data and was not calibrated to raingauges. Reflectivity conversion to rain rates is sensitive to the relationship used. A tropical relationship was tested using the approach described by Vieux & Bedient (1998). Applying a scalar multiplier is analogous to systematic bias removal when radar is gauge-adjusted using mean-field techniques. The scalars applied to the spatially variable radar rainfall maps range from 0.5, 1.0, 1.5 and 2.0. The scalar multiplier for radar rainfall rate, R multiplied by 2.0 is shown as: *R_2.0*. The range in response is sensitive to the rainfall scalar as shown in Fig. 3. The slope of the best fit line for *R_1.0* is 0.9184 indicating an 8% bias towards underprediction in this case. The broad range of response in stage, from 10 to over 30 m, to rainfall rates is in marked contrast to infiltration, which affects stage on the order of only several metres. As is generally accepted, rainfall rates and spatial distribution have dramatic impacts on runoff. The sensitivity shown to radar rainfall confirms that radar-based QPE is a critically important factor for accurate streamflow prediction. As more typhoon and tropical storm events become available, further calibration and validation will be possible.

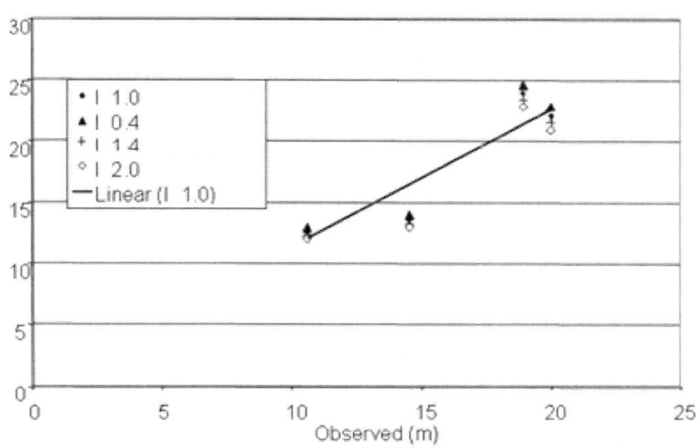

Fig. 2 Stage sensitivity to infiltration scalars with linear best fit line applied to the *I_1.0* data only.

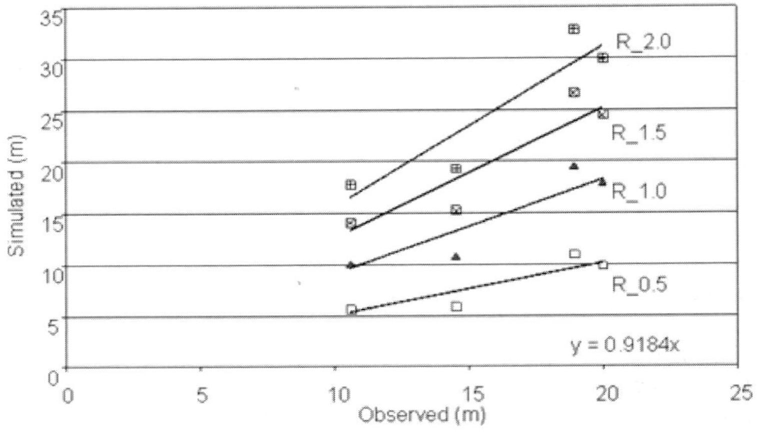

Fig. 3 Stage sensitivity to radar rainfall scalars.

Simulated and observed stage hydrographs at the Wu Du Bridge located along the Keelung River are shown in Fig. 4. The model closely matches the observed stage in the rising limb and for the second peak. The third observed peak occurred after the end of the radar rainfall processed for this post-analysis study. The shape of the recession limb is important from the standpoint of judging the adequacy of the model representation. Except for the third peak, which was not simulated, the recession limb of the model tends to follow the observed recession. Table 1 shows the comparison between simulated and observed stage at locations along the Keelung. The statistics for several points along the river where observed hydrographs may be compared with simulated results indicate that the mean absolute percentage error (MAPE) is 8.6%. These are calibration results and remain to be validated as more events occur and data are recorded.

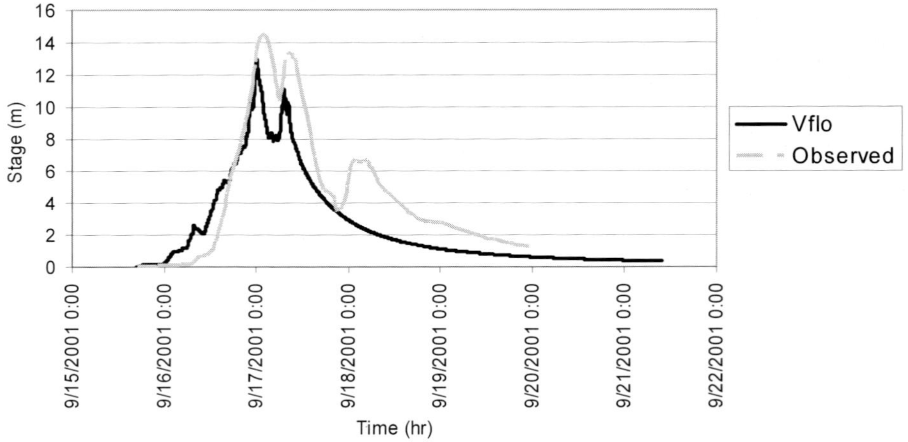

Fig. 4 Simulated and observed stage hydrographs at the Wu Du Bridge.

Table 1 Simulated and observed maximum stages for Typhoon Nari at select locations.

Keelung River location	Simulated stage (m)	Observed stage (m)	Absolute percentage error
Zhang An Bridge	17.01	18.94	10.2
Wu Du	12.93	14.54	11.1
Da Hua Bridge	10.97	10.61	3.4
Jie Shou Bridge	7.94	7.22	9.9
			MAPE=8.6

SUMMARY

Close collaboration between Taiwan government agencies, the United States government, and private industry has resulted in the creation of new monitoring and prediction techniques and technologies to assist Taiwan in monitoring and predicting the occurrence of flooding. The coupling of advanced multiple-sensor techniques for QPE and a physics-based fully distributed hydrological model provides improved accuracy for streamflow prediction in a critical social and economic region of Taiwan. The coupling of advanced techniques in rainfall monitoring with distributed hydrological modelling allows prediction to occur on smaller time and space scales. The terrain and hydroclimatic conditions in Taiwan cause a complex rainfall structure to evolve during typhoon landfall. Sensitivity demonstrated for this region shows that accurate rainfall rates are critically important to operational flood forecasts.

REFERENCES

Gourley, J. (1998) Multiple sensor precipitation estimation over mountainous terrain. MS thesis, School of Meteorology, University of Oklahoma, USA.

Howard, K. W., Gourley, J. J. & Maddox, R. A. (1997) Uncertainties in WSR-88D measurements and their impacts on monitoring life cycles. *J. Weath. Forecasting* **12**, 167–174.

Sui, Chung-Hsiung, Huang, Ching-Yuang, Tsai, Yi-Ben, Chen, Ching-Sen, Lin, Pay-Liam, Shieh, Shinn-Liang, Li, Ming-Hsu, Liou, Yuei-An, Wang, Tai-chi Chen, Wu Ray-Shyan, Liu Gin-Rong & Chu, Yen-Hsyang (2002) Meteorology-hydrology study targets typhoon Nari and Taipei. *EOS, Trans. Am. Geophys. Union* **83**(24), June.

Vieux, B. E. (2001) *Distributed Hydrological Modelling Using GIS.* Water Science Technology Series. Kluwer Academic Publishers, Norwell, Massachusetts, USA.

Vieux, B. E. (2002) Predictability of flash floods using distributed parameter physics-based models. Report of a Workshop on Predictability and Limits to prediction in Hydrological Systems, 77–82. Committee on Hydrological Science, National Research Council, National Academy Press, Washington, DC, USA.

Vieux, B. E. & Moreda, F. G. (2002) Ordered physics-based parameter adjustment of a distributed model. In: Chapter 20 of *Advances in Calibration of Watershed Models* (ed. by Q. Duan, S. Sorooshian, H. V. Gupta, A. N. Rousseau & R. Turcotte), 267–281. Monograph Series on Water Resources, Trans. Amer. Geophys. Union, Washington, DC, USA.

Vieux, B. E. & Vieux, J. E. (2002) V*flo*™: a real-time distributed hydrological model (Proc. 2nd Federal Interagency Hydrological Modelling Conference, 28 July–1 August 2002). Las Vegas, Nevada, USA.

Vieux, B. E. & Bedient, P. B. (1998) Estimation of rainfall for flood prediction from WSR-88D reflectivity: a case study, 17–18 October 1994. *J. Weath. Forecasting* **13**, 126–134.

Weather Radar Information and Distributed Hydrological Modelling (Proceedings of symposium HS03 held during IUGG2003 at Sapporo, July 2003). IAHS Publ. no. 282, 2003.

Applying radar rainfall data in basin hydrological modelling

OSCAR KALINGA, THIAN YEW GAN & JIANCANG XIE

Department of Civil and Environmental Engineering, University of Alberta, Edmonton, Alberta T6G 2G7, Canada

tgan@civil.ualberta.ca

Abstract To assess the advantage of basin-scale hydrological modelling gained from using radar rainfall data as input over the standard gauge measurements, this study applied the WSR-88D level III hourly radar rainfall data of the National Weather Service, USA, to simulate the basin hydrology of the Blue River basin using the physics-based hydrological model, DPHM-RS. To reduce the errors of radar rainfall data, the radar data are first re-sampled with respect to the gauge rainfall data of Oklahoma Mesonet by the statistical objective analysis (SOA) approach. Preliminary results show that adjusting the radar rainfall data by SOA improved the simulated runoff hydrograph.

Key words gauge rainfall data; hydrological modelling; radar rainfall data; statistical objective analysis

INTRODUCTION

To overcome the limitation of point measurements from raingauges, there is a growing interest in utilizing radar data for basin-scale hydrological modelling (Sun *et al.*, 2000). Since radar rainfall data are derived from radar echoes, there is also a need to analyse errors associated with precipitation estimated from different radar processing techniques (Borga, 2002). This study uses the WSR-88D level III hourly rainfall estimates supplied by the Hydrology Laboratory (HL) of the National Weather Service (NWS) office, USA, meteorological data of Oklahoma Mesonet, and a semi-distributed hydrological model, DPHM-RS developed by Biftu & Gan (2001).

Major sources of errors affecting the WSR-88D level III hourly rainfall estimates are radar electronic calibration, the *Z–R* relationship, rainfall attenuation, radar range effect, anomalous propagation, ground cluster, beam filling, and the radar sampling strategy (Austin, 1987), and techniques used to produce stage III data from stage II products, such as averaging the overlapping radar bins. In general, more spatially and temporally variable spring and summer convective rainfall is better represented by radar data than the sparse raingauges, while the less variable stratiform winter precipitation could be well captured by sparse raingauges (Stellman *et al.*, 2001). This study focuses on the following objectives:

(a) To assess the advantage of basin-scale hydrological modelling gained from using radar rainfall data as input over the standard input of gauge measurements in terms of basin subdivision and radar data resolution.

(b) To integrate radar rainfall fields with gauged rainfall data via a statistical objective analysis scheme as an effort to reduce errors in terms of storm type (i.e. convective and stratiform storms), and storm size.

DESCRIPTION OF STUDY SITE

The study site is the Blue River basin (BRB) of the South Central Oklahoma, where its stream gauging station is located at latitude 33°59′49″N and longitude 96°14′27″W, 1.6 km west of Blue, Oklahoma, USA (see Fig. 1). Its elevation ranges from 153 to 350 m a.m.s.l. and there is no impoundment on the Blue River. Woody savannah is the dominant vegetation cover, occupying almost 80% of BRB. The major soil group of BRB is clay loam mixed with sand or silt in some portions. The average annual precipitation ranges from about 400 mm in the extreme western panhandle to 1420 mm in the southeastern corner of the State (NOAA, 1977). In winter, Oklahoma lies in the southern range of the polar jet stream and the northern range of the subtropical jet stream, leading to extremely variable temperature, precipitation (rain, ice and snow), strong winds, and blizzard conditions. The January temperature ranges from a daytime high of 20°C to a nighttime low well below zero. The primary source of moisture is from the Gulf of Mexico. The Pacific Ocean off the coast of Mexico is a source of moisture under certain airflow patterns. Much of the spring precipitation results from large thunderstorms, many of which produce tornadoes and large hail. These severe storms occur as surface low-pressure and frontal systems develop when a transient upper-air trough approaching from the west interacts with warm, moist air from the Gulf of Mexico (Tortorelli, 1991). Further, important moisture sources include local and upwind land surfaces. Convective storms generally move eastward across the state to provide most spring and early summer rainfall. If intense thunderstorms repeatedly develop over the same terrain for several hours or days, localized flooding can be massive.

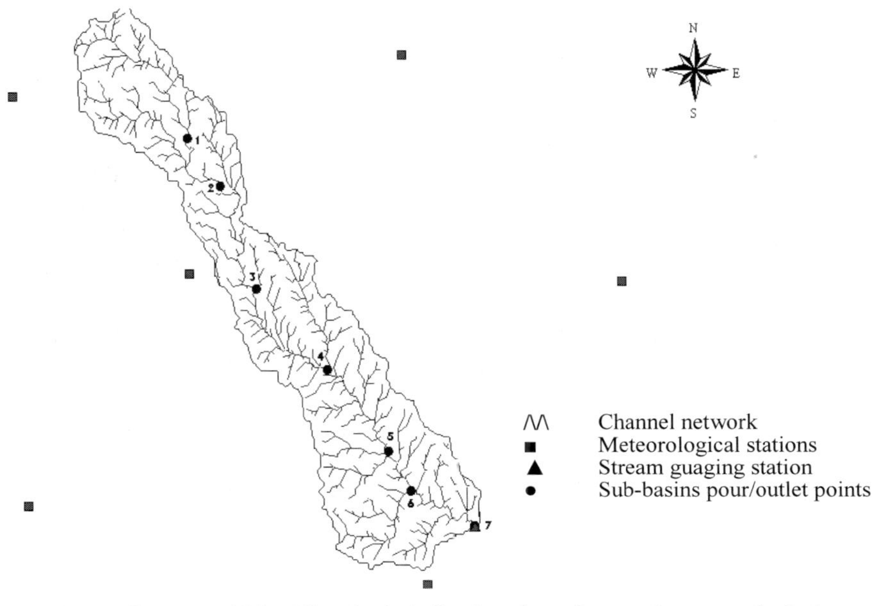

Fig. 1 Location map of Blue River basin indicating channel network, meteorological stations, streamflow gauging station, and sub-basins pour/outlet points.

RESEARCH METHODOLOGY

The research methodology comprises of the following:

(a) Assess the difference between the observed and the simulated runoff hydrograph of BRB by DPHM-RS model driven by: (a) gauge rainfall data; (b) radar rainfall data; (c) mean aerial precipitation (MAP) of gauged data; and (d) MAP of radar data, for convective (spring and summer) and stratiform (winter) storms.

(b) For the WSR-88D hourly stage III data produced via the global, conceptual processing technique of Krajewski (1997), a variation of the statistical objective analysis (SOA) scheme of Preira Fo *et al*. (1998) was applied to refine the radar rainfall data to better reflect local conditions by statistically re-sampling the WSR-88D stage III data (see Fig. 2) with respect to the tipping bucket measurements of Oklahoma Mesonet, and the adjusted radar data assessed as described in (a).

(c) At 4 × 4 km resolution, BRB can be covered by about 1024 cells of radar data (see Fig. 2). To gain an idea of the optimum radar data resolution based on storm types, repeat the tests of (a) at coarser resolutions, such as 8 × 8, etc., to 128 × 128 km.

(d) To study the effect of basin subdivision on the simulated hydrograph, BRB was subdivided into a number of sub-basins (e.g. see Fig. 3) and tests described in (b) were repeated to determine an optimum basin subdivision.

DESCRIPTION OF DATA

Radar rainfall estimates

The techniques used to derive operational WSR-88D hourly products are divided into three stages (Seo *et al*., 1999). Stage I involves producing the Hourly Digital Precipitation (HDP) from individual radars of the National Weather Service (NWS) at a spatial resolution of approximately 4 km × 4 km, on a polar stereographic projection called the Hydrological Rainfall Analysis Project (HRAP) grid. Then, depending on what is the prevailing weather condition, a standard Z–R relationship ($Z = 300R^{1.4}$) or a tropical Z–R relation ($Z = 250R^{1.2}$) is used to convert the radar measured reflectivity (HDP) into a radar rainfall rate estimate, called a stage I product.

Stage II deals with the data quality control that consists of correcting the mean-field bias, the filling of missing values, and the removal of erroneous values of the hourly radar product (stage I). Bias is estimated from the difference between the rain-gauge-measured rainfall and the radar-derived values at the gauge's nine surrounding HRAP cells. This bias is computed for the non-zero gauge-radar pairs, with a minimum of three pairs used to calculate the mean field bias. The computed mean field bias is applied for the respective hour over the entire radar coverage area until a new set of gauge-radar pairs is obtained (Seo *et al*., 1999). In the absence of at least three gauge-radar pairs for a given hour, the previous hour mean field bias is applied to the whole coverage.

If a gauge reports rainfall for a given hour while the radar data are missing, a maximum value at the gauge's site is assigned to all HRAP cells falling under the gauge's radius of influence. The gauge's radius of influence under the radar's umbrella

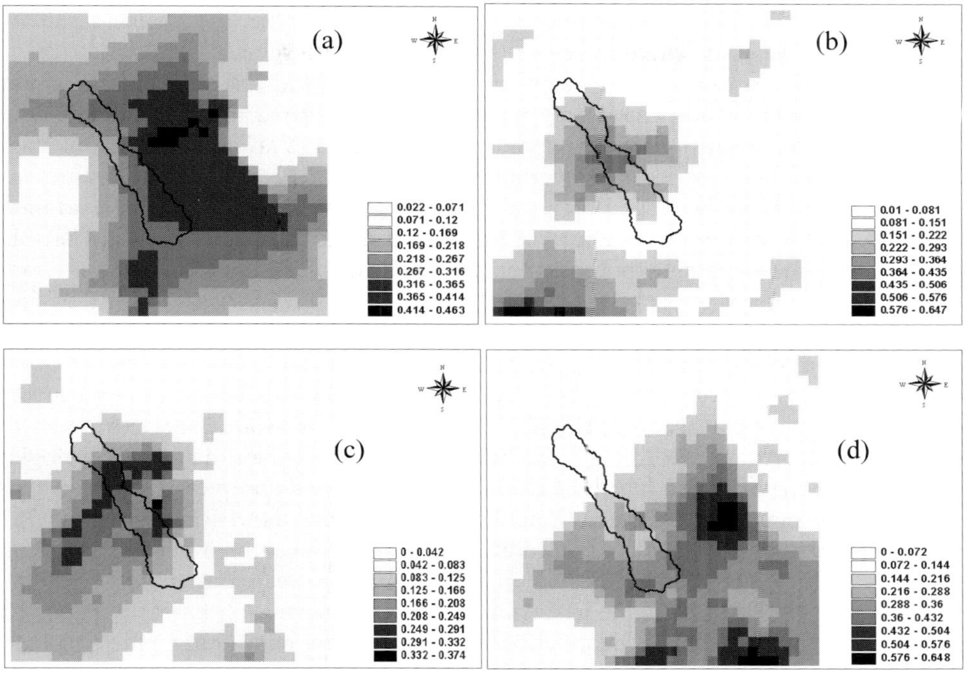

Fig. 2 Four episodes of radar rainfall (cm h^{-1}) occurring over BRB on 7 January 1998 at z-times (a) 00z, (b) 01z, (c) 02z, and (d) 03z.

Fig. 3 Blue River basin sub-divided into seven sub-basins.

is typically two to three HRAP bins. On the other hand, for the case where radar data are available, but show rainfall over a gauge that reports zero, then the respective HRAP bin will be assigned a zero value. This process results to a stage II product.

In stage III, any erroneous gauge or radar data due to bright banding, ground clutter, or anomalous propagation are removed, followed by combining the individual stage II products from the respective radars within the area of coverage into a single mosaic. Specifically, for areas of overlapping radar coverage, the multiple values at a given HRAP grid are averaged. Pereira Fo *et al.* (1998) showed that this averaging, among other factors, leads to errors in the final stage III products of the WSR-88D network used for operational purposes (e.g. streamflow forecasting).

Land-use and terrain data

To drive the physics-based, DPHM-RS model of Biftu & Gan (2001) to BRB, the input data required are the topographic data, land-use/cover, soil data, meteorological data, channel cross-section, depth of groundwater table, and observed streamflow data. The land-use classes determined from the Arc/Info GIS for BRB consist of 78% woody savannah, 14% deciduous needle leaf forest, 4% evergreen needle leaf forest, 2% grasslands, 1% croplands, and 1% urban and built-up areas. Based on the 100 m × 100 m resolution DEM data, the basin was divided into seven sub-basins (Fig. 3). The slope, flow direction, flow accumulation, mean elevation, and topographic soil index of each sub-basin was also determined. Four dominant soil types were identified for BRB: loam, clay, sandy clay loam, and silty clay loam. Hourly hydroclimatic data such as air temperature, global solar radiation, relative humidity, wind speed, soil temperature, and gauge measured rainfall data were taken from the Oklahoma Mesonet. The hourly ground heat flux and net solar radiation were estimated within DPHM-RS.

STATISTICAL OBJECTIVE ANALYSIS (SOA) AND DPHM-RS

SOA scheme

As an effort to reduce errors in the stage III radar rainfall data described in the previous section, the radar precipitation data were re-sampled with respect to raingauge measurement using the SOA scheme originally developed by Gandin (1963) and modified by Pereira Fo *et al.* (1998). The SOA scheme exploits the strength of both systems, such that weather radar maps out the spatial details more accurately while raingauges measure the mean value of precipitation field more accurately than radars. By combining these two data sources we can potentially improve the rainfall estimates and consequently the estimated flow hydrograph.

SOA will be applied to the WSR-88D stage III radar rainfall data of BRB on a seasonal time series since we expect the correlation structure of rainfall fields to differ between seasons, e.g. winter in BRB is dominated by stratiform storms, while spring and summer are more subjected to convective storms. The proposed analysis equation used in SOA is:

$$R_a(x_j, y_j) = R_r(x_j, y_j) + \sum_{n=1}^{N} w_{jn}\left[R_g(x_n, y_n) - R_r(x_n, y_n)\right] \quad (1)$$

where $R_a(x_j,y_j)$ is analysed rainfall (mm) at grid point j; $R_r(x_j,y_j)$ is radar estimated rainfall (mm) at j; $R_g(x_n,y_n)$ is raingauge rainfall measurement (mm) at station n; $R_r(x_n,y_n)$ is radar estimated rainfall (mm) at station n; w_{jn} is *a posteriori* weight; N is the number of raingauge stations; and (x, y) are coordinates (km).

As a necessary condition to determine the weights, we need to assume that the raingauge measurement (observation) and radar measurement (background) errors are uncorrelated and unbiased. The expected analysis error variance is minimized by differentiating equation (1) with respect to each of the weights, w_{jn}.

Hydrological model (DPHM-RS)

A physics-based hydrological model (DPHM-RS) (Biftu & Gan, 2001), designed to assimilate data in a semi-distributed approach, is used to simulate streamflow of BRB. The model is divided into six components: interception, evapotranspiration (ET), soil moisture, subsurface flow, surface flow, and channel routing (see Fig. 4). ET from land and vegetation is modelled separately, exchanges of energy and water fluxes at the land–atmosphere interface are considered to be one-dimensional (vertical), and a kinematic response function is used to route surface and sub-surface runoff from each sub-basin to the stream and then via a network of stream channels to the basin outlet.

DISCUSSION

For any hydrological modelling based on gauge rainfall data, we expect some discrepancies between the simulated and observed runoff hydrographs because gauge data are point measurements that cannot adequately account for the spatial variability of rainfall fields. On the other hand, the use of spatial radar rainfall data alone could still lead to simulated hydrograph that usually suffer in terms of an underestimation in volume/peak runoff than in phase shift because radar captures the spatial variability of rainfall events well but tends to underestimate the mean rainfall in comparison to raingauges. We therefore expect the integration of radar and gauge rainfall data to improve the basin rainfall estimation and hence the resultant simulated hydrograph to agree better with the observed both in phase and in runoff volume. As this point, the effect of radar data resolution on the resultant simulated hydrograph is difficult to predict, but we may find the optimum data resolution to be something other than 4 km, and dependent on the climatic and hydrological regime of the river basin. We would expect more accurate results from using radar rainfall than gauge rainfall data, especially if the basin is dominated by convective (more highly variable in space and time) than stratiform (less variable) storms.

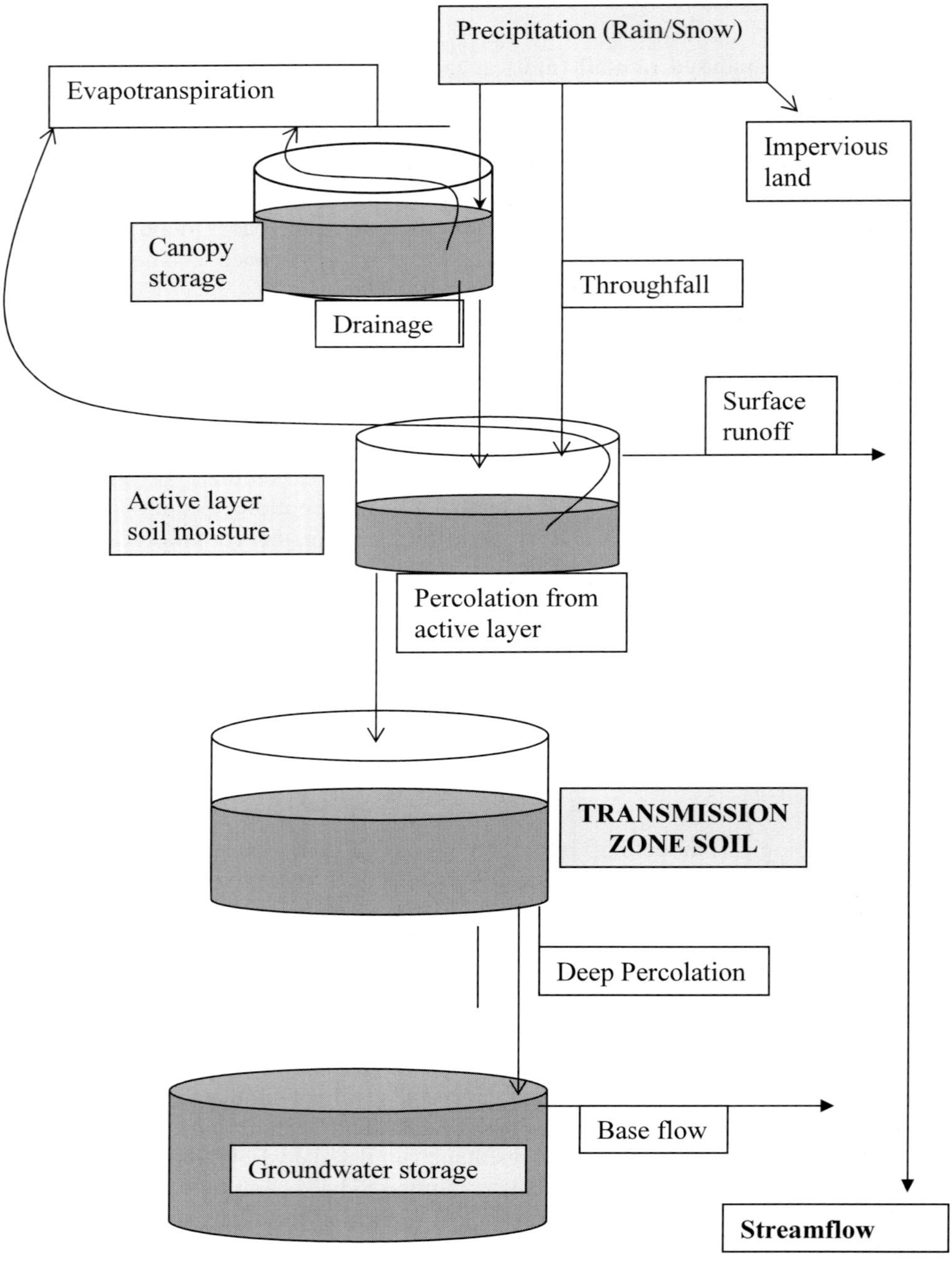

Fig. 4 A schematic diagram showing the major components of DPHM-RS.

PRELIMINARY RESULTS

The response function from a 1 cm effective rainfall (Fig. 5) for each of the seven sub-basins (Fig. 3) was derived using a kinematic wave equation. These response functions have a major influence on the final shape of the estimated basin outflow hydrograph because they are used to compute the actual surface runoff for each sub-basin. With 57 grids of radar rainfall data falling within the BRB, each sub-basin had an average of eight grids (ranging from 5 to 10 grids per sub-basin). The efficiency index (*Ef*) of Nash and Sutcliffe (1970) was used for statistical evaluation of model estimated hydrographs.

Generally the result (Fig. 6) shows that the model estimated hydrographs with gauge, radar, and radar adjusted (SOA) rainfall input forcing both failed to predict the second peak of the observed hydrograph. This can partly be attributed to the sub-division of the basin as evidenced by: (a) the sub-basin's response functions (Fig. 5);

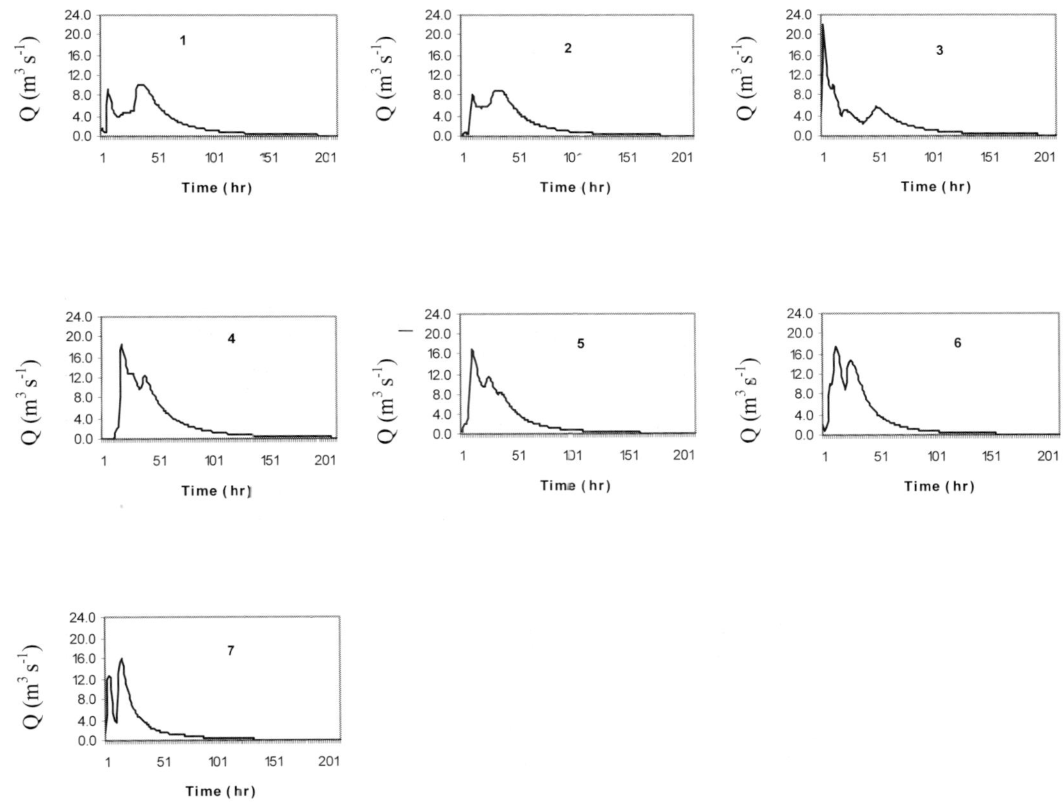

Fig. 5 Kinematic response (routing) function for each of the seven sub-basins of the Blue River basin (BRB).

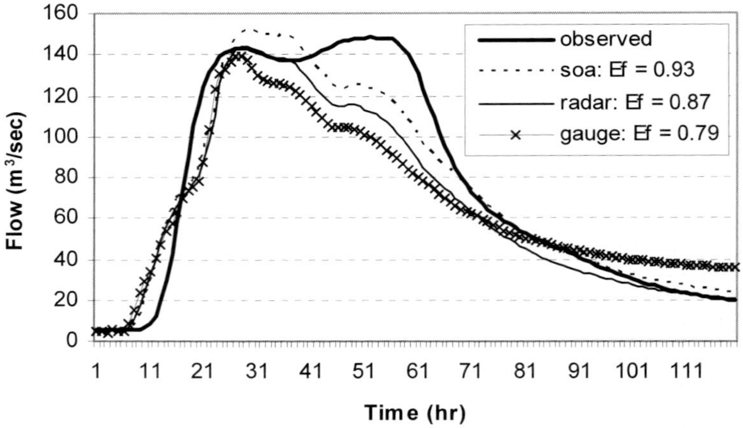

Fig. 6 Comparison of observed and estimated hydrographs for the period of 13–17 March 1995 using different rainfall estimation methods (*Ef* is the simulation efficiency of Nash & Sutcliffe, 1970).

(b) the DPHM-RS model structure, particularly the surface runoff component (Fig. 4); (c) spatial rainfall distribution for this particular rainfall event; and (d) magnitude and duration of the storm/flood event. This will be explored further by: (a) comparing the amount of average rainfall for each sub-basin to see if more rainfall was concentrated on sub-basins with a dominating single peak response function, namely 3, 4, and 5 (Fig. 5); (b) simulating a hydrograph for medium and small storm/flood events; and (c) combining sub-basins with similar response function among each other (e.g. 1 and 2) and with the observed hydrograph to allow them to have a large influence on the final outflow hydrograph which can lead us into obtaining an optimum subdivision of the basin.

Comparing the hydrographs simulated by the three rainfall forcings, namely, gauge, radar, and SOA shows that, SOA produced a better overall outflow hydrograph (*Ef* = 0.93), even though it overestimated the first peak of the observed hydrograph. The radar input rainfall predicted well the first observed peak but it systematically under-estimated the observed hydrograph thereafter, and hence its simulation efficiency (*Ef* = 0.87) was lower. The gauge estimated hydrograph (*Ef* = 0.79) was relatively poor, partly because not one raingauge (out of six) falls within BRB. The adjusted radar rainfall data produced a better hydrograph in both total volume and phase as compared to radar or gauge data alone because it combined the capability of radar to capture the spatial distribution of rainfall (resulting in better phase prediction of the hydrograph) and the more accurate mean rainfall recorded by raingauge. This is evident particularly at the tail part of the estimated hydrographs (Fig. 6) where the radar estimated hydrograph is much more in-phase with the observed hydroraph, followed by SOA and finally by the gauge data alone.

Acknowledgements This research is partly funded by the NSERC of Canada, and the first author is also supported by a F.S. China scholarship of the University of Alberta.

The Oklahoma Meson programme provided the meteorological data, and the Hydrology Laboratory of the National Weather Service of USA provided the WSR-88D level III hourly Radar rainfall data.

REFERENCES

Austin, P. M. (1987) Relation between measured radar reflectivity and surface rainfall. *Mon. Weath. Rev.* **115**, 1053–1070.

Biftu, G. F. & Gan, T. Y. (2001) Semi-distributed, physically based, hydrological modelling of the Paddle River Basin, Alberta, using remotely sensed data. *J. Hydrol.* **244**, 137–156.

Borga, M. (2002) Accuracy of radar rainfall estimates for streamflow simulation. *J. Hydrol.* **267**, 26–39.

Frederick, R. H., Miller, J. F. & Tracey, R. . (1977) *Five to Sixty Minute Precipitation Frequency for the Eastern and Central US.* NOAA Technical Memorandum NWS Hydro-35, Silver Springs, Maryland, USA.

Gandin, L.S. (1963) *Objective Analysis of Meteorological Fields.* Translated by R. Harding, Israel Program for Scientific Translation.

Krajewski, W. F. (1997) Rainfall estimation using weather radar and ground stations. In: *Weather Radar Technology for Water Resources Management* (Ed. by B. Braga Jr & O .Massambani). http://www.unesco.org.uy/phi/libros/radar/content.html

Nash, J. E. & Sutcliffe, J. V. (1970) River flow forecasting through conceptual models, 1. A discussion of principles. *J. Hydrol.* **10**, 282–290.

Preira Fo, A. J., Crawford, K. C. & Hartzell, C. L. (1998) Improving WSR-88D hourly rainfall estimates. *Weath. Forecasting* **13**, 1016–1028.

Seo, D. -J., Briedenbach, J. P. & Johnson, E. R. (1999) Real-time estimation of mean field bias in radar rainfall data. *J. Hydrol.* **223**, 131–147.

Stellman, K. M., Fuelberg, H. E., Garza, R. & Mullusky, M. (2001) An examination of radar raingauge-derived mean arial precipitation over Georgia watersheds. *Weath. Forecasting* **16**, 133–144.

Sun, X., Mein, R. G., Keenan, T. D. & Elliott, J. F. (2000) Flood estimation using radar and raingauge data. *J. Hydrol.* **239**, 4–18.

Tortorelli, R. L. (1991) Floods and Droughts: Oklahoma, National Water Summary 1988-89: US Geol. Survey, *Water Supply Paper 2375*, 451–452.

Weather Radar Information and Distributed Hydrological Modelling (Proceedings of symposium HS03 held during IUGG2003 at Sapporo, July 2003). IAHS Publ. no. 282, 2003.

A distributed hydrological model applied to Heihe mountainous basin in western China

XIA JUN[1,2], WANG GANGSHENG[1] & TAN GE[1]

1 *Institute of Geographical Sciences and Natural Resources Research, Chinese Academy of Sciences, Beijing 100101, China*

2 *Wuhan University, Wuhan 430072, China*

xiaj@igsnrr.ac.cn

Abstract: The development of western China is confronted with serious challenges on water resources due to the ecosystem's vulnerability and the increasing demand on water. Water cycle modelling to a changing environment is the base of sustainable water management. The main aim of this paper is to develop a distributed hydrological modelling coupled physical process grid with the Time Variant Gain Model (TVGM) that was developed by the first author during the 1990s at the University College of Galway, Ireland, as a nonlinear system approach applied to river flow forecasting. It can be shown that the integrated hydrological system approach (TVGM), with distributed physical process or conceptual units, could obtain a flexible result for hydrological modelling given the scarcity of fully physical process data. By preliminary application and examination in the Heihe mountainous basin of northwest China with an area of about 10 000 km^2 in arid and semiarid regions, it was found that this model is successful, with more advantages than the conditional distributed model applied in China. Further research on this approach is addressed in this paper.

Key words coupling; distributed hydrological modelling; system approach; China

INTRODUCTION

The problem of water resources has become more urgent in western China because of global change and socio-economic development, which bring new challenges to hydrological sciences. Research on the water cycle process and its space-time change are innovation issues in the field of international geography.

A distributed watershed hydrological model is an effective way to represent the time–space changes of the hydrological cycle, which are based on DEM/GIS coupled with a unit hydrological model. At present, there are three modelling approaches: (a) fully distributed physical models, such as SHE (Beven, 2001) and similar physically based rainfall–runoff models, establishing the temporal and spatial relationship between grid elements applying numerical analysis; (b) distributed conceptual models, such as distributed Xinanjiang model and SWAT (Soil and Water Assessment Tool) (Arnold *et al.*, 1997), integrating existing lumped conceptual models with each grid element or sub-basin; (c) semi-distributed models, such as TOPMODEL (Beven, 2001; Takeuchi, *et al.*, 1999), simulating hydrological processes by applying topographic information.

The physically-based rainfall–runoff models, represented by SHE, have the advantage that they are based on physical theory. However, they are difficult to apply because of their demands of both input date, much of which is not directly measurable, and computational resources. This has led to the development of simplified distributed models by a combination of physical or conceptual processes with a system approach.

This paper develops such distributed hydrological modelling, called the Distributed Time Variant Gain Model (DTVGM). It will be shown that DTVGM can be applicable in the case of input information imperfections such as the lack of enough precipitation and evaporation observations for distribution modelling, and it has few parameters, which can be estimated in terms of system identification approach to reduce uncertainty. Using a case study of distributed hydrological modelling in the Heihe mountainous basin in west China, it was shown that the DTVGM can obtain higher efficiency than that of the present SWAT model, and that hydrological modelling can satisfy the requirement of water resources management.

MODEL DEVELOPMENT

Introduction to TVGM

The first author developed the Time Variant Gain Model (TVGM) during the 1990s in the University College of Galway, Ireland, as a nonlinear system approach applied to river flow forecasting (Xia *et al.*, 1989, 1995, 1997, 2002). It can be shown that a complex nonlinear system process formulated by the Volterra functional series could be replaced by a simple time variant gain model (TVGM), which can also significantly improve the efficiency of rainfall–runoff forecasting for semiarid or monsoon influenced basins related to present linear system approaches.

Observed data of many basins show that hydrological system gain G is time-variant and has a relation to basin wetness that could be represented by the soil moisture content (S) or the antecedent precipitation index (API):

$$G(t) = g_1 \cdot S(t)^{g_2}, \quad \text{or } G(t) = g_1 \cdot API(t)^{g_2} \tag{1}$$

The linear expression of equation (1) can be written as:

$$G(t) = g_1 + g_2 \cdot S(t) \text{ or } G(t) = g_1 + g_2 \cdot API(t) \tag{2}$$

where, g_1 and g_2 are parameters in the time-variant gain function.

The effective rainfall calculation in runoff generation component can be written as:

$$R(t) = G(t) \cdot X(t) = g_1 X(t) + g_2 API(t) X(t) \tag{3a}$$

$$\text{or} \quad R(t) = g_1 X(t) + \int_0^t g_2 U_0(t-\sigma) X(\sigma) X(t) \, \mathrm{d}\sigma \tag{3b}$$

where $X(t)$, $R(t)$, $U_0(t-\sigma)$ are precipitation, effective rainfall and response function of the API model, respectively.

It could be proved that the time-variant gain in equations (2)–(3a) and flow routing process in the basin are represented as a simple relationship, i.e.:

$$Y(t) = \int_0^m U(\tau) R(t-\tau) \, \mathrm{d}\tau \tag{4}$$

where, $Y(t)$ is runoff, $U(\tau)$ is a linear response function of the flow routing sub-system and m is finite memory length of response function.

Owing to the introduction of time-variant gain component, basin system behaviour could be formulated equivalent to a Volterra nonlinear system, given by:

$$Y(t) = \int_0^m H_1(t-\tau)X(\tau)d\tau + \int_0^m \int_0^t H_2(t-\sigma, t-\tau)X(\sigma)X(\tau)d\tau \quad (5)$$

where $H_1(\tau) = g_1U(\tau), H_2(\sigma,\tau) = g_2U_0(\sigma)U(\tau)$.

It can be verified that TVGM has higher efficiency of hydrological modelling than the present simple hydrological system approach (see Xia, 1997, 2002).

Model structure of DTVGM

DTVGM includes two components: one is the runoff generation process on grid elements; the other is the flow routing process on ranked grids. At present, the runoff generation process is divided into two layers in the vertical direction: the upper layer is the surface flow based on the concept of TVGM; the lower layer is the subsurface flow

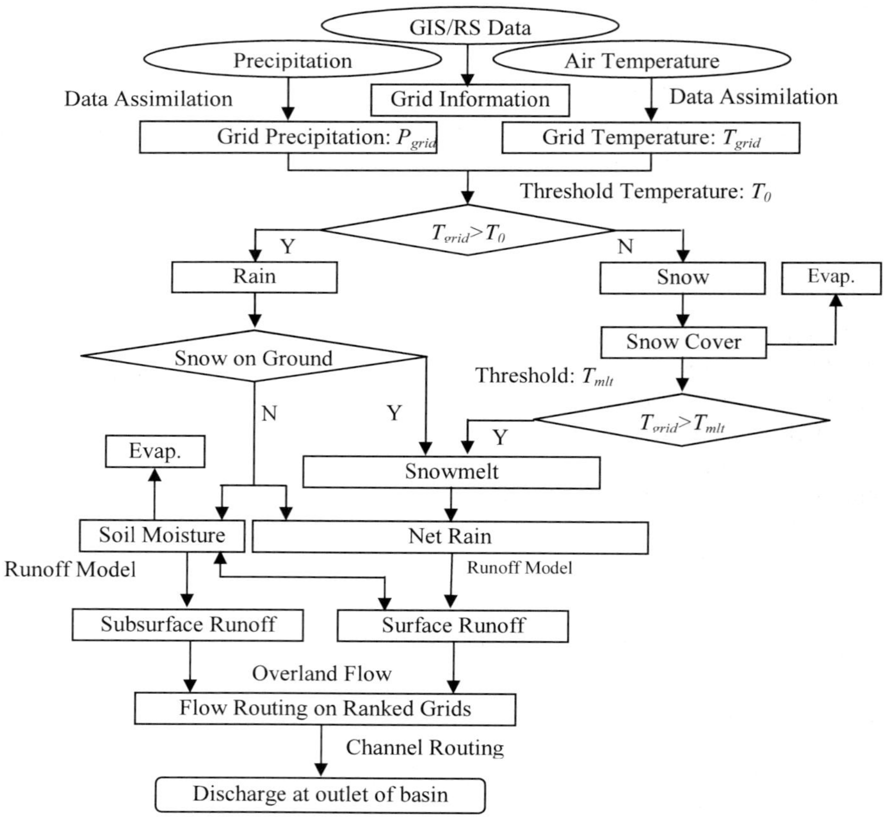

Fig. 1 Model structure of DTVGM. (Y: yes, N: no, Evap: evapotranspiration, T_0: a threshold temperature to separate rain and snow, T_{mlt}: a threshold temperature of snowmelt).

developed through a control volume approach, including a water balance equation and a storage–outflow equation. DTVGM combines surface runoff and subsurface runoff processes by soil moisture content. In contrast, the kinematic wave models are applied to simulate the flow routing process. The model structure of DTVGM is shown in Fig. 1.

Runoff generation model

To calculate runoff generation on grids, the first step is to separate rain from snow according to the local temperature T_{av} (Yang & Zeng, 1999). The total runoff on each grid consists of surface runoff and subsurface runoff. The surface runoff on grids is based on the concept of TVGM; and the subsurface runoff model is developed through a control volume approach, including a water balance equation and a storage–outflow equation. In addition, the widely used degree-day method (Beven, 2001) is applied to simulate the snowmelt process.

Moreover, the evapotranspiration estimation is also an important process. Many equations have been developed to estimate potential evapotranspiration E_p. These equations use combinations of meteorological variables in addition to plant information to estimate E_p, such as the Hargreaves method, the Priestley-Taylor method and the Penman-Monteith method (Arnold *et al.*, 1997). The Hargreaves method is the simplest and uses air temperature as the primary independent variable for determining E_p. Modelling actual evapotranspiration typically involved calculating potential evapotranspiration by one of the above equations and then modifying it by some function of soil moisture content (Thompson, 1999).

Flow routing model

The flow routing model includes two issues: the flow route and the routing method. In DTVGM, the flow route can be described as ranked grids according to the flow directions of grid elements. We define the outlet grids as the first rank, and the grids from which water flows into the first ranked grids as the second rank, the rest may be deduced by analogy, so the most upstream grids are defined as the highest rank. Flow routing is undertaken from those grids having higher rank onto the grids having lower rank. The kinematic wave routing is adopted as the routing method in DTVGM, which is a more physically-based modelling approach (Thompson, 1999).

CASE STUDY ON HEIHE MOUNTAINOUS BASIN

Heihe mountainous basin

The Heihe River is one of the three inland rivers in the Hexi Corridor Region of north-west China, which has an area of about 143 000 km^2 and consists of 35 rivers. The mainstream area is 116 000 km^2. The upstream above Yingluo Gorge, with a drainage area of 10 009 km^2 and river length of 303 km, is the main runoff generation region, the so-called Heihe mountainous basin, with the outlet Yingluo Gorge hydrology station (38°48′N, 100°11′E). The basin altitude ranges from 1737 to 5010 m (average 3670 m).

Because of its arid and high-cold traits, the Heihe River basin has few hydrological and meteorological stations to collect data, which makes it difficult for hydrological modelling. The complex hydrological models are not suitable for such a basin because the necessary data and parameters can't all be collected. However, simple models like DTVGM are advantageous for dealing with the problem.

The DEM of the Heihe mountainous basin (see Fig. 2) is extracted and divided into 38 277 grids. The area of each grid (500 m × 500 m) is 250 000 m^2, so the extracted area is 9569.25 km^2. In addition, the grids of the basin are partitioned into 456 ranks (see Fig. 3).

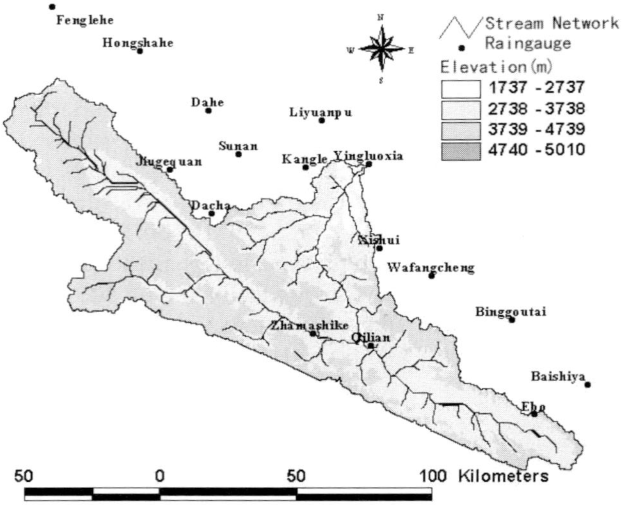

Fig. 2 DEM and stream network of the Heihe mountainous basin.

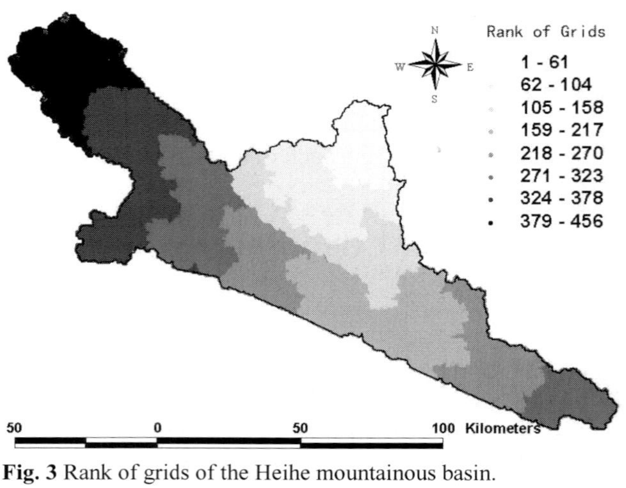

Fig. 3 Rank of grids of the Heihe mountainous basin.

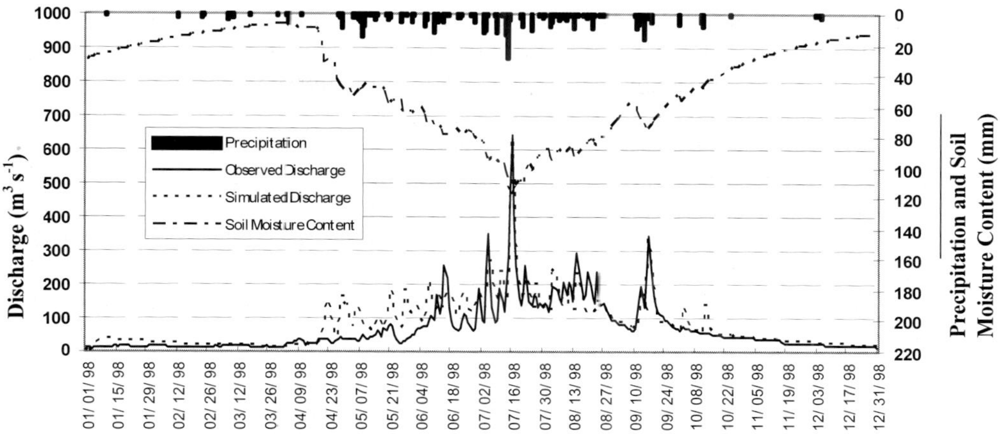

Fig. 4 Comparison between observed and simulated runoff of the Heihe mountainous basin.

Table 1 Simulated result of streamflow peaks.

Observed:		Simulated by DTVGM:		Error:	
Date	Discharge ($m^3 s^{-1}$)	Date	Discharge ($m^3 s^{-1}$)	Relative error of discharge (%)	Time error of peaks (day)
1998-6-15	257	1998-6-16	183	–28.8	1
1998-7-5	348	1998-7-6	274	–21.3	1
1998-7-16	645	1998-7-16	640	–0.8	0
1998-7-22	254	1998-7-23	225	–11.4	1
1998-8-15	293	1998-8-15	240	–18.1	0
1998-9-13	192	1998-9-14	185	–3.6	1
1998-9-17	344	1998-9-17	338	–1.7	0
Model efficient coefficient: $R = 0.75$					

Model efficient coefficient: $R = \left[1 - \frac{\Sigma(Q_C - Q_O)^2}{\Sigma(Q_O - \overline{Q_O})^2}\right] \times 100\%$ where $Q_O, Q_C, \overline{Q_O}$ represent observed discharge simulated discharge and observed mean discharge, respectively.

RESULTS

The distributed input module includes data assimilation (the spatial distribution calculation of precipitation and air temperature) and the spatial distribution initialization of soil moisture content, etc. (Luo, 1997; Wang, 1999; Weisse *et al.*, 2001). The computing module consists of runoff generation and flow routing processes. Daily streamflow of 1998 is simulated and the results are shown in Fig. 4 and Table 1.

The application results by DTVGM indicate the following points: (a) the simulated discharge process is similar to that observed, the model efficienvy coefficient reaches 0.75, which is more successful in distributed hydrological modelling; (b) the simulated streamflow peaks, especially the larger discharge, have relatively high precision;

(c) the distributed inputs and variables in DTVGM ensure that the model can produce a positive response to spatial variations; (d) compared with the simulation by SWAT (the model efficiency coefficient is less than 0.60), the simulation by DTVGM is more successful; and (e) DTVGM can be coupled with other modules according to the basin characteristics; for example, it is coupled with a snowmelt model to enhance its physical mechanism in the case of the Heihe mountainous basin because of it being a high-cold region. Moreover, research is going on to build a linkage of time variant gain factor, $G(t)$, in the TVGM with land remote sensing, to analyse impact of land use and cover change on hydrological response.

CONCLUSIONS

Distributed hydrological models have a higher requirement for data than lumped models. DTVGM could overcome some of the present difficulties of space–time modelling of hydrological processes due to the scarcity of fully physical process data.

Further research needs to be done on the following: (a) quantity and quality of observed data and the data assimilation method have a great impact on simulation by case study; (b) snowmelt estimation has been considered in the application study, but the initial conditions related to snow cover are absent, which affected the snowmelt simulation; (c) the runoff mechanism of frozen earth should be involved in DTVGM to improve the subsurface runoff model; and (d) establishing the relationship between land use and land cover change. The Gain factor, G, is a key issue of this model application in further research.

Acknowledgements The work described in this paper was supported by the Hundred Talents Programme & Key Project of CAS (KZCX1-09-02, CX10G-E01-08), and the Natural Sciences Foundation of China (50279049).

REFERENCES

Arnold, J. G., Williams, J. R., Srinivasan, R. & King, K. W. (1997) Model theory of SWAT. US Dept of Agriculture, Agricultural Research Service, Grassland, Soil and Water Research Laboratory, USA.

Beven, K. J. (2001) *Rainfall–Runoff Modelling*. John Wiley & Sons Ltd, Chichester, West Sussex, UK.

Luo, Q. L. (1997) Data processing in meteorological data assimilation. In: *Systems Engineering - Theory and Practice* (in Chinese with English abstract), 98–103.

Takeuchi, K., Ao, T. Q. & Ishidaira, H. (1999) Introduction of block-wise use of TOPMODEL and Muskingum-Cunge method for the hydro-environmental simulation of a large ungauged basin. *Hydrol. Sci. J.* **44**(4), 633–646.

Thompson, S. A. (1999) *Hydrology for Water Management*. A. A. Balkema, Rotterdam, The Netherlands.

Wang, Y. S. (1999) Data assimilation—its cause, its meaning and main procedures. *Marine Forecasts* (in Chinese with English abstract) **16**, 11–19.

Weisse, A., Michel, C., Aubert, D. & Loumagne, C. (2001) Assimilation of soil moisture in a hydrological model for flood forecasting. In: *Soil–Vegetation–Atmosphere Transfer Schemes and Large-Scale Hydrological Models*. (ed. by A. J. Dolman, A. J. Hall, M. O. Kavvas, T. Oki & J. W. Pomeroy) (Proc. Maastricht Symp., July 2001), 249–256. IAHS Publ. no. 270.

Xia, J. (1989) Part 2-Nonlinear system approach. Research Report of the Third International Workshop on River Flow Forecasting. University College Galway, Ireland (unpublished) .

Xia Jun, O'Connor, K. M., Kachroo, R. K. & Liang, G .C. (1997), A nonlinear perturbation model considering catchment wetness and its application in river flow forecasting. *J. Hydrol.* **200**, 164–178.

Xia, J. (2002) A system approach to real time hydrological forecasts in watersheds. *Water Int.* **27**(1), 87–97.

Yang, Z. N. & Zeng, Q. Z. (1999) *Glacier Hydrology* (in Chinese), 159–203. Chongqing Press, Chongqing, China.

Weather Radar Information and Distributed Hydrological Modelling (Proceedings of symposium HS03 held during IUGG2003 at Sapporo, July 2003). IAHS Publ. no. 282, 2003.

Development of distributed red soil runoff model using radar data

YUTAKA ICHIKAWA
Graduate School of Global Environmental Studies, Kyoto University, Sakyo-ku, Kyoto 606-8501, Japan
ichikawa@wr.kuciv.kyoto-u.ac.jp

KATSUHIRO NAKAGAWA
Okinawa Subtropical Environment Remote-Sensing Centre, Communications Research Laboratory, 4484, Aza-Onna, Onna, Kunigami, Okinawa 901-2401, Japan

KAZUKI FUJIWARA, MICHIHARU SHIIBA
Graduate School of Global Environmental Studies, Kyoto University, Sakyo-ku, Kyoto 606-8501, Japan

SHUICHI IKEBUCHI
Disaster Prevention Research Institute, Kyoto University, Gokasho, Uji 611-0011, Japan

Abstract Okinawa region (the south part of Japan) has been suffering from problems caused by red soil runoff. This region is often attacked by heavy rainfall, which causes overland flow and soil erosion. The soil particles of this region are very fine and will not sink once they are involved in water flow. The muddy water pollutes the shores and coasts, and causes heavy damage to ecosystems of the sea, farming and tourism. The final goal of this research is to develop a distributed red soil runoff model combined with the rainfall data observed by the bistatic radar, which the Communications Research Laboratory (CRL) is now installing at Okinawa main island. In this paper, we developed a distributed red soil runoff model based on a kinematic wave flow model and a digital topographic model, as a preliminary step to the research. The model was applied to Taiho River basin (24 km^2), which is located at the north part of Okinawa main island. The application did not show favourable results. The model should be improved so that it takes into account the change of land cover, which has considerable influence on soil erosion processes.

Key words distributed model; Japan; Okinawa region; red soil; runoff; slope erosion

INTRODUCTION

Okinawa region (the south part of Japan) has been suffering from problems caused by severe red soil runoff (Yoshinaga *et al.*, 1996). This soil is distributed over Okinawa region and is red-yellow. It is quite erosive because it consists of very fine particles and has medium permeability (Noborikawa & Terasawa, 1982; Onaga & Yoshinaga, 1988). When the soil is eroded and involved in water flow, the soil particles will not sink and will stay in the water for a long time. Since the courses of rivers in Okinawa region are generally short and steep, the muddy water containing red soil particles easily reaches the shores and coasts, and it causes heavy damage to ecosystems of the sea, farming and tourism.

In order to mitigate the red soil runoff, it is essential that the volume of the soil runoff be predicted and the effects of countermeasures be evaluated by using a physically-based model. The final goal of this research is to develop a distributed red soil runoff model combined with rainfall data observed by the radar which Communications Research Laboratory (CRL) is now installing at Okinawa Island. This radar will be able to observe spatial distribution of rainfall with high resolution as well as the size distribution of rain drops, which has considerable influence on the soil detachment. In this paper, we developed a distributed red soil runoff model based on a rainfall–runoff model developed by the authors (Shiiba *et al.*, 1998, 1999; Ichikawa *et al.*, 2001), as a preliminary step to the research.

METHODOLOGY

The volume of the red soil runoff has strong relation to the rainfall–runoff processes. Therefore, in order to precisely evaluate the volume of the soil runoff, a rainfall–runoff model which can physically describe the rainfall–runoff processes is needed. As mentioned above, we have been developing a physically based rainfall–runoff model. This model represents a hill slope as a set of small rectangular planes (slope units), and routes water flow by applying a kinematic wave model to each slope unit (Shiiba *et al.*, 1999; Ichikawa *et al.*, 2001). In this study, we develop a red soil runoff model based on this rainfall–runoff model. However, we have not yet applied it to Okinawa region and do not know whether it can reproduce rainfall–runoff phenomena of the Okinawa region. We then examine the ability of this model before developing a red soil runoff model.

EXAMINATION OF THE APPLICABILITY OF THE RAINFALL–RUNOFF MODEL

Outline of the rainfall–runoff model

The rainfall–runoff model used in this study represents a hill slope as a set of slope units, and routes water flow by using a kinematic wave model considering the field capacity of the soil layer. This model assumes that a part of the water which infiltrates into a soil layer is captured by the capillary force and lateral water flow occurs when water content exceeds the field capacity. The kinematic wave equations considering field capacity (Shiiba *et al.*, 1998) are as follows:

$$h = \begin{cases} h_f + h_c\left\{1 - \left(d - h_f / d\right)^N\right\}^{1/N} & (0 \le h_f < d) \\ h_f + h_c & (h_f \ge d) \end{cases} \quad (1)$$

$$q = \begin{cases} a h_f & (0 \le h_f < d) \\ \alpha(h_f - d)^m + a h_f & (h_f \ge d) \end{cases} \quad (2)$$

where q is discharge per unit width, h is water content, h_f is free water content, $a(= k\sin\theta/\gamma_e)$ is water flow velocity in a soil layer, k: permeability, γ_e: effective porosity

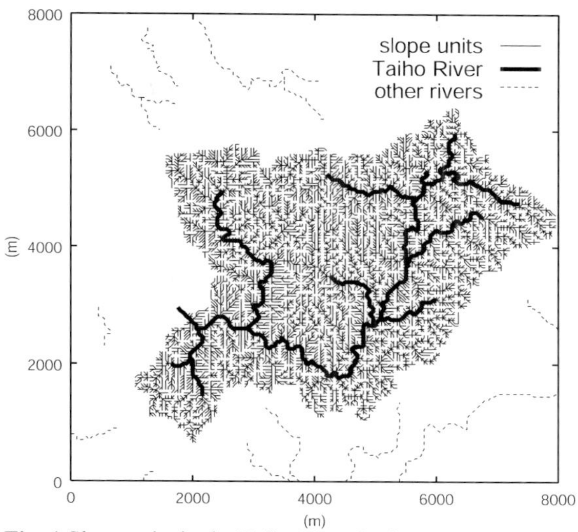

Fig. 1 Slope units in the Taiho River basin.

of a soil layer, $d(= \gamma_e D)$ is depth of effective pore contained in a soil layer, D: depth of a soil layer, $\alpha = \sqrt{\sin\theta}/n$, θ is slope gradient, n is Manning's roughness coefficient, m is a constant, N is a constant, $h_c = \gamma_c D$, and γ_c is the field capacity of a soil layer. This kinematic wave model is applied to each slope unit and is solved by using a finite difference scheme.

Application of the rainfall–runoff model

The rainfall–runoff model was applied to Taiho River basin (24 km^2) which is located in the northern part of Okinawa main island. Figure 1 shows the slope units in the basin. They were generated from a grid based DEM whose grid spacing was 50 m. The period of application was from 16 to 18 August 1997. First, we conducted a runoff simulation with the parameter values which we have usually used in the application to Honshu Island (the main island of Japan). The parameter values were given as: $n = 0.3$ (m$^{-1/3}$ s^{-1}), $m = 1.667$ [-], $k = 0.015$ [m s^{-1}], $\gamma = 0.4$ [-], $\gamma_c = 0.1$ [-], $D = 1.0$ [m], and $N = 6.0$ [-].

Figure 2 shows the simulation result with the above values. The solid line, dotted line and dot-dash line are the calculated discharge, observed discharge and observed rainfall intensity, respectively. We can see a considerable difference between the calculated and observed discharge around the time of peak. The observed discharge responds to the temporal variation of the rainfall quickly, while the response of the calculated one is relatively slow. This simulation means that the parameter values used in the application to Honshu Island are not available in the application to the Okinawa region.

Then we revised the values of k ($0.015 \rightarrow 3.9 \times 10^{-5}$), γ ($0.4 \rightarrow 0.5$), γ_c ($0.1 \rightarrow 0.4$) and D ($1.0 \rightarrow 0.35$) on the basis of the information collected from a field survey and the documents (Noborikawa & Terasawa, 1982; Onaga & Yoshinaga, 1988). Figure 3

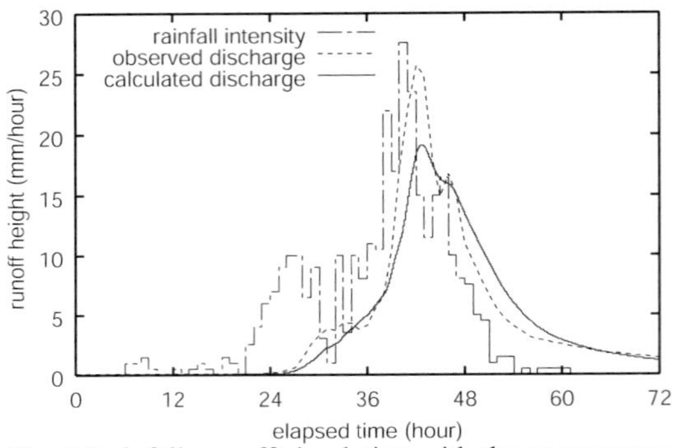

Fig. 2 Rainfall–runoff simulation with the parameter values used in the application to Honshu Island. The period of simulation is 16–18 August 1997.

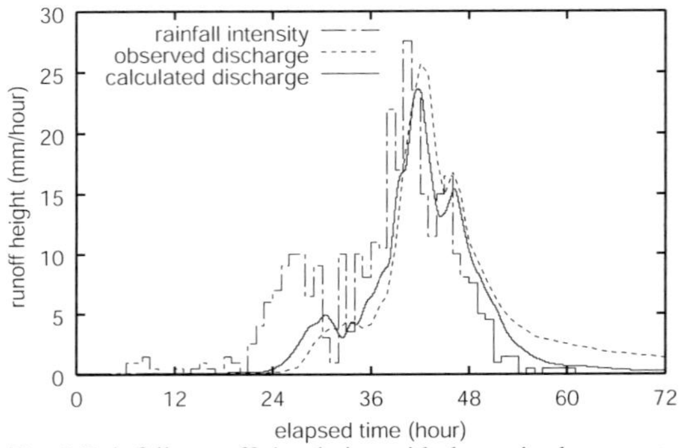

Fig. 3 Rainfall–runoff simulation with the revised parameter values (16–18 August 1997).

shows the simulation result with the revised values. Although there are differences between the calculated and observed discharge around the rising and recession parts of the hydrographs, the degree of agreement of the hydrographs was generally improved. Figure 4 shows the simulation result of another period (1–2 October 1997) with the revised parameter values. This simulation also shows a reasonable result. These

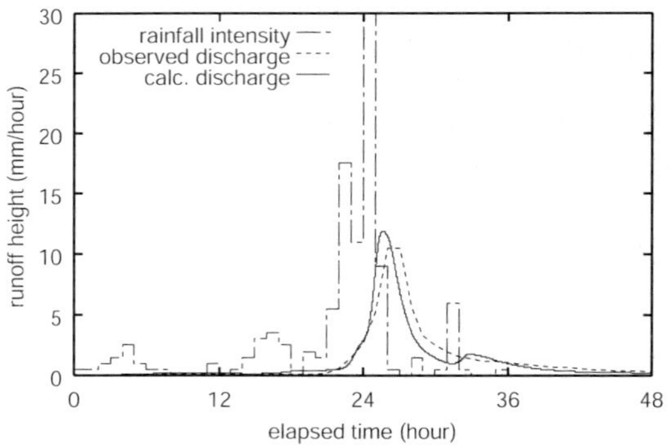

Fig. 4 Rainfall–runoff simulation with the revised parameter values (1–2 October 1997).

simulations mean that the rainfall–runoff model has sufficient applicability to Okinawa region if the parameter values are properly determined. In the next section, we develop a red soil runoff model based on this rainfall–runoff model.

DEVELOPMENT OF RED SOIL RUNOFF MODEL

The weight of eroded fine particles per unit area and unit time, m_f, is given by the following equation (Ashida *et al.*, 1980):

$$m_f = \rho_s E(1-\gamma)p_f \tag{3}$$

where ρ_s is density of the soil (kg m^{-3}), E is erosion rate (m s^{-1}), γ is porosity of the soil [-], and p_f is percentage of fine particles in the soil by weight. ρ_s, γ and p_f are the physical properties which are known by using soil tests. In order to evaluate E, we employ a relationship between the erosion rate and the friction velocity which was obtained by Ashida & Tanaka (1974).

Ashida & Tanaka (1974) conducted an experiment of clay erosion and examined a relationship between E/u_* (erosion rate/friction velocity) and τ/τ_c (tractive force/critical tractive force). From the experiment, they found that the erosion rate increased considerably with increasing tractive force when τ/τ_c was nearly equal to 1, while E/u_* was almost constant when τ/τ_c exceeded 5, and that the constant value ranged between 10^{-4} and 10^{-5}. In the field test of bare ground erosion, Ashida *et al.* (1980) also detected that the relationship E/u_* became constant when the tractive force increased considerably. The constant value was observed to be 10^{-3}–10^{-4}. In this study, we assume that the tractive force becomes sufficiently large when the surface flow occurs, and evaluate the erosion rate as the product of the friction velocity and a constant value. This constant value is determined by calibration in this study, although it should be obtained from an experiment or a field test.

From the above discussion, we give the basic equations of the red soil runoff model as:

$$\frac{\partial(CA)}{\partial t} + \frac{\partial(CQ)}{\partial x} = R_E \tag{4}$$

$$R_E = \rho_s E(1-\gamma)p_f S \tag{5}$$

$$E = cu_* \tag{6}$$

where C is concentration of suspended soil, A, Q and S are cross-sectional area, discharge and wetted perimeter of surface flow, respectively, R_E is rate of soil entrainment, and c is a constant.

The discharge and cross-sectional area of surface flow can be calculated by using the rainfall–runoff model. Also, S in equation (5) and u_* in equation (6) can be calculated from the water depth of surface flow. These three equations are applied to each slope unit, and C is calculated by using a finite difference scheme. This model assumes that the erosion of the red soil due to surface flow occurs on the slope units of bare ground including the cultivated land.

APPLICATION OF THE RED SOIL RUNOFF MODEL

The red soil runoff model was applied to Taiho River basin. The same parameter values of the rainfall–runoff model as the application of the rainfall–runoff model were used. The soil density, ρ_s, and the percentage of fine particles, p_f, which are the parameters of the red soil runoff model, were determined as 2750 (kg m^{-3}) and 1.0 [-], respectively. Although the constant, c, in equation (6) should be determined from an experiment or a field test, we have not yet conducted either of them. Then we used the following three values as c: (a) 10^{-7}, (b) 5.0×10^{-8}, and (c) 10^{-8}, in the application of the model. The periods of applications were 16–18 August 1997 and 1–2 October 1997.

Figures 5 and 6 show the simulation results of 16–18 August 1997 and 1–2 October 1997, respectively. The marks (×) in both figures show the observed soil flux and the three lines in the figures are the calculated soil flux hydrographs. Figure 5 shows that the calculated hydrograph of case (B) ($c = 5.0 \times 10^{-8}$) agrees well with the observation. However, Fig. 6 shows that none of the calculated hydrographs agree with the observation. The observed soil flux at the time of peak shown in Fig. 5 is smaller than the one shown in Fig. 6, though the total rainfall amount of the former event (August) is larger than that of the latter event (October) as shown in Figs 3 and 4 (the peak rainfall intensities are almost same). This finding means that the volume of

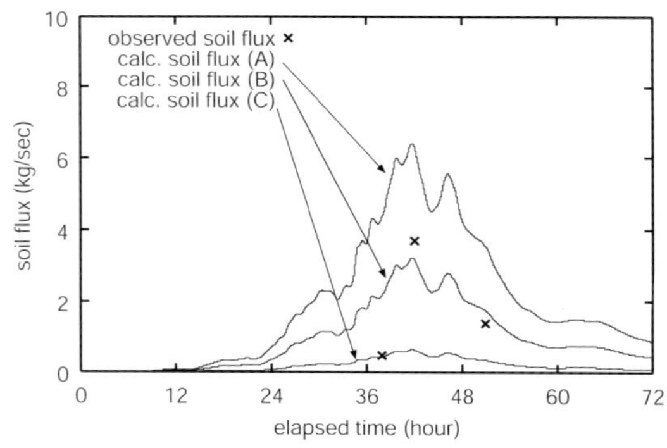

Fig. 5 Red soil runoff simulation (16–18 August 1997).

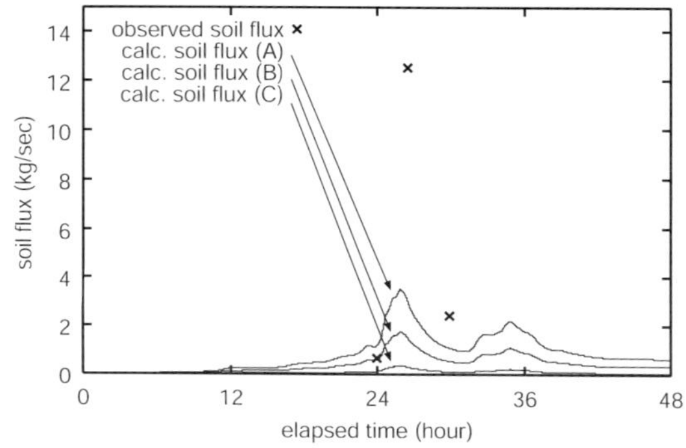

Fig. 6 Red soil runoff simulation (1–2 October 1997).

soil erosion does not always correspond with the magnitude of rainfall event in this basin. The soil erosion processes are considerably affected by the state of land cover, especially in the cultivated land. However, the soil runoff model developed in this study does not take the change of land cover into consideration. The soil runoff model should be improved in this phase.

CONCLUSIONS

In this study, we developed a red soil runoff model for Okinawa region on the basis of the physically-based rainfall–runoff model. We first examined the applicability of the rainfall–runoff model to Okinawa region. The model showed favourable applicability when the parameter values were properly determined according to the soil properties of Okinawa region. Then we developed a red soil runoff model by combining the rainfall–runoff model with the relationship between the erosion rate and the friction velocity, and applied this model to Taiho River basin. Since we have not yet conducted an experiment or a field test to determine the ratio of the erosion rate to the friction velocity, we used the three values as the ratio in the application. We applied the model to two events, and could not find the ratio which was available for both events. The model should be improved so that it takes into account of the change of land cover, which has considerable influence on the soil erosion processes, as well as the soil detachment due to rain drops, whose size distribution will be detected by the radar.

Acknowledgements We are thankful to North Dams Office of Okinawa General Bureau, Cabinet Office, who provided the hydrological data of Taiho River basin. We also wish to thank Mr Tatsuo Omija of Okinawa Prefectural Institute of Health and Environment and Mr Kiyoshi Kuniyoshi of Okinawa Prefectural Agricultural Experiment Station, for providing valuable research papers and documents on red soil erosion and runoff.

REFERENCES

Ashida, K., Egashira, S. & Kanayashiki, T. (1980) On the yield of wash load by slope erosion. (Proc. 24th Japanese Conference on Hydraulics), 135–141. JSCE (in Japanese).

Ashida, K. & Tanaka, K. (1974) Erosion and sediment transport mechanism on the sandy bed with some clay content (in Japanese with English abstract). *Ann. Disaster. Prev. Res. Inst.* Kyoto University **17**(B), 571–584.

Ichikawa, Y., Murakami, M., Tachikawa, Y. & Shiiba, M. (2001) Development of a basin runoff simulation system based on a new digital topographic model (in Japanese with English abstract). *J. Hydraul., Coastal Environ. Engng JSCE* **691**, 43–52.

Noborikawa, S. & Terasawa, S. (1982) Physical properties of the great soil groups of Okinawa island (in Japanese with English abstract). *Soil Phys. Cond. Plant Growth* **46**, 2–11.

Onaga, K. & Yoshinaga, A. (1988) Physical properties of major Okinawan upland soils (in Japanese with English abstract). *Soil Phys. Cond. Plant Growth* **58**, 17–29.

Shiiba, M., Ichikawa, Y., Sakakibara, T. & Tachikawa, Y. (1999) A new numerical representation form of basin topography (in Japanese with English abstract). *J. Hydraul., Coastal Environ. Engng JSCE* **642**, 1–9.

Shiiba, M., Tachikawa, Y., Ichikawa, Y., Hori, T. & Tanaka, K. (1998) Development of slope runoff models which consider field capacity and pipe flows (in Japanese with English abstract). *Ann. Disaster Prev. Res. Inst.* Kyoto University **41**(B-2), 229–235.

Yoshinaga, A., Onaga, K. & Sakai, L. (1996) Preventive measures for red clay outflow on farmland in Okinawa and prevention treaty of outflow of red clay (in Japanese). *Water Sci.* **228**, 1–13.

Weather Radar Information and Distributed Hydrological Modelling (Proceedings of symposium HS03 held during IUGG2003 at Sapporo, July 2003). IAHS Publ. no. 282, 2003.

A flash flood forecast model for the Three Gorges basin using GIS and remote sensing data

YANGBO CHEN

Department of Water Resources and Environment, Sun Yat-Sen University, Guangzhou 510275, China
eescyb@zsu.edu.cn

JIAXI HU

Department of Management, Three Gorges University, Yichang, Hubei 443002, China

JIE YU

Cascade Dispatching Centre, CTGPC, Yichang, Hubei 443002, China

Abstract This paper presents a physically-based, distributed hydrological model for flash flood forecasting of the Three Gorges basin, which uses GIS techniques and remote sensing data. This model is divided into several sub-models, which include a Basin Digitization Model (BDM) to divide the whole basin into a number of square grid cells, Precipitation Feeding Model (PFM) to give grid-cell based precipitation, an Evapotranspiration Model (EM) to calculate cell-based evapotranspiration, a Runoff Production Model (RPM) to determine the runoff produced at cell scale, and a Runoff Routing Model (RRM) to route runoff to a basin outlet. This model mainly relies on remotely sensed data to physically derive model parameters, which is useful for ungauged or poorly gauged basins.

Key words digital elevation model; distributed model; flash flood forecast; GIS; remote sensing; weather radar

INTRODUCTION

The Three Gorges basin, with a drainage area of 58 000 km^2, is at the middle of the Yangtze River, China's largest river. The Three Gorges Dam, the largest dam in the world, is built at the end of the Three Gorges basin. The development of a better flood forecasting system is needed for the effective operation of the Three Gorges Dam. This paper presents a physically-based, distributed hydrological model for flash flood forecasting for the Three Gorges basin, which uses GIS techniques and remote sensing data. This model relies mainly on remotely sensed data to physically derive model parameters, which is a strong point for applications to ungauged or poorly gauged basins.

MODEL STRUCTURE

The presented model uses GIS techniques and remote sensing data, including digital weather radar estimated precipitation, satellite sensed DEMs and vegetation information. The model is divided into several sub-models, which include a Basin Digitization Model (BDM), a Precipitation Feeding Model (PFM), an Evapotranspiration Model (EM), a Runoff Production Model (RPM) and a Runoff Routing Model (RRM).

In the BDM, a square-grid network DEM is employed to divide a study basin into a number of square grid cells, and every cell is considered as a sub-basin (called unit-basin), which consists of three layers: a canopy layer (CL), a soil layer (SL) and an underground layer (UL). For every grid cell, unique vegetation type and leaf area index are assigned. The flow direction, flow accumulation, slope, and flow length for every grid cell are calculated by ArcView based on the DEM.

The PFM treats precipitation information measured by digital weather radar and gauged precipitation. The digital weather radar rainfall reflectivity is first transferred into rainfall intensity and assigned to each grid cell. The precipitation measured by raingauges is given to each grid cell using the Theissen Polygon method. The whole Three Gorges basin is not covered by the digital weather radar, therefore the spatial distributions of precipitation are created from weather radar and the raingauge information.

The EM calculates the grid-cell based evapotranspiration, which uses vegetation types and leaf area index as key parameters. The RPM determines runoff production that is formed by the Xinanjiang model at the grid-cell scale. Runoff is produced after the soil is saturated, and then divided into surface and subsurface runoff according to the steady infiltration rate.

The RRM routes the runoff to the basin outlet. The runoff routing is divided into hillslope routing that routes the surface runoff at each grid cell and river routing that routes the surface runoff through the river channel network. Hillslope routing is controlled by the flow velocity derived from the Manning equation, and river routing is governed by the flow velocity derived from the Darcy-Weisbach equation.

BASIN DIGITIZATION MODEL (BDM)

Whole basin division

The whole basin is divided into a number of unit-basins by using a Digital Elevation Model (DEM; EROS Data Center, 2001). The DEM used in this model is a square-grid network model, which divides the whole basin into a number of square grids. The grid is called a unit-basin that is treated as a uniform basin in which elevation, vegetation type, soil characteristics, and thus model parameters are considered to take the same values.

The unit-basin is then divided into three layers: the canopy layer, the soil layer and the underground layer. The canopy layer is from the terrain surface to the top of the vegetation. The evaportranspiration takes place in this layer, and the Evapotranspiration Model is used to determine the evapotranspiration at the unit-basin scale. In the soil layer, soil water is filled by the precipitation and depleted via evapotranspiration. The underground layer is beneath the soil layer, where water stored in the layer is assumed to move as steady flow.

Deriving parameters

After the basin division, flow direction, flow accumulation, slope and flow length are derived from the DEM, which are important parameters for the Runoff Routing Model.

Flow direction and runoff routing network Runoff produced at each unit-basin is assumed to flow to only one of its eight neighbouring unit-basins (Moore *et al.*, 1991). According to the flow direction of each unit-basin, a flow network is extracted. The network is referred to as a runoff routing network, which describes the runoff flow routes of every unit-basin from its centre to the basin outlet.

Slope The slope in each unit-basin is determined as:

$$\tan(\theta_i) = \frac{H_i - H_j}{D} \tag{1}$$

where θ_i is the slope of unit-basin; H_i is the elevation of the *i*th unit-basin; H_j is the elevation of the destination unit-basin of the *i*th unit-basin, and D is the projected length of the unit-basin.

Flow length The flow length L_i in each unit-basin is calculated as follows:

$$L_i = \frac{D}{\cos(\theta_i)} \tag{2}$$

PRECIPITATION FEEDING MODEL (PFM)

The Precipitation Feeding Model (PFM) is used to convert the precipitation measured by radar or raingauges to grid-cell based precipitation. The resolution of radar measured precipitation is usually 1 km × 1 km. The precipitation of each unit-basin simply takes the precipitation of the nearest radar grid. To give gauged precipitation to each unit-basin, the Thiesson polygons method is adopted. The Thiesson polygon consists of several polygons, and each polygon has one raingauge inside it. In this case, one polygon covers several unit-basins, and the precipitation of unit-basins inside the polygon is assumed to the same as that of the raingauge in it.

EVAPOTRANSPIRATION MODEL (EM)

The land cover of the whole basin is classified into three types: water surface, soil surface (with no vegetation) and vegetated surface. According to the land cover classification, different evapotranspiration schemes are applied.

The water surface evaporation is calculated as the follows:

$$E_{i,t} = EC_i \tag{3}$$

where $E_{i,t}$ is the evaporation of the *i*th unit-basin at time *t*, and EC_i is potential evaporation of water surface. The soil evaporation is calculated as follows:

$$\begin{aligned} &E_{i,t} = ES_i \quad \text{if} \quad P_{i,t} > ES_i \\ &E_{i,t} = P_{i,t} + \left(ES_i - P_{i,t}\right)\left(\frac{W_{i,t}}{WM_i}\right)^2 \quad \text{if} \quad P_{i,t} < ES_i \end{aligned} \tag{4}$$

where $E_{i,t}$ is the evaporation of the *i*th unit-basin at time *t*, ES_i is potential evaporation of soil surface, $P_{i,t}$ is the precipitation of the *i*th unit-basin at time *t*, $W_{i,t}$ is the storage of the *i*th unit-basin at time *t*, and WM_i is the storage capacity of the *i*th unit-basin.

The vegetated surface evapotranspiration includes two parts: the canopy evaporation and transpiration. Canopy evaporation takes place during wet stages. In the period the canopy intercepts precipitation that is depleted via evaporation. The canopy evaporation is calculated as follows:

$$E_{i,t} = \min\left[U_{i,t}, EP_i\left(\frac{U_{i,t}}{UM_i}\right)^{\frac{2}{3}}\right] \tag{5}$$

where EP_i is potential evaporation over the canopy, $U_{i,t}$ is the intercepted water at time *t*, and UM_i is the interception capacity, which is calculated as:

$$UM_i = k \cdot LAI_i \tag{6}$$

where LAI_i is the leaf area index of the *i*th unit-basin, and *k* is a coefficient that is taken as 0.2 mm (Xu *et al.*, 1994). The transpiration takes place when intercepted water is depleted, which is calculated as:

$$E_{i,t} = EC_i\left(\frac{W_{i,t}}{WM_i}\right)^2 \tag{7}$$

where EC_i is potential transpiration.

RUNOFF PRODUCTION MODEL (RPM)

The Runoff Production Model calculates the runoff produced in each unit-basin. The runoff is classified into three types corresponding to the land cover types: runoff produced on water surface, runoff produced on soil surface and runoff produced on vegetated surface.

Runoff produced on water surface is calculated as:

$$R_{i,t} = \max(0, P_{i,t} - E_{i,t}) \tag{8}$$

where $R_{i,t}$ is the runoff produced in the *i*th unit-basin due to the precipitation of $P_{i,t}$. Runoff production on a soil surface unit-basin is controlled by the saturation module. Net precipitation first fills out the soil layer, and only after the soil layer is saturated, runoff is produced. The runoff produced on a soil surface unit-basin is calculated as:

$$R_{i,t} = \max(0, P_{i,t} - E_{i,t} + W_{i,t} - WM_i) \tag{9}$$

The water balance equation of the soil layer is expressed as:

$$W_{i,t+1} = W_{i,t} + P_{i,t} - E_{i,t} - R_{i,t} \tag{10}$$

The method for calculating the runoff produced on a vegetated land surface is the same as a soil surface. First net precipitation $P'_{i,t}$ is calculated as:

$$P'_{i,t} = \max(0, P_{i,t} + U_{i,t} - E_{i,t} + UM_i) \tag{11}$$

The precipitation first fills the soil layer, and only after the soil layer is saturated, the runoff is produced, which is calculated as:

$$R_{i,t} = MAX(0, P'_{i,t} + W_{i,t} - WM_i) \tag{12}$$

The runoff produced on the water surface is considered as surface runoff. The runoff produced on the soil and vegetated unit-basins is divided into surface runoff and subsurface runoff. Subsurface runoff $RG_{i,t}$ is controlled by the infiltration, which is calculated as:

$$RG_{i,t} = I_i \tag{13}$$

where I_i is steady infiltration rate. The difference between the total runoff and subsurface runoff is surface runoff $RS_{i,t}$:

$$RS_{i,t} = R_{i,t} - RG_{i,t} \tag{14}$$

RUNOFF ROUTING MODEL (RRM)

The runoff routing includes surface runoff and subsurface runoff routing. The surface runoff routing is divided into river routing and hillslope routing. The river routing is the runoff routing along the river channel, while the hillslope routing is the runoff routing along the land surface, mostly along a slope. The division between hillslope routing and river routing is made through the flow accumulation threshold value. If the flow accumulation value is larger than the threshold value, then the surface routing within the unit-basin is river routing, otherwise it is hillslope routing. In this study, the threshold value is set to 200.

The most important parameter for surface runoff routing is the velocity of the runoff along the routing network. If the routing velocity on a unit-basin is known, then the routing time along the unit-basin is determined as:

$$\Gamma_i = \frac{L_i}{v_i} \tag{15}$$

where Γ_i is the surface runoff routing time along the ith unit-basin, v_i is the surface runoff routing velocity along the ith unit-basin. Summing up Γ_i, the total routing time of the ith unit-basin is determined.

If the river routing is considered as the one-dimensional open channel steady flow, the cross-section of the river channel is assumed as rectangular, and river bed slope is constant within a unit-basin, the river channel routing velocity can be derived as:

$$v = \sqrt{\frac{2gs}{fB}} \tag{16}$$

according to the Darcy-Weisbach equation, where g is gravity velocity, s is the river bed slope, B is the width of river channel, and f is Darcy-Weisbach constant.

The hillslope routing is described by the Manning equation and the continuity equation. Assuming the flow is distributed uniformly on the hillslope, the average hillslope flow velocity is derived as:

$$v = 2.3031 \frac{s^{0.2994} R^{0.4012}}{n^{0.5988} (DL)^{0.4012}} \tag{17}$$

where s is the slope, R is the runoff, and n is the Manning coefficient.

Acknowledgements This study is supported by the National Natural Science Foundation of China (Grant no. 50179019), and the Cascade Dispatching Centre of CTGPC for providing research data. The authors thank Professor Eric Wood and James Smith at Princeton University for their beneficial discussions.

REFERENCES

EROS Data Center (2001) HYDRO1K Documentation, downloaded from USGS website at: http://www.edcdaac.usgs.gov/gtopo30/hydro/readme.html

Moore, I. D., Grayson, R. B. & Ladson, A. R. (1991) Digital terrain modeling: a review of hydrological, geomorphologic and biological applications. *Hydrol. Processes* January–March, 3–30.

Xu, L., Lettenmaier, D. P., Wood, E. F. & Burges, S. J. (1994) A simple hydrologically based model of land surface water and energy fluxes for GSMs. *J. Geophys. Res.* **99**(D7), **14**, 415–428.

Weather Radar Information and Distributed Hydrological Modelling (Proceedings of symposium HS03 held during IUGG2003 at Sapporo, July 2003). IAHS Publ. no. 282, 2003.

Performance assessment of the BTOPMC model in a Nepalese drainage basin

NAWA RAJ PRADHAN
Disaster Prevention Research Institute, Kyoto University, Uji, Kyoto 611-0011, Japan
pradhan@rdp.dpri.kyoto-u.ac.jp

RAGHUNATH JHA
Department of Civil Engineering, Institute of Engineering, Pulchowk Campus, Katmandu, Nepal

Abstract BTOPMC, a physically-based distributed hydrological model based on the block-wise use of TOPMODEL with Muskingum-Cunge flow routing method, was selected to evaluate the applicability in a Nepalese basin. Model performance was judged by a range of quantitative and qualitative measures of fit applied to both the calibration and validation periods. BTOPMC can be successfully used as a tool for integrated water resources investigation in a large watershed, the mountain physiographic region, Nepal.

Key words BTOPMC model; Nepalese river basin; performance assessment; Sun Kosi River; water balance

INTRODUCTION

Nepal, with an area of 147 181 km^2, has more than 6000 rivers stretching from the Great Himalayas with an elevation of more than 8800 m, to the southern plain called "Terai" with an elevation of 100 m, within a 200 km width. This provides a dense network of rivers with steep topographical conditions and 80% of the total annual rainfall concentrates in a relatively short period of about four months. These peculiar characteristics lead to a set of hazards, including frequent floods. The country has been regarded as one of the richest nations of the world in water resources, having annual average surface water availability of 1 521 934 m^3 km^2 and 11 200 m^3 per capita. Nepal is a country with vast hydroelectric resources. The total hydropower potential of Nepal is estimated to be 83 000 MW. Moreover, Nepal is an agriculture-based country.

In view of the important role that the water resource can play in Nepal and the need to harness it in a sustainable manner, it is essential to know the various facets of the water resources. Depending on the situation, different approaches to its use is needed to be designed, sometimes to "keep the water away from the people", and at other times "keep the people away from the water". Suitably selected models can help to identify issues which are useful for planning various development projects with minimum risk of flood disasters, alerting people to keep away from such disasters in time, and providing planners with information about water for use in various development projects like hydropower, irrigation, etc. Until now, quite a few such models have been tested in Nepal. With the advancement of technology, GIS and satellite images have played a vital role for the easy access of a wide range of data. The model, which can incorporate such data, is very fruitful for a country like Nepal, where only a few scattered and less reliable data sources are available.

The inability of lumped type models to explicitly handle land-use changes and water-use systems modification make them unsuitable, without extensive modification, for the present requirements. More suitable types of models include distributed models, but most of them are data demanding and not yet practical for large ungauged basins. Thus, to overcome all the complexities and difficulties stated above, BTOPMC (Ao *et al.*, 1999b; Takeuchi *et al.*, 1999)—a physically-based distributed hydrological model based on the block wise use of TOPMODEL (Beven & Kirkby, 1979) with Muskingum-Cunge flow routing method—was selected to evaluate the applicability in a Nepalese catchment. In contrast to other models, in TOPMODEL the catchment is not divided into homogeneous units, but natural heterogeneity (i.e. topography) and its effects on hydrological processes are represented by distribution functions (Seibert, 1999).

MODEL DESCRIPTION

The component models for BTOPMC are as follows:

(a) Runoff Generation Model: The block-wise use of TOPMODEL is adopted. The model parameters and the average soil-topographic index are calibrated or computed for each block, respectively.

(b) Routing Model: The variable coefficient Muskingum-Cunge routing method is used by which the river reach characteristics can be considered, and the velocity and discharge for any grid can be obtained simultaneously.

(c) Terrain Model: A river stream network is produced from the DEM with an automatic sink removal algorithm developed by the authors.

(d) Precipitation Model: the Thiessen method is adopted to give rainfall for each grid cell in which rainfall is given the value of the nearest raingauge. Basin-averaged rainfall can also be used.

(e) Evapotranspiration Model: The Penman method is used to estimate monthly evapotranspiration and the estimated values are provided to BTOPMC model.

STUDY AREA

The study area, the Sun Kosi River basin, lies within the Kosi basin. The study area is located in the eastern part of Nepal. The Kosi basin is one of Nepalese three major river systems and lies between latitudes of 26°50′N and 28°N and longitudes of 85°30′E and 88°12′E. The drainage basin area within Nepal is estimated to be 33 000 km^2. The Kosi basin is divided into three major topographic and ecological areas: the Mountain area in the north, the hill area in the midland and the Terai area in the south. The Sun Kosi basin lies between the mountain area and the hill area. The mountain area is situated at an altitude exceeding 3000 m, with the highest point reaching more than 8000 m. The hill area lying between 3000 and 300 m consists of high ridges and steep slopes including the Mahabharat range and Siwalik Hills. The Terai area, lying below 300 m, forms the southern belt extending along the Indian border.

Rainfall distribution in the Sun Kosi basin is in a concentric circular pattern. Maximum annual rainfall in the Sun Kosi is 3500 mm with a minimum of 2000 mm

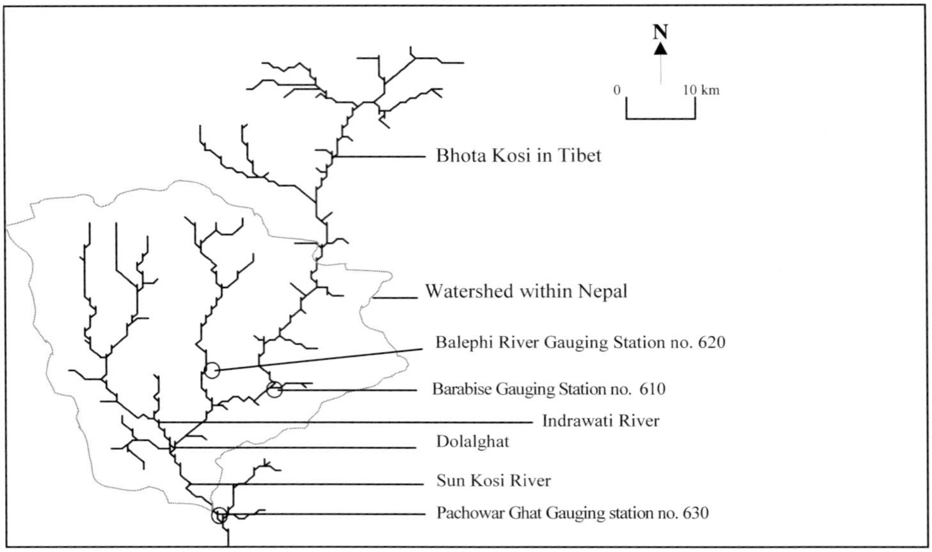

Fig. 1 Sun Kosi drainage basin above Pachowar Ghat gauging station no. 630.

and rainfall is particularly heavy in the upper reaches of the Sun Kosi along the Bhote Kosi basin where annual rainfall exceeds 3500 mm.

The total length of the Sun Kosi River is approximately 330 km, of which 280 km lies in Nepalese territory (Japan International Co-operation Agency, 1985). The river gradient is approximately 1/210 throughout the entire length of its course in Nepal and 1/450 between Tribeni and Dolalghat. The Sun Kosi, the Tamur and Arun rivers meet at Tribeni and the Indrawati River joins the Sun Kosi River at Dolalghat. The Sun Kosi study area was taken above Pachowar Ghat gauging station no. 630 (see Fig. 1).

DISCUSSION

Of the 330 km total length of the Sun Kosi River, 280 km lies in Nepalese territory and the rest lies in the Tibetan Plateau where data information was not available (see Fig. 1). To cope with this problem, the observed discharge is given as a boundary condition so that the drainage basin above that gauging station can be removed. Thus the observed discharge of Barabise gauging station no. 610 (see Fig. 1) is given as input for the calibration and validation periods.

The whole drainage basin area under study was approximately 5000 km^2. This study area was divided into eight blocks. The basin is described by a digital terrain model of resolution 30 arc second downloaded from GTOPO30 (USGS, US Geological Survey internet site). Despite of the coarse resolution, topographic data and inconsistency in observed hydro-meteorological data, the simulation results of the model were promising. The simulation results from calibration for the year 1991 are shown in Table 1. The results were good enough with 80% Nash efficiency. Figure 2 shows the comparison of hydrographs for this calibration period in 1991.

Table 1 Calibrated simulation results for year 1991.

Simulated average discharge (m^3 s^{-1})	211.3
Observed average discharge (m^3 s^{-1})	229
Nash efficiency (%)	80.0
Volume ratio of simulated and observed average discharge	0.9
Simulated water balance (m)	–0.2
Peak time of simulation (day)	229
Peak time of observation (day)	222
Simulated maximum peak discharge (m^3 s^{-1})	1162.6
Observed maximum peak discharge (m^3 s^{-1})	1360

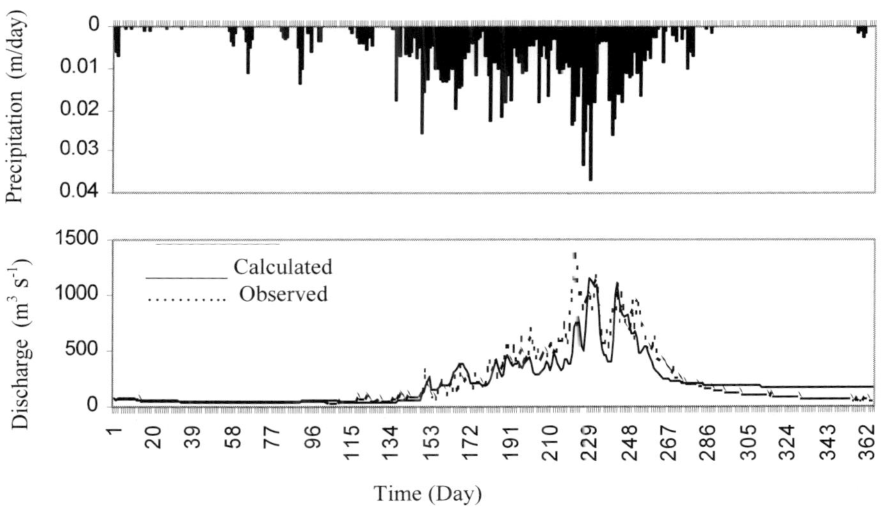

Fig. 2 BTOPMC Model calibration for Sun Kosi River in 1991 at Pachowar Ghat gauging station no. 630.

The same parameters were used for verification for the year 1990. The result was good enough with 75.8% Nash efficiency. The simulation results from validation year 1990 are shown in Table 2. Figure 3 shows the comparison of hydrographs for this validation period in 1990.

In the calibration period, the rate of recession for the groundwater flow was too low, which gave poor water balance for the later dry season of the year (refer Fig. 2). In TOPMODEL, the saturation deficit controls the discharge from the local area. The groundwater discharge, $q_b(i,t)$, is determined by the local saturation deficit, $SD(i,t)$, the gradient of the local surface, $\tan\beta i$, and the parameters relating to transmissivity of the block, T_o and m, as follows:

$$q_b(i,t) = T_o\, e^{-SD(i,\,t)/m} \tan\beta i \qquad (1)$$

Thus saturation deficit variation with respect to time was analysed. In Fig. 4, the saturation deficit gradually decreased and from day 229 onwards the soil is completely

Table 2 Validated simulation results for year 1990.

Simulated average discharge ($m^3 s^{-1}$)	284.1
Observed average discharge ($m^3 s^{-1}$)	253.7
Nash efficiency (%)	75.8
Volume ratio of simulated and observed average discharge	1.1
Simulated water balance (m)	–0.2
Peak time of simulation (day)	226
Peak time of observation (day)	199
Simulated maximum peak discharge ($m^3 s^{-1}$)	1597.9
Observed maximum peak discharge ($m^3 s^{-1}$)	1410

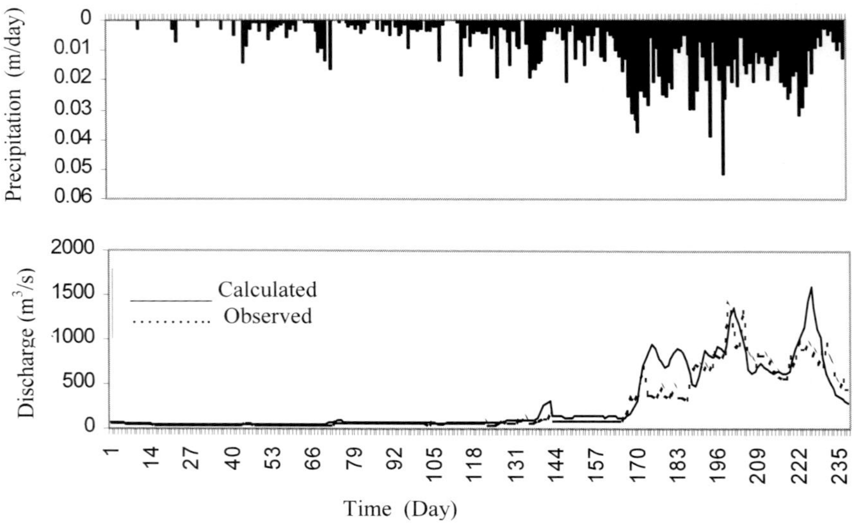

Fig. 3 BTOPMC Model verification for Sun Kosi River in 1990 at Pachowar Ghat gauging station no. 630.

saturated leading to low recession rate of groundwater flow. This is why the model failed to give satisfactory water balance for the later dry season of the year shown in Fig. 2. Thus, in validation period, the later dry season of the year was not taken into account (see Fig. 3).

All the parameter calibration was done by the trial-and-error method. Apart from the distributed soil topographic index, eight sets of parameters were estimated for the drainage basin divided into eight blocks shown in Table 3. Regarding the volumetric analysis, lateral transitivity played the most important role whose increment highly changed the base flow during the rainy season. On the other hand lateral transitivity and saturation deficit of the soil influenced each other in water balance analysis. The rate of recession of the groundwater flow in the later dry season is very low (see Fig. 2).

Hence, in the calibration by trial and error process the values of lateral transitivity were kept very low, as shown in Table 3. The increment of Manning's roughness coefficient decreased the peaks by shifting them backward.

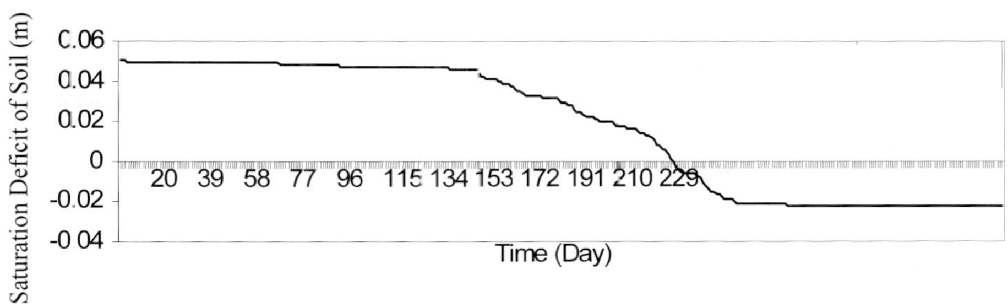

Fig. 4 Variation of saturation deficit of soil with respect to time for block 8.

Table 3 BTOPMC calibrated parameters for year 1991 and used in verification for year 1990.

Block Number	*To* ($m^2 h^{-1}$)	*m* (m)	Sr_{max} (m)	$S_{bar}0$ (m)	*n* ($m^{-1/3} s^{-1}$)
1	0.51	0.03	0.008	0.501	0.01
2	0.10	0.013	0.004	0.21	0.01
3	0.15	0.015	0.003	0.06	0.01
4	0.03	0.01	0.002	0.021	0.05
5	10.05	0.02	0.002	0.82	0.02
6	10.02	0.02	0.003	0.81	0.01
7	2.50	0.023	0.002	0.01	0.01
8	0.01	0.05	0.002	0.051	0.01

T_o: saturated transmissivity of soils;
m: decay factor of T_o;
Sr_{max}: capacity of root zone storage;
$S_{bar}0$: initial saturation deficit of soil;
n: Manning roughness coefficient.

On overall analysis of the performance, the BTOPMC model can be successfully used as a tool for integrated water resources investigation in a large watershed, mountain physiographic region, in Nepal.

REFERENCES

Ao, T. Q., Takeuchi, K. & Ishidaira, H. (1999) BTOPMC-1.0: User's guide to the computer programs (unpublished).

Beven, K. J. & Kirkby, M. J. (1979) A physically based, variable contributing area model of hydrology. *Hydrol. Sci. Bull.* **24**(1), 43–69.

Japan International Co-operation Agency (1985) *Master Plan Study of the Kosi River Water Resources Development.*. JICA, Kathmandu, Nepal.

Seibert, J. (1999) On TOPMODEL's ability to simulate groundwater dynamics. In: *Regionalization in Hydrology* (ed. by B. Diekkrugger, M. J. Kirkby & U. Schroder) (Proc. Braunschweig Conf., March 1999), 211–220. IAHS Publ. no. 254.

Takeuchi K., Ao, T. Q. & Ishidaira, H. (1999) Introduction of block-wise use of TOPMODEL and Muskingum-Cunge method for the hydro–environmental simulation of a large ungauged basin. *Hydrol. Sci. J.* **44**(4), 633–646.

Weather Radar Information and Distributed Hydrological Modelling (Proceedings of symposium HS03 held during IUGG2003 at Sapporo, July 2003). IAHS Publ. no. 282, 2003.

Coupling of MODFLOW and WATFLOOD in hydrological modelling of a small watershed

SAYED-FARHAD MOUSAVI

College of Agriculture, Isfahan University of Technology, Isfahan 84154, Iran

mousavi@cc.iut.ac.ir

NICHOLAS KOUWEN

Department of Civil Engineering, University of Waterloo, Waterloo, Ontario N2L 3G1, Canada

Abstract In water resources assessment, coupling of watershed and groundwater models integrates surface water and groundwater components. In this research, the coupling was performed by applying the recharge component of the WATFLOOD/SPL hydrological model to the MODFLOW groundwater model in the Laurel Creek watershed, Ontario, Canada. Net groundwater leakage computed by MODFLOW was combined with WATFLOOD/SPL surface flow and interflow. The flows obtained were compared with flows computed by the WATFLOOD/SPL model. Several runs were performed using radar-rainfall data and the hourly measured streamflow for the period 1993–1999. The results showed that use of adjusted radar precipitation improved the predicted streamflows. Running the MODFLOW model for 28 years of consecutive data showed that it takes about 20 years for the system to reach equilibrium, regardless of the starting heads. A simple lower-zone function (in WATFLOOD/SPL) was substituted for the MODFLOW to predict the streamflows.

Key words coupling; Laurel Creek, Canada; MODFLOW model; radar precipitation; WATFLOOD/SPL model

INTRODUCTION

Mathematical modelling of watersheds is a tool for assessing water resources. Current watershed models generally represent the surface water component adequately, but they simplify the dynamics of groundwater. On the other hand, groundwater models ignore the dynamics of surface water. VanderKwaak & Sudicky (1996) developed a fully coupled, surface-subsurface flow and solute transport model. A mathematical model for coupling the surface and subsurface flow was presented by Gandolfi & Savi (2000).

Groundwater forms an important source of municipal, industrial, agricultural and residential water supply in Ontario (MacRitchie *et al.*, 1994). The Regional Municipality of Waterloo and Grand River Conservation Authority are currently involved in detailed hydrological studies throughout the Laurel Creek watershed (part of the Grand River basin). The purpose of this paper is to couple WATFLOOD/SPL hydrological model with MODFLOW groundwater model in the Laurel Creek watershed. The study sought to determine whether replacing detailed aquifer properties by a simple function would affect simulation of streamflows.

MATERIALS AND METHODS

Hydrological model

The coupling process was performed by applying the WATFLOOD/SPL hydrological model to the Laurel Creek watershed (Fig. 1), taking the recharge component of this model and using it as an input to the MODFLOW groundwater model. The net groundwater leakage to the Laurel Creek is combined with surface flow and interflow computed by WATFLOOD/SPL and routed down the creek.

The surface hydrological processes are represented by the WATFLOOD/SPL modelling system, which consists of two parts: WATFLOOD is the data management system that includes a number of data pre-processing programs, and SPL is the hydrological simulation model (Kouwen & Mousavi, 2002). For subsurface hydrological processes, the MODFLOW groundwater model is used to simulate groundwater flow in three dimensions (Harbaugh & McDonald, 1996a,b).

Study area

The Laurel Creek watershed in southern Ontario, Canada (Fig. 1), with a drainage area of 70.45 km^2, was chosen. This watershed is located at longitude of 80°39′58″–80°28′46″W and latitude of 43°26′12″–43°30′24″N. Mean annual precipitation in the Grand River basin varies from 813 to 965 mm, and mean annual temperature of the region is about 7.4°C.

The square-grid (1 × 1 km) approach was used to allow for an easy interface with remotely sensed data, specially the radar rainfall measurements. Hourly rainfall

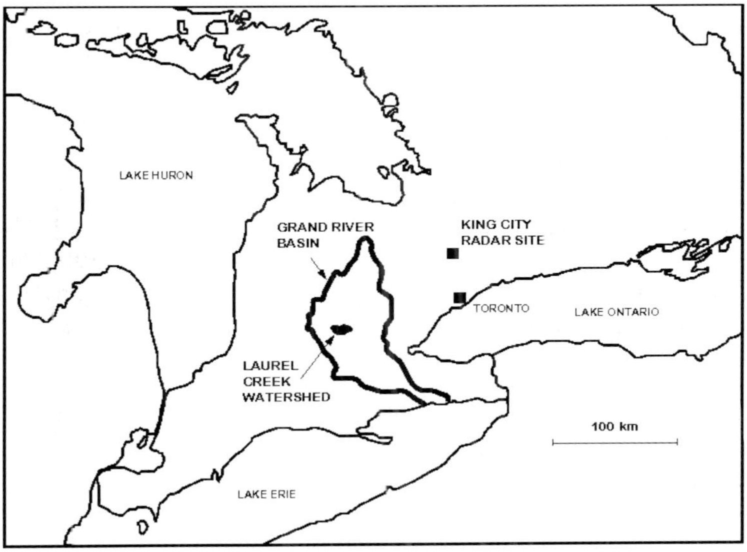

Fig. 1 Location map of Laurel Creek watershed.

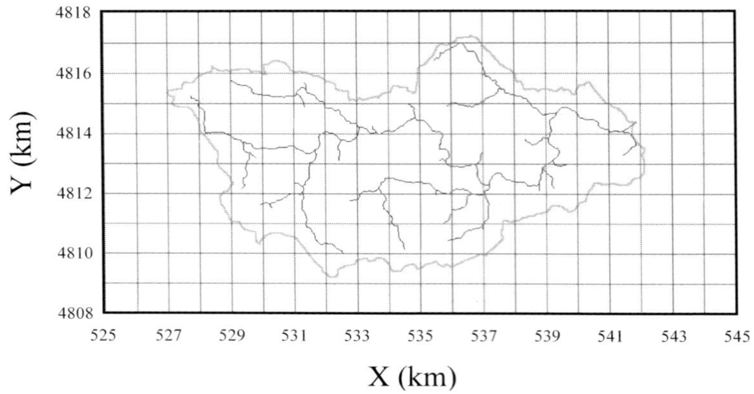

Fig. 2 Grid pattern for Laurel Creek watershed.

and temperature data of 1993–1999 were used. The Universal Transverse Mercator (UTM) coordinates were adopted for the surface and ground water models. The X-coordinates are from 525 to 545 km and Y-coordinates are from 4808 to 4818 km (Fig. 2).

Modelling scenarios

Three approaches were compared. First, WATFLOOD/SPL was applied and surface flow, interflow and groundwater outflow were computed. Next, the groundwater recharge was computed in WATFLOOD/SPL and used in MODFLOW. Two cases were considered in MODFLOW: (a) the groundwater system was considered as one unconfined aquifer, and (b) the groundwater system consisted of one unconfined and three confined aquifers.

Since actual water table depth (as the starting head) of each cell was not known in MODFLOW, three different positions of water table were assumed: (a) full saturation, or starting head equal to the average surface elevation of the cell; (b) starting head equal to average elevation minus 10% of the aquifer thickness; and (c) starting head equal to average elevation minus 20% of the aquifer thickness. Thirteen wells are discharging groundwater at a rate of minimum 0.0108 to maximum 0.0415 $m^3 s^{-1}$. It was assumed that the wells are running 12 h a day for the whole year. The MODFLOW model was run for 7 years (1993–1999) using the above starting heads. Then, another 7 and 14 years of consecutive synthetic data were incorporated into the model to attain the equilibrium.

RESULTS AND DISCUSSION

Behaviour of the coupled models

Several runs were performed using the hourly-measured streamflow at the Weber gauge station (located at X and Y coordinates of 539 and 4813 UTM, respectively) for

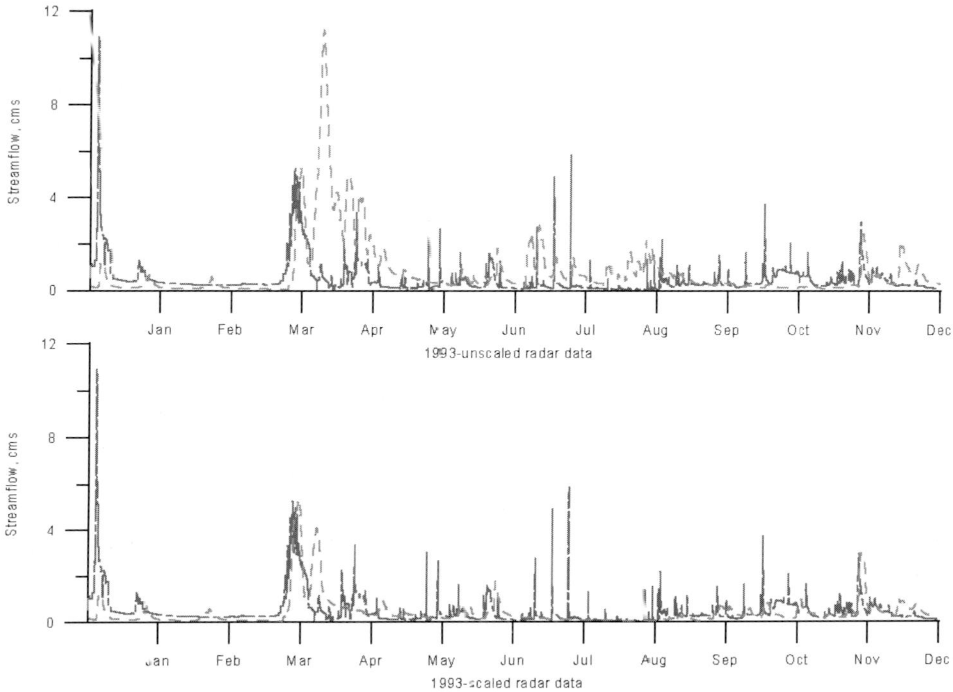

Fig. 3 Computed (broken line) *vs* measured (solid line) streamflow in 1993. Radar rainfall data were used to compute the hydrographs. The lowER hydrograph was obtained by using the calibrated radar rainfall data.

the period 1993–1999. Figure 3 shows the computed *vs* measured streamflow for 1993. Although there are large differences between gauge and radar accumulation for many of the individual events (because of underestimation or overestimation) (Fassnacht *et al.*, 1998), nevertheless, radar–rainfall data for this year are quite reliable. As seen, the computed curve does not fit the measured curve very well, especially in April. Snowmelt in March and April has produced a lot of runoff. In April, the highest predicted streamflow by WATFLOOD/SPL model was about 10.9 $m^3 s^{-1}$, while the highest measured flow was around 5.5 $m^3 s^{-1}$. Changing the scale factors for conversion of radar rainfall (Fassnacht *et al.*, 1999) improved the fit and decreased the discrepancies. After running the MODFLOW for 7 years (1993–1999), the final heads were obtained in each grid.

Comparison of the results (not shown here) showed that there is not much difference between head changes of the one-layer and four-layer models in many grids. This result shows that detailed information of different layers may not be necessary to improve the final-head prediction by MODFLOW. Since the exact position of the water table was not available, three arbitrary positions were assumed and MODFLOW was run for 28 years of consecutive data. Figure 4 shows that, at the beginning of the simulation, there is a large difference between the scenarios. But, as time passes, these differences diminish. On the basis of this figure, it takes about 20 years for the system to reach equilibrium, regardless of the starting heads.

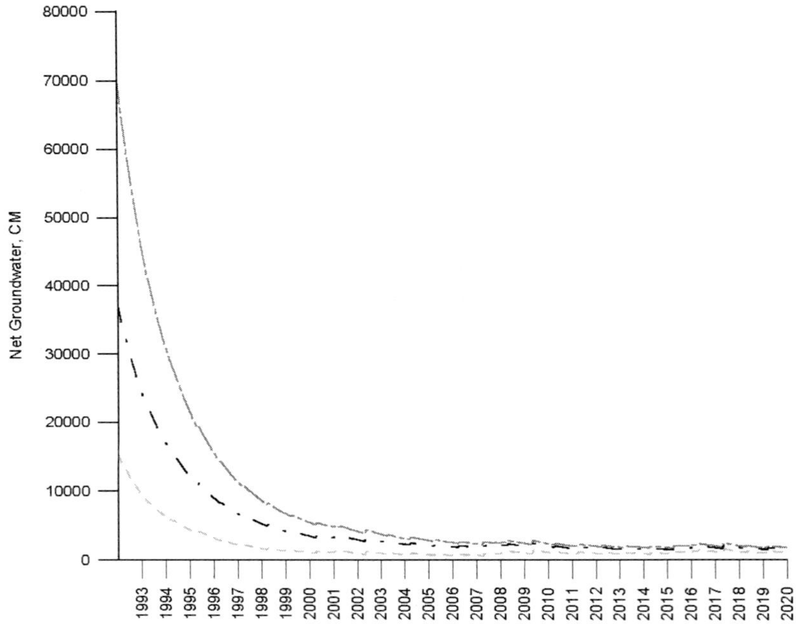

Fig. 4 Net groundwater leakage to the Laurel Creek. Starting head scenarios (top, middle, and bottom lines) are the starting head equal to the average surface elevation, the average elevation minus 10%, and the average elevation minus 20% of the aquifer thickness, respectively.

Comparison of stand-alone WATFLOOD and coupled WATFLOOD–MODFLOW

In WATFLOOD/SPL, an exponential groundwater depletion function is used to gradually deplete the lower-zone storage:

$$QLZ = LZF.LZS^{PWR} \tag{1}$$

Where *QLZ* is lower-zone discharge, *LZF* is a lower-zone function, *LZS* is lower-zone storage, and *PWR* is an exponent. The LZF value of equation (1) was set to a very low value (e.g. 10^{-15}) to produce only surface flow and interflow. The recharge computed by WATFLOOD/SPL was used to drive MODFLOW. The leakage computed by MODFLOW was then combined with the WATFLOOD/SPL surface runoff and interflow to produce inflow to the stream system. The combined flows were routed in the creek. Figures 5 and 6 compare the flows computed with the stand-alone WATFLOOD/SPL and the combined model for 1993–1996. Initially, the two models show different results. However, as time passes, the differences between the two hydrographs diminish and eventually, both methods result in the same flow. The simulation results for 1997–1999 are not shown because they are similar to 1996.

The conclusion from comparing the hydrographs in Figs 5 and 6 is that the simple lower-zone function can substitute the MODFLOW model in this particular watershed, and detailed aquifer properties are not needed to forecast the streamflow.

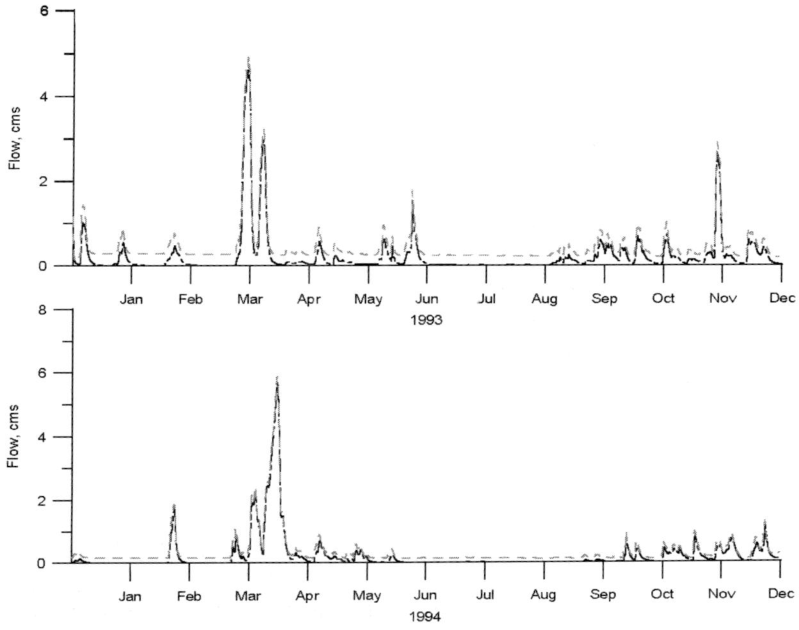

Fig. 5 Comparison of the simulated hydrographs with the stand alone WATFLOOD (solid line) and the combined WATFLOOD–MODFLOW (broken line) for 1993 and 1994.

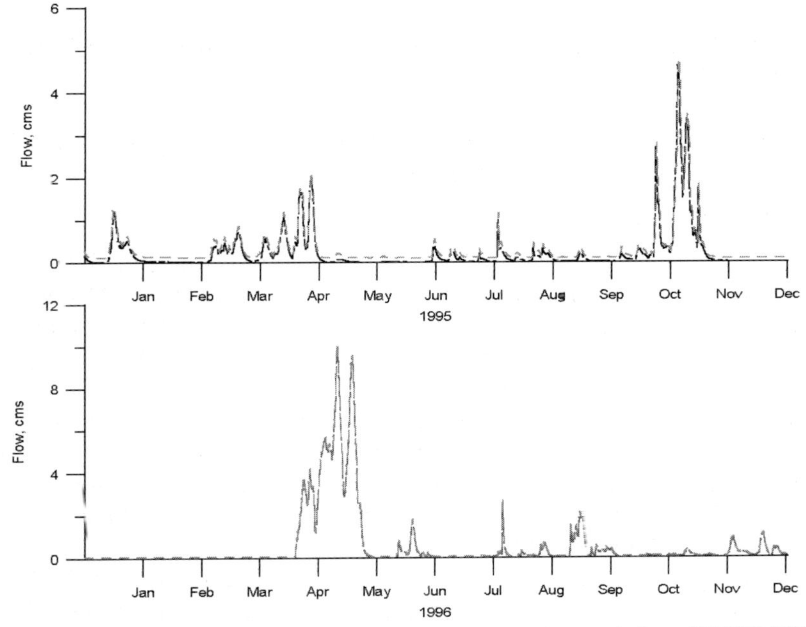

Fig. 6 Comparison of the simulated hydrographs with the stand-alone WATFLOOD (solid line) and the combined WATFLOOD –MODFLOW (broken line) for 1995 and 1996.

Acknowledgements The authors would like to thank the University of Waterloo for supporting the development of the WATFLOOD model, Environment Canada and Grand River Conservation Authority for providing the data, and Mr Frank Seglenieks for preparing the watershed files and some of the figures.

REFERENCES

Fassnacht, S. R., Soulis, E. D., Snelgrove, K. R. & Kouwen, N. (1998) Application of weather radar to model the snow hydrology of southern Ontario. (Proc. 55th Eastern Snow Conf.), 115–123. Jackson, New Hampshire, USA.

Fassnacht, S. R., Soulis, E. D. & Kouwen, N. (1999) Algorithm application to improve weather radar snowfall estimates for winter hydrological modelling. *Hydrol. Processes* **13**, 3017–3039.

Gandolfi, C. & Savi, F. (2000) A mathematical model for the coupled simulation of surface runoff and infiltration. *J. Agric. Engng. Res.* **75**, 49–55.

Harbaugh, A. W. & McDonald, M. G. (1996a) User's documentation for MODFLOW-96, an update to the US Geological Survey Modular Finite Difference Ground-Water Flow Model. *US Geol. Survey Open-File Report 96-485*, Reston, Virginia, USA, 56 p.

Harbaugh, A. W. & McDonald, M. G. (1996b) Programmer's Documentation for MODFLOW-96, an update to the US Geological Survey Modular Finite Difference Ground-Water Flow Model. *US Geol. Survey Open-File Report 96-486*, Reston, Virginia, USA.

Kouwen, N. & Mousavi, S. F. (2002) WATFLOOD/SPL hydrological model and flood forecasting system. In: *Mathematical Models of Large Watershed Hydrology* (ed. by V. P. Singh & D. K. Frevert). Water Resources Publications, LLC, Highlands Ranch, Colorado, USA.

MacRitchie, S. M., Pupp, C., Grove, G., Howard, K. W. F. & Lapcevic, P. (1994) Groundwater in Ontario: hydrogeology, quality concerns and management. NHRI Contribution no. CS-94011, Environment Canada, Burlington, Saskachewan, Canada.

VanderKwaak, J. E. & Sudicky, E. A. (1996) Simulating flow and transport in coupled surface subsurface hydrological systems. *EOS: Trans. Am. Geophys. Union* **77**(46), F264.

Prediction of evapotranspiration and streamflow with a distributed model over the large Wuding River basin

XINGGUO MO[1], SUXIA LIU[1,2], ZHONGHUI LIN[1] & WEIMIN ZHAO[3]

1 *Institute of Geographical Sciences and Natural Resources Research, Chinese Academy of Sciences, Beijing 100101, China*
moxg@igsnrr.ac.cn

2 *School of Mathematics, University of New South Wales, Sydney 2052, Australia*

3 *Bureau of Hydrology, Yellow River Conservancy Commission, Zhengzhou 450004, China*

Abstract A GIS-based distributed hydrological model (LISFLOOD) is used to simulate the evapotranspiration and streamflow over the large Wuding River basin of China, with LAI being retrieved from remote sensing NDVI. The model runs over 1995 for calibration and 1996 for validation. It is shown that the model can simulate the rainfall–runoff processes well in the wet season and is poor when simulating the baseflow in winter. The evapotranspiration is the principal component of water budget, the annual amounts of which are different over the land use types. Soil evaporation is greater than transpiration and sensible heat flux is 1.5 times higher than latent heat flux at the basin scale. About 78% of the rainfall amount is infiltrated into soil and 6% for surface runoff. The annual amount of recharge to groundwater is small.

Key words distributed hydrological model; evapotranspiration; streamflow; Wuding River basin

INTRODUCTION

Land-use/cover changes (LUCC), such as desertification, deforestation and reforestation in semiarid and arid regions are significantly altering the hydrological cycle and surface energy balance over a basin. The interactions between topography, vegetation, soil and climate in a basin determine the hydrological regime in a complex way, resulting in strong spatio-temporal variability of hydrological cycling processes. Prediction of the resulting changes on water cycles at the basin scale requires a distributed hydrological model that takes into account the relationships between hydrology, vegetation and atmosphere.

By embedding a soil–vegetation–atmosphere transfer (SVAT) model into a distributed hydrological model, the spatial and temporal variability of the involved processes described by SVAT models can be transferred to basin scales under the support of detailed physical–geographic properties (Strasser & Mauser, 2001; Beven, 2002). Remote sensing information, such as Normalized Difference of Vegetation Index (NDVI) which provides reliable phenological evolution of vegetation greenness over a large area, has been used widely to retrieve biophysical parameters (Maisongrande *et al.*, 1995; Gao, 1995; Sellers *et al.*, 1996) for such models.

The purpose of this study is to simulate the daily evolution of evapotranspiration and streamflow over the Wuding River basin, China, from 1995 to 1996, with a distributed model (LISFLOOD) coupled with a SVAT model. First, the study area and the data are briefly described. Then a section outlining the distributed model follows. Next, it describes how to retrieve vegetation leaf area index (LAI) from NDVI. Finally, simulation results in river discharge, evapotranspiration, surface energy budget, infiltration, surface runoff and recharge are presented.

STUDY AREA AND DATASET

The Wuding River basin, one of the largest sub-basins of the Yellow River, is located in the Loess Plateau of China at the range of 107–111°E and 37–39.5°N, covering an area of 30 261 km^2 (Fig. 1). The altitude ranges from 600 to 1800 m with the main flatland being about 1200 m above sea level in the northwestern part of the basin. The length of the main stream is 491.1 km with a mean slope of 1.97‰. Annual precipitation is about 400 mm, while most of the rainfall events occur in the summer monsoon season, especially from July to August. The yearly mean stream flux is about 30 $m^3 s^{-1}$. Mainly due to human disturbance, the Wuding River basin has been severely threatened by water erosion in summer and wind erosion in winter.

The land-use/cover is classified into six types in the basin, namely, agricultural field, deciduous broadleaf forest, mixture forest, dwarf shrub, grassland and desert with fractions of 22.4%, 1%, 0.3%, 10%, 45% and 21.4%, respectively. The soil texture is distinguished as sandy loam, sandy silt, sand, silt and coarse sand, among which 55% is sandy silt. The basin is divided into 29 370 grids with 1 km resolution. A

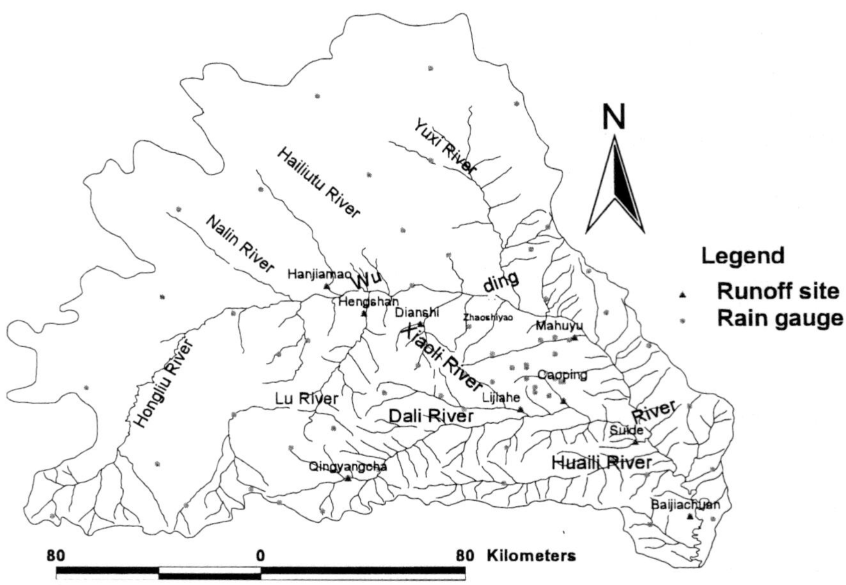

Fig. 1 The channel network of the Wuding River basin.

DEM from US Geological Survey is used to derive the stream channel pattern and the slope of each grid in the basin.

Daily precipitation records at 82 raingauges over the whole basin and meteorological records at 18 climatological stations in and around the basin are available from 1995 and 1996. The precipitation and meteorological measurements are interpolated to each grid with the distance weight least squares method.

MODEL

The distributed hydrological model is mainly based on LISFLOOD designed for large basin water cycle and flood prediction (De Roo *et al.*, 2000). It is a GIS-based model with PCRaster programming. In the model frame, the influences of topography, precipitation, land use and soil types are taken into account. A two-layer canopy energy balance SVAT scheme (Shuttleworth & Wallace, 1985) and the variable infiltration capacity scheme (Liang & Lettenmaier, 1994) describing the subgrid heterogeneity of infiltration are embedded into the original LISFLOOD model (Mo *et al.*, 2002). Both overland and channel runoff are routed with the kinematic wave equations. The model runs over 1995 and 1996 with a 1-km^2 grid size and daily time step.

LEAF AREA INDEX, LAI

LAI is an important biophysical variable that regulates the land surface energy partitioning between canopy and the underneath soil surface, as well as canopy rainfall interception. Uncertainty of LAI may result in bias on evapotranspiration and thus on water balance estimation. Numerical experiments showed that LAI changes of 1 resulted in 5% annual evapotranspiration variation (De Roo *et al.*, 2001).

In this study, monthly maximum composite NOAA–AVHRR NDVI with 1 km resolution in 1996 is utilized to retrieve vegetation canopy LAI. Annual evolution of LAI is interpolated to daily values for simulation purposes. A linear scaling between LAI and NDVI is applied as follows (Chen & Cihlar, 1996):

$$LAI = LAI_{max} \frac{NDVI}{NDVI_{max}} \tag{1}$$

where LAI_{max} is the maximum leaf area index, $NDVI_{max}$ is the maximum vegetation index. The LAI_{max} and $NDVI_{max}$ for the six vegetation types are set as Table 1. In the

Table 1 Ratio to the whole basin area, maximum values of *LAI* and *NDVI*, annual precipitation and evapotranspiration for the land cover types (A: Agricultural fields; B: Deciduous broadleaf forest; C: Mixed forest; D: Dwarf shrub; E: Grassland; F: Desert).

	A	B	C	D	E	F
LAI_{max}	4	5	4	4	4	0
$NDVI_{max}$	0.674	0.721	0.721	0.674	0.674	--
Precipitation (mm)	389	397	326	352	361	312
Evapotranspiration (mm)	400	335	371	333	386	191

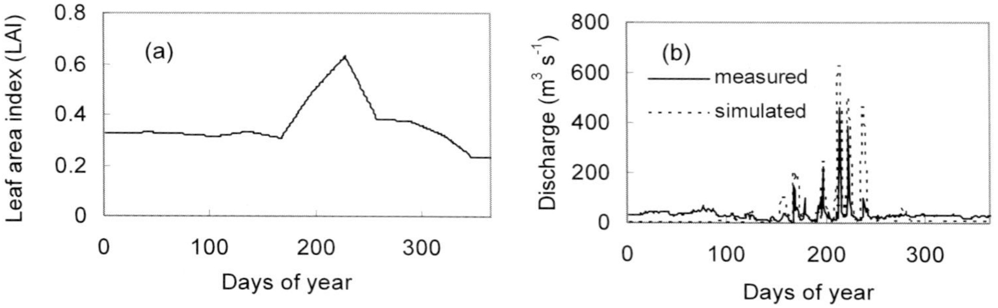

Fig. 2 (a) Daily evolution of the averaged leave area index. (b) Comparison of the simulated and observed daily discharge at the basin outlet.

Wuding River basin, the average LAI is small most of the year and the maximum value of 0.6 occurred in July and August as shown in Fig. 2(a).

RESULTS

River discharge

The model runs over 1995 and 1996 with rainfall and meteorological input at daily time step. The observed discharge at the outlet Baijiachuan in 1995 is used to calibrate the model. The most sensitive parameters to discharge simulation are river channel Manning roughness coefficient N and the infiltration parameter B, which describes the partitioning of rainfall into infiltration and surface runoff. Parameter B is sensitive to total discharge, whereas N is sensitive to the temporal phase of the hydrograph. The two parameters are adjusted manually to calibrate the discharge. On the annual time scale, the difference between the simulated and measured discharge is 5% and the Pearson correlation coefficient of daily discharge is 0.64. The model is able to simulate the rainfall–river flow process well in two of three peaks in summer as shown in Fig. 2(b). Comparing with the basin average rainfall, it is obvious that the simulated discharge highly corresponds to the rainfall processes (Fig. 3). However, since there are some reservoirs and flood-lagging dams in the basin to reduce the flood peaks and discharge volume, considerable differences between the simulated and measured discharges exist as the present model has not taken into account reservoir effects for the streamflow prediction. It is also noticed that the daily average observed discharge is about 30 $m^3 s^{-1}$ in winter, which is much higher than the simulated discharge.

Evapotranspiration

The results in 1996 show that the evapotranspiration amount (342 mm $year^{-1}$), which is very close to the averaged precipitation (356 mm $year^{-1}$), is undoubtedly the principal term of water balance over the Wuding River basin. The residual of precipitation minus evapotranspiration (14 mm) is lower than the simulated runoff depth, which can be explained by the difference compensated from soil moisture and groundwater

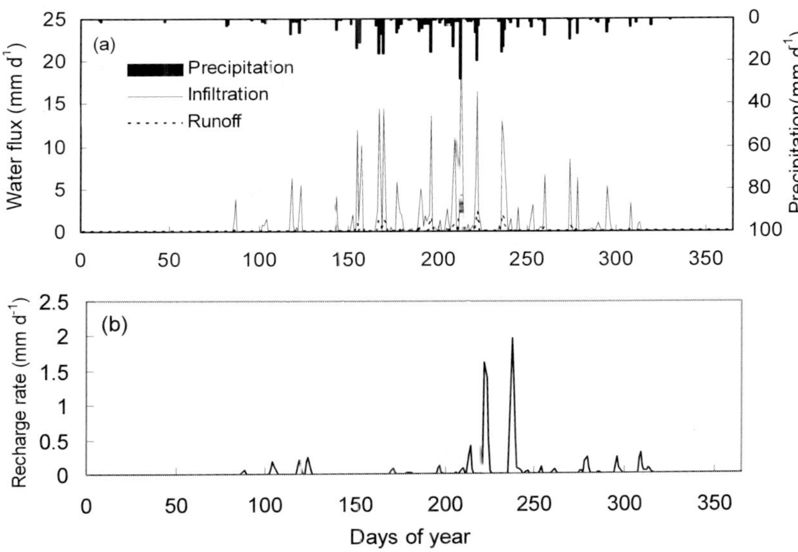

Fig. 3 Daily precipitation, infiltration and surface runoff (a) and recharge rate to groundwater (b) over the basin.

storage. Usually it takes about five years for soil water storage to reach the equilibrium (Strasser & Mauser, 2001).

As shown in Table 1, the yearly value of evapotranspiration from the agricultural field is the greatest, the next being that from the grassland. In the desert, the evapotranspiration is quite small, consuming only 61% of total precipitation. As a consequence, a large amount of soil water is recharged to the groundwater. This may explain why the groundwater level is much shallower in the desert part of the basin. Evapotranspiration from some other vegetation types exceeds the precipitation supplied from soil moisture storage.

As shown in Fig. 4(a), there is no remarkable difference between the annual evolution patterns of transpiration *TR* and soil evaporation *ES*. As expected, the

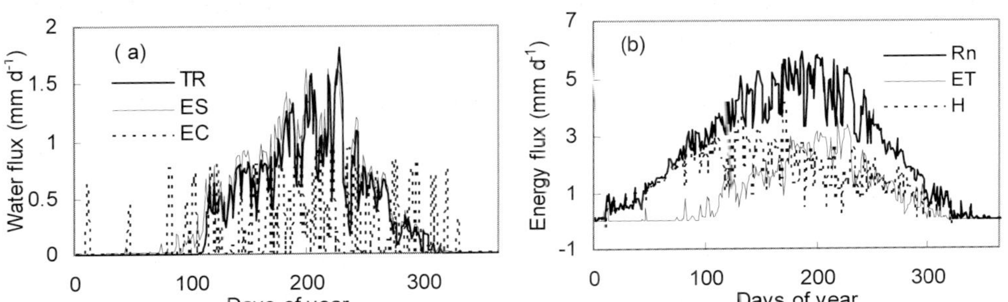

Fig. 4 (a) Daily evolution of transpiration (*TR*), soil evaporation (*ES*) and canopy intercepted water evaporation (*EC*) over the basin. (b) Daily net radiation (*Rn*), latent heat flux (*ET*) and sensible heat fluxes (*H*) over the basin.

evaporation from vegetation canopy intercepted water *EC* only occurred on raining days. Annually, the ratios of *TR*, *ES* and *EC* to *ET* are 0.38, 0.46 and 0.16, respectively, which means that the soil evaporation exceeds transpiration in this arid and semiarid basin. This may be due to the high proportion of desert with a low vegetation cover fraction and LAI.

Surface energy budget

Assuming that soil heat flux is a small term of the surface energy budget on a daily scale, we only concentrate on net radiation *Rn*, latent heat flux *ET* and sensible heat *H* for surface energy budget. It is shown that *ET* (with water depth equivalence in mm) was very small in the early spring and winter period when precipitation was scarce, and then reached its high values stage from Days of Year (DOY) 170 to 230 when the rainy season arrived and the vegetation was well developed (Fig. 4(b)). Annually *Rn*, *ET* and *H* are 914, 337 and 485 mm, respectively, and the Bowen ratio is 1.43, indicating that the sensible heat flux is almost 1.5 times higher than latent heat flux in this basin.

Infiltration, surface runoff and recharge to groundwater

In the Wuding River basin, most of the rainfall that reaches the ground surface is infiltrated into the ground (78%), and only a small part is assigned to surface runoff (6% or 22 mm), as shown in Fig. 3(a). However, there are always two to three heavy storms in summer that may cause considerable flooding and inundation conditions. Since annual precipitation is inadequate in the basin, soil moisture is usually low. As a result, percolation is also small. It is found from daily simulation that the maximum daily recharge is about 2 mm day^{-1} after a heavy rainfall and the yearly recharge amount is about 14 mm (Fig. 3(b)).

Acknowledgements Financial support for this work was from the Chinese Academy of Sciences innovative project (kzcx2-310), NSFC project (90211007) and the Fellowship of the Royal Society, UK. The first author is indebted to Dr Ad de Roo at Space Application, Institute DG Joint Research Centre, European Commission and Prof. Keith Beven at the University of Lancaster for providing the LISFLOOD model.

REFERENCES

Beven, K. (2002) Towards an alternative blueprint for a physically based digitally simulated hydrologic response modelling system. *Hydrol. Processes* **16**, 189–206.

Chen, J. M. & Cihlar, J. (1996) Retrieving leaf area index of boreal conifer forests using Landsat TM images. *Remote Sens. Environ.* **5**,153–162

De Roo, A. P. J., Wesseling, C. G. & van Deursen, W. P. A. (2000) Physically-based river basin modelling within a GIS: the LISFLOOD model. *Hydrol. Process.* **14**, 1981–1992.

De Roo, A. P. J., Dojik, M., Schmuck, G., Koster, E. & Lucjeer, A. (2001) Assessing the effects of land use changes on floods in the Meuse and Oder basin. *Phys. Chem. Earth (B)*, **26**(7–8), 593–599.

Gao, W. (1995) Parameterization of subgrid-scale land surface fluxes with emphasis on distributing mean atmospheric forcing and using satellite-derived vegetation index. *J. Geophys. Res.* **100**(D7), 14 305–14 317.

Liang, X. & Lettenmaier, D. (1994) A simple hydrologically based model of land surface water and energy fluxes for general circulation models. *J. Geophys. Res.* **99**(D7), 14 415–14 428.

Maisongrande, P., Ruimy, A., Dedieu, G. & Saugier, B. (1995) Monitoring seasonal and interannual variations of gross primary productivity, net primary productivity and net ecosystem productivity using a diagnostic model and remotely-sensed data. *Tellus* **47B**, 178–190.

Mo, X., Beven, K. & Pappenberger, F. (2002) Water budget estimation and prediction uncertainty with the distributed model-LISFLOOD over the Lushi basin, China. *J. Hydrol.* (submitted)

Sellers, P. J., Los, S. O., Tucker, C. J., Justice, C. O., Dazlich, D. A., Collatz, G. J. & Randall, D. A. (1996) A revised land surface parameterization (SiB2) for atmospheric GCMs. Part II: The generation of global fields of terrestrial biophysical parameters from satellite data. *J. Climate* **9**, 706–737.

Shuttleworth, W. J. & Wallace, J. S. (1985) Evaporation from sparse crops: An energy combination theory. *Quart. J. Roy. Met. Soc.* **111**, 1143–1162.

Strasser, U. & Mauser, W. (2001). Modelling the spatial and temporal variations of the water balance for the Weser basin 1965–1994. *J. Hydrol.* **254**, 199–214.

Weather Radar Information and Distributed Hydrological Modelling (Proceedings of symposium HS03 held during IUGG2003 at Sapporo, July 2003). IAHS Publ. no. 282, 2003.

A catchment surface runoff simulation for land surface model study

SHUFEN SUN & HUIPING DENG
Institute of Atmospheric Physics, LASG, PO Box 9804, Beijing 100029, China
ssf@lasg.iap.ac.cn

Abstract TOPMODEL has become attractive in a large-scale hydrological study, as it is a conceptual and distributed model with clearly physical bases to present the topographic control on hydrology and spatial distributions of soil moisture in a catchment. In this paper a numerical hydrologic simulation of 28 years (1960–1987) in the Soumou River catchment in west China was conducted by using TOPMODEL implemented with a simple land surface model. In order to estimate topography index, a key parameter to TOPMODEL, a digital elevation model (DEM) was applied for the elevation data with 100 m resolution in the catchment. The numerical results from the model were in good agreement with the 28-year observed data, which indicates the potential of TOPMODEL use in a large-scale land surface process model. A sensitivity study of the effect of the channel initiation threshold (CIT) on the simulated runoff result was also conducted.

Key words physical meaning; runoff simulation; Soumou River catchment, China; TOPMODEL; topography index

INTRODUCTION

The hydrological process is an important process on the land surface (Sellers *et al.*, 1986), which regulates the partition of precipitation on the land surface and affects the spatial distribution pattern of soil moisture, and in turn influences the water cycle and energy balance on the land surface. The hydrology model is not well established in the most current climate models for correctly predicting surface water conflux process and its spatial redistribution either horizontally or vertically. Besides, due to the low spatial resolution in current GCM, it is not able to provide many important hydrological aspects in detail. There are two kinds of traditional hydrology models, empirical models and physically-based models. The empirical models are black-box models which can only predict water efflux at the exit mouth of a catchment but are not able to provide any information about how soil moisture is distributed and in turn where the saturation zone is to produce surface runoff. Besides, some parameters in the models have no clear physical meaning, and all the parameters in the models need long-term runoff data to fix them. Since the 1980s, physically-based and distributed models, including very detailed physics such as the SHE model (Abbott *et al.*, 1986), have been developed. Compared with the empirical models, they can project a detailed history of many hydrological processes and they do not need long-term historical data for model parameter identification. However, this kind of model includes too many physical coefficients, which are also very difficult to obtain, and they consume a lot of computer resources. It makes the models unsuitable for extensive use in the climate study. In order to meet the requirement from the land surface modelling study in GCM, TOPMODEL (Topography hydrology model), a partly distributed and partly physically-based model, has recently been studied. The model has the advantages of good physical meaning, a few parameters, reduced computer time, and more importantly, takes soil moisture spatial distribution into consideration. It is hoped that the description of water cycle and energy balance in the current land surface model can be improved for climate study (Randal *et al.*, 2000) by applying TOPMODEL. This paper will briefly introduce the principal physical base first. Then, based on TOPMODEL, the river discharge from 1960 to 1987 in the Soumou River catchment, a tributary of the Yangtze River basin in China, is simulated. Results show TOPMODEL can reproduce the observation data quite well.

BASIC FEATURES OF TOPMODEL

Basic principle and structure of TOPMODEL

Numerous field studies have indicated that surface runoff in wet regions is mainly produced by saturation excess runoff (Dunne flow). It means a spatial variation of soil properties and antecedent soil moisture will result in different surface runoff production. For a large area (e.g. a grid size in GCM), the saturation excess runoff will occur in a certain portion of the large area where there is no soil moisture deficit, or where groundwater reaches the ground surface. Even though there are several surface runoff models based on a saturation excess runoff mechanism, only a few models take the topography influence on spatial distribution pattern of soil moisture into consideration. In fact it is clearly shown that the topography is an essential factor to decide the spatial distribution of soil moisture and water table in a hillslope catchment. TOPMODEL is one of the models which fully takes the influence of topography issue on the soil moisture and groundwater table in the catchment into consideration.

By using the prescribed topography index of a catchment and average water storage deficit calculated in the catchment, the model can directly predict the variations of groundwater table and special distributions of soil moisture in the root zone, and in turn predict the area portion in a catchment where saturated excess runoff occurs. The model has the advantage of a few parameters with good physical meaning and saving computer time (Beven & Kirkby, 1979; Beven, 2000). TOPMODEL is based on the following assumptions (Beven & Kirkby, 1979; Sivapalan *et al.*, 1987): (a) the water table is approximately parallel to the soil surface; (b) the saturated soil hydraulic transmissivity exhibits an exponential decline with the depth below the surface; and (c) under quasi steady-state conditions, the groundwater table is recharged with a special uniform rate. From the model, variations of the groundwater table and total baseflow from the catchment can be derived as follows (Beven, 2000):

$$D_i = \overline{D} + m\left(\lambda - \ln\frac{a}{k_0 \tan\beta}\right) \quad (1)$$

$$Q_b = Ak_0 e^{-\lambda} e^{-\overline{D}/m} \quad (2)$$

$$Q_o = Ak_0 e^{-\lambda} \tag{3}$$

where D_i is local storage deficit in the unsaturation zone, $\overline{D}$ is mean of storage deficit in whole catchment, and λ is the average topography index of the catchment equal to:

$$\frac{1}{A}\int_0^A \ln\frac{a}{\tan\beta}\,\mathrm{d}A \tag{4}$$

A is the entire catchment area, Q_b is total base flow from the drainage; k_o is local saturated transmissivity at the surface, m is a parameter to adjust the decreasing rate of local transmissivity with increasing water storage, and $\tan\beta$ is the hydraulic gradient equal to the rate of catchment surface elevation change. Q_o and m can be determined by using the river runoff data. In order to complete TOPMODEL to calculate other components such as evaporation, vegetation interception, infiltration to soil, surface runoff production and so on, a simple land surface model is implemented into TOPMODEL.

Calculation of topography index

How to calculate the topography index is a key issue (Quinn *et al.*, 1995; Deng & Li, 2002) for TOPMODEL performance. Currently, digital topography models (DTMs) or digital elevation models (DEMs) are extensively used to calculate the spatial distribution of topography index in a catchment (Quinn *et al.*, 1995; Wolock & Price, 1994; Wolock & McCabe, 1995). Using different resolution elevation data and different methods will affect the final result of the topography index distribution. If grid cells in the catchment having a river channel segment are considered as grid cells without a river channel segment and become water collecting areas, the number of grid cells with the high topography index increases. It changes the distribution of topography index and in turn enlarges average topography index in the catchment. In order to reduce the effect of the uncertainty of defining a grid with or without a river channel, an upper limit area, named Channel Initiation Threshold, CIT is set up. If a grid cell with the draining area is greater than the CIT, the grid cell should have a river segment and its draining area is equal to CIT. The curves of topography index distributions and its accumulated frequencies have been calculated with different CITs based on the DEM with 100-m resolution (Fig. 1). The maximum topography index in the catchment and average topography index increases with the CIT value increment, and there is a small difference in the accumulated frequencies for CITs greater than 0.5-km^2 when 100-m resolution elevation data are used.

MODEL VALIDATION

To evaluate the TOPMODEL used here, the long-term runoff data of the Soumou River catchment 1960–1987 were simulated by the model together with a simple land surface model. The catchment is located in mountainous northwest China and has an area of 2538 km^2. There are only two rainfall data sets from two raingauge stations.

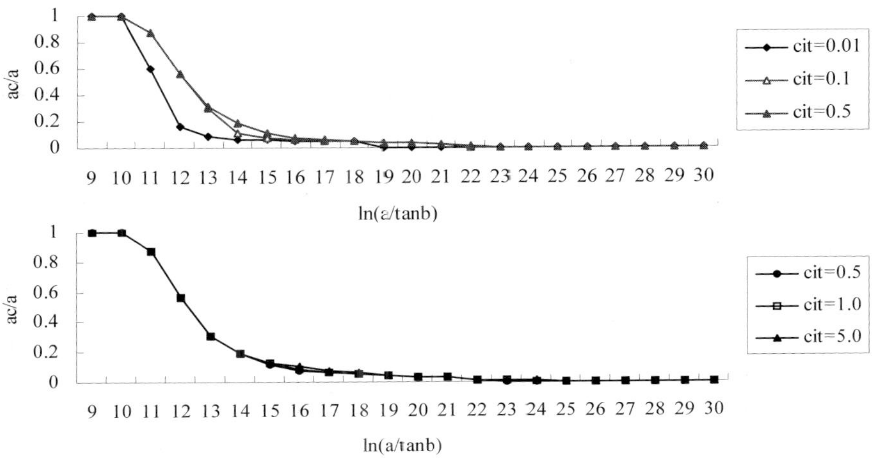

Fig. 1 Distribution functions of ln(a/tanβ) with different upslope catchment area limited threshold

One set of data is from the Marcon County meteorology station from 1954 to 1987 and the other is from the Hongyuan County meteorology station from 1961 to 1987. The input forcing data of precipitation from 1961 to 1987 was obtained by area proportional average of data from these two stations. By using the river runoff recession curve in the dry season from 1980 to 1984, Q_o and m are determined. Q_o and m are equal to 3.5 mm day^{-1} and 39 mm from May to September and 2.0 mm day^{-1} and 60 mm from October to April, respectively. According to the distribution information of vegetation type for each grid and the estimation of maximum intercepted water storage S_{Di} of ith specific vegetation (4 mm for forest, 3 mm for shrub, 2 mm for sparse forest, 1.5 mm for grass and 0.5 mm for sparse grass) in the catchment, the average maximum intercepted water storage S_D for each grid can be obtained based on area proportional average. For example, the average S_D for the whole catchment was 2.0 mm in early 1970, 1.7 mm in 1986 and 1.5 mm in 1999. Potential evaporation E_o was obtained based on Penman's equation by using the data from the Marcon County meteorology station. Radiation data were obtained from empirical formulae (Gao & Lu, 1982). By using the above parameters, the river runoff of the Soumou River catchment from 1860 to 1987 is simulated. In general, the model simulates the observation data of the river runoff pretty well in both magnitude and change trend. A large difference may occur in some seasons and some years. This may be due to the input forcing data of precipitation because there are only two raingauge stations used to estimate the whole catchment precipitation, which may greatly deviate from the real situation.

Figures 2 and 3, as examples, show two years of results from 28 years simulation based on the CIT equal to 0.5 km^2. Figure 2 shows the comparison of daily averaged river runoff between the model simulation and observations in the two years. Figure 3 shows the comparison of monthly averaged river runoff between the model simulation and observations in the same two years. The model can simulate observation runoff pretty well in many years but does not agree well in some years such as in 1978, as

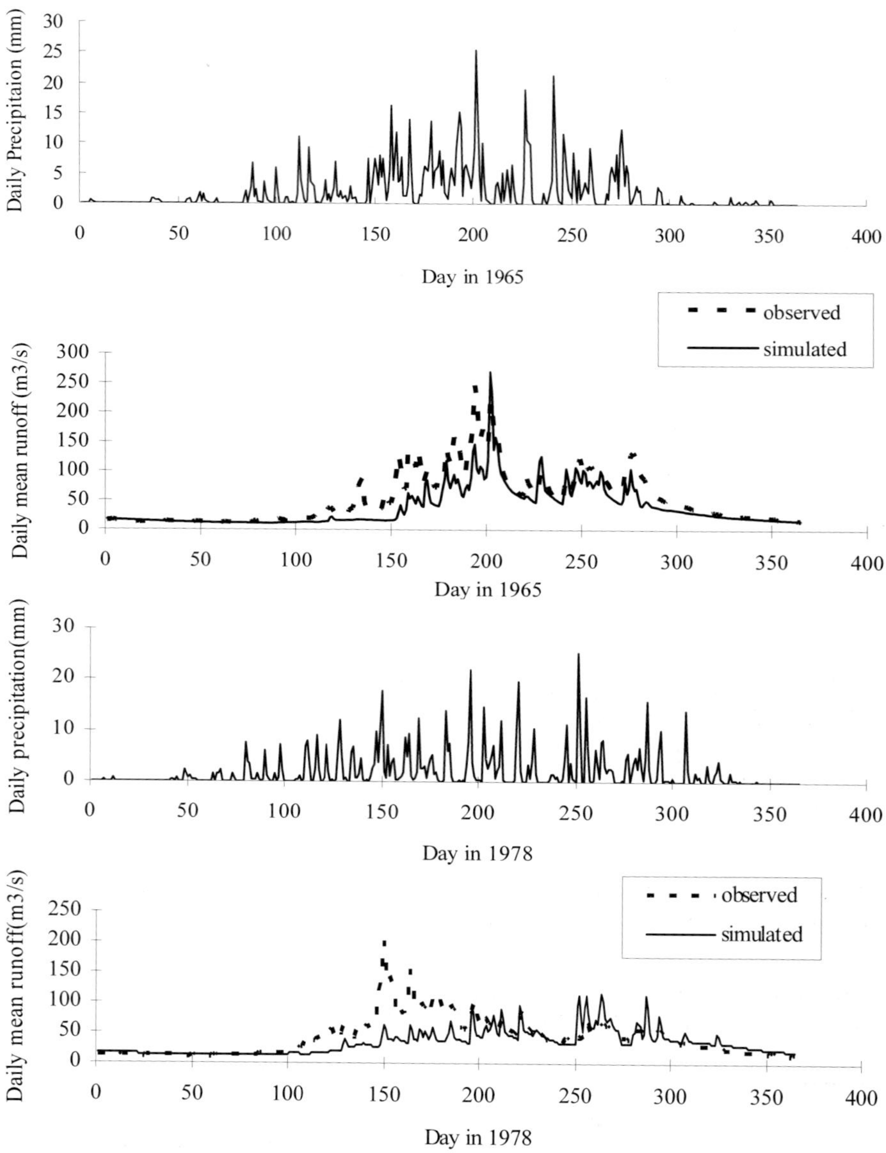

Fig. 2 Comparison of daily observed and simulated runoff in 1965 and 1978.

shown in the figures in either magnitude or change trend. From daily averaged runoff and precipitation of 1978 in Fig. 2, it clearly shows the inconsistency between the observed rainfall input and the observed river runoff, i.e. less precipitation and more observed river runoff in days 150–190 but more precipitation and less runoff in days 250–300. It is obviously unreasonable. The only possible explanation is that using two raingauge station data sets to estimate the precipitation in the whole catchment may cause big errors.

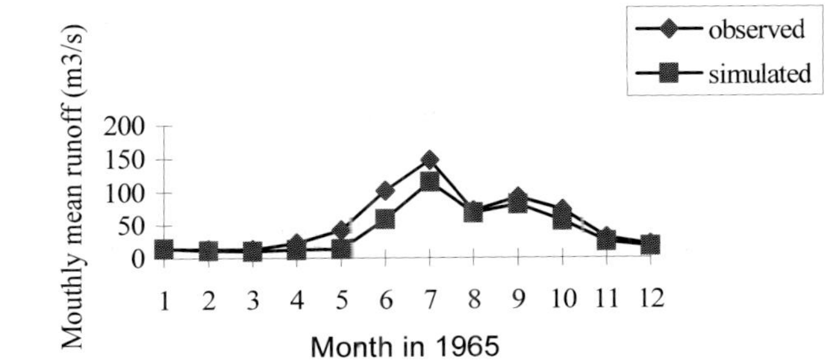

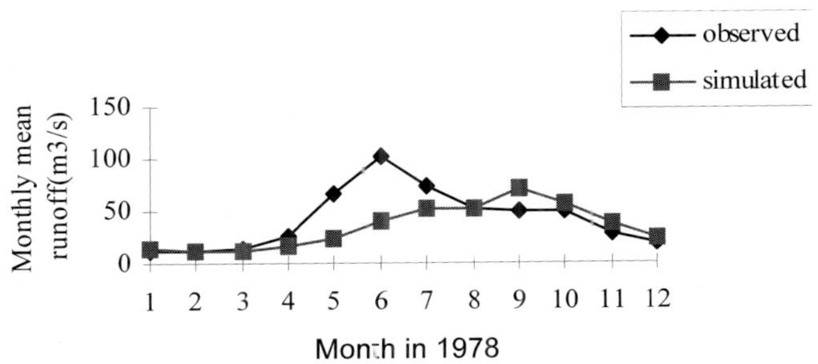

Fig. 3 Comparison of monthly observed and simulated runoff in 1965 and 1987.

In order to evaluate general quality of the model, Table 1 shows the percentage efficiency E of each year from 1960 to 1987 with CIT equal to 0.01, 0.1, 0.5, 1 and 5 km^2, respectively. E is defined as:

$$E = 1 - \frac{\sigma_{\varepsilon}^2}{\sigma_0^2} \tag{4}$$

where σ_o is the variance of the observation data and σ_ε the variance of difference between the observation and predicted results by the model. The coefficients E show

Table 1 Percentage efficiency with different CIT values (km^2).

Year	CIT (km^2)					Year	CIT (km^2)				
	0.01	0.1	0.5	1	5		0.01	0.1	0.5	1	5
1960	0.53	0.66	0.70	0.72	0.72	1974	0.79	0.79	0.79	0.79	0.78
1961	0.43	0.46	0.52	0.53	0.54	1975	0.57	0.61	0.64	0.65	0.67
1962	0.69	0.73	0.73	0.72	0.72	1976	0.76	0.75	0.73	0.73	0.74
1963	0.64	0.67	0.71	0.72	0.73	1977	0.70	0.67	0.65	0.64	0.63
1964	0.66	0.66	0.66	0.65	0.65	1978	0.28	0.33	0.38	0.38	0.38
1965	0.77	0.77	0.72	0.71	0.71	1979	0.53	0.56	0.60	0.60	0.60
1966	0.54	0.58	0.63	0.64	0.65	1980	0.82	0.83	0.83	0.83	0.83
1967	0.68	0.68	0.68	0.67	0.67	1981	0.84	0.86	0.85	0.85	0.86
1968	0.62	0.62	0.64	0.64	0.65	1982	0.86	0.85	0.84	0.83	0.83
1969	0.80	0.80	0.81	0.81	0.81	1983	0.80	0.83	0.85	0.85	0.86
1970	0.29	0.29	0.33	0.33	0.36	1984	0.68	0.72	0.76	0.76	0.77
1971	0.68	0.68	0.68	0.68	0.68	1985	0.52	0.52	0.53	0.53	0.55
1972	0.71	0.72	0.71	0.71	0.72	1986	0.72	0.72	0.74	0.74	0.74
1973	0.69	0.69	0.70	0.69	0.68	1987	0.65	0.67	0.69	0.70	0.70

some difference between CITs less than 0.5 km^2 and greater than 0.5 km^2. But for CIT = 0.5, 1 and 5 km^2, E is almost the same because their curves of the topography index distribution and corresponding accumulated frequency shown in Fig. 1 are very close. From the table, the values of E for most years are greater than 0.5, especially E for CITs greater than 0.5 km^2 are around 0.70–0.80. It means the model works well in simulating the runoff in the Soumou River catchment.

CONCLUSIONS

Generally, the TOPMODEL implemented with a simple land surface model can predict the runoff in a hillslope catchment well and can determine soil moisture distribution. When using DTMs or DEMs, the effect of CIT, Channel Initiation Threshold, should be taken into consideration. Its effect on the simulation result is related to the surface elevation data resolution and river catchment characteristics. It is important to keep in mind that TOPMODEL can work well for the hillslope catchments with moist soil and shallower groundwater.

Acknowledgements This work was conducted under the support of three Chinese projects of NSF grant 40233034, NSF 40075019 and KZCX2-SW-210.

REFERENCES

Abbott, M. B., Bathurst, J. C. & Cunge, J. A. (1986) An introduction to the European Hydrological System–System Hydrologique Européen (SHE). *J. Hydrol.* **87**, 45–59.

Beven, K. J. & Kirkby, M. J. (1979) A physically based, variable contributing area model of basin hydrology. *Hydrol. Sci. Bull.* **24**(1), 43–69.

Beven, K .J. (2000) *Rainfall-Runoff Modelling*. John Wiley & Sons Inc., New York, USA.

Deng, H. P. & Li, X.B. (2002) Relationship of upslope contribution area and soil water content in TOPMODEL. *Progr. Geogr.* **21**(2), 103–110 (in Chinese).

Gao, G. D. & Lu, Y. R. (1982) *Radiation and Heat Balance of Land Surface in China.* Science Press, Beijing, China (in Chinese).

Quinn, P. F., Beven, K. J. & Culf, A. (1995) The ln(a/tanβ) index: how to calculate it and how to use it within the TOPMODEL framework. *Hydrol. Processes* **9**(2), 161–182.

Randal, D. K., Max, J. S., Agnes, D., Stieglitz, M. & Praveen, K. (2000) A catchment-based approach to modelling land surface processes in a general circulation model 1. Model structure. *J. Geophys. Res.* **105**(No.D20), 24808–24822.

Sellers, R. J., Mintz, Y., Sud, Y. C. & Dalcher, A. (1986) A simple biosphere model (SiB) for use within general circulation models. *J. Atmos. Sci.* **43**(6), 505–531.

Sivapalan, M., Wood, E. F. & Beven, K. J. (1987) On hydrologic similarity, 2: A scaled model of storm runoff production. *Water Resour. Res.* **23**(12), 2266–2778.

Wolock, D. M. & McCabe, Q. J. (1995) Comparison of single and multiple flow direction algorithms for computing topographic parameters in TOPMODEL. *Water Resour. Res.* **31**(5), 1315–1324.

Wolock, D. M. & Price, C. V. (1994) Effects of digital elevation model and map scale and data resolution on a topography–based watershed model. *Water Resour. Res.* **30**(11), 3041–3052.

Approaching realistic soil moisture status with an improved mesoscale numerical weather prediction model

SUXIA LIU[1,2], LANCE LESLIE[1], MILTON SPEER[3], REES BUNKER[4] & RUSSEL MORISON[1]

1 *School of Mathematics, The University of New South Wales, Sydney, New South Wales 2052, Australia*
suxia@maths.unsw.edu.au

2 *Institute of Geographical Sciences and Natural Resources Research, Chinese Academy of Sciences, Beijing 100101, China*

3 *Bureau of Meteorology, Sydney, New South Wales, Australia*

4 *Rural Fire Office, New South Wales, Australia*

Abstract An advanced soil moisture scheme (Richards) is coupled to a high resolution numerical weather prediction model (HIRES) replacing the original Force–Restore scheme and changing other related processes. The new scheme now provides HIRES with an upper and lower bound for soil moisture through a process of checking model precipitation that is used to calculate soil moisture. A comparison of HIRES model results is presented using the new and original soil moisture schemes applied to the Goulburn River catchment in southeastern Australia. It is shown that precipitation is the most important factor to be considered before introducing an advanced scheme to correctly simulate soil moisture. It is also shown that, based on the correct precipitation input, the Richards scheme provides a more realistic soil moisture profile. Improving model soil moisture will ultimately provide better estimates of forest fire danger indices used as guidance by weather forecasters in assessing bushfire risk.

Key words bushfire; forest fire danger index; Goulburn River catchment, Australia; numerical modelling; soil moisture

INTRODUCTION

Although widely recognized for long-term climatic simulations (Eltahir, 1998) with GCMs, the importance of soil moisture is still less emphasized in short-range forecasting (Lin *et al.*, 2002; Cassardo *et al.*, 2002) in numerical weather prediction (NWP) models. Simulating soil moisture with a NWP model presents two main challenges. Firstly, NWP models generally have relatively simple land-surface schemes, which treat soil moisture processes poorly, and generally soil moisture is not output as a model predicted variable when using these schemes. Such schemes include, for example, the Force–Restore method (Deardorff, 1977; Noilhan & Planton, 1987). Secondly, the model forecast validity period in short to medium range weather forecasting (from 48 h to several days), is not long enough for values of model output variables such as precipitation, temperature and relative humidity to affect soil moisture processes. For soil moisture processes to be assessed using values of model predicted variables, the time period should be no less than about one month.

One aim of this study is to improve soil moisture calculation in a high resolution numerical weather prediction model (HIRES). Soil moisture calculation is assessed over a period covering approximately six weeks (November/December, 1997) for the Goulburn River catchment in southeastern Australia prior to and following a large bushfire that occurred in the catchment. Given the lack of a long-term soil moisture data set that could be used to initialize HIRES, we present a methodology to develop such a data set.

The ultimate aim of this study is to improve the model prediction of Forest Fire Danger Index (FFDI), which in Australia is the main index used as guidance by fire weather forecasters to issue fire weather warnings. FFDI and most schemes that calculate a fire danger index are based on values of air temperature, relative humidity, wind speed, and a factor that represents the dryness of the vegetation based on the number of days since precipitation has occurred. Few schemes, if any, have yet considered routinely soil moisture status, which is a key factor in assessing the dryness of vegetation and hence bushfire risk. Therefore, it is anticipated that enhancing the way soil moisture is calculated in HIRES will lead to improve values of FFDI for assessing bushfire risk.

In this paper, the HIRES model is first briefly described. This is followed by a section outlining adjustments to HIRES to include more soil moisture processes. Third, the data set developed for the simulation is described. The final section presents the modelling results prior to and after the soil moisture improvements are made.

HIRES MODEL

The University of New South Wales HIgh RESolution (HIRES) model is a limited-area numerical weather prediction model developed by one of the authors, and many versions have been tested extensively in a wide range of applications. It is a primitive equation model on a Lambert–Conformal projection and utilizes the sigma coordinate system on an Arakawa C grid. A four-dimensional 6-hourly cycled statistical interpolation scheme is employed to update boundary conditions. The semi-implicit semi-Lagrangian approach (splitting the advective and gravity wave terms) is used for horizontal advection. The Mellor–Yamada boundary layer scheme (Level 2.25) is used to resolve the boundary layer meteorology. Leslie *et al.* (1985) embedded a detail physical process parameterization scheme into HIRES, which includes a stability dependent boundary layer with eddy diffusivity functions of bulk Richardson number; the vertical diffusion above boundary layer based on mixing length hypothesis, a surface heat budget with prognostic equation for surface temperature, a large-scale precipitation scheme, and a modified Kuo convection scheme. More recently, options utilizing the Kain-Fritsch cumulus parameterization scheme (Kain & Fritsch, 1990), and an explicit physics version (Wang, 1999) have been included.

For the hydrological cycle in HIRES, relative humidity is held to the value specified in the initial state by the Global Assimilation and Prognosis system (GASP) (Bourke *et al.*, 1995; Seaman *et al.*, 1995) at a horizontal resolution of 200 km and/or the Limited Area Prediction System (LAPS) (Puri *et al.*, 1998) at 75 km. If surface temperature and dew point analyses are available, then those specify the surface humidity. Otherwise the surface humidity is diagnosed as the weighted average of saturated humidity and other variables. The Force–Restore method was used in the original version of HIRES to calculate soil moisture. This two-layer soil model prognostically represents the rapid response of near-surface moisture to forcing by precipitation or evaporation and also provides a source of moisture restored from the deep soil to the surface when there is no precipitation.

REVISION TO THE HIRES MODEL

In the original version of HIRES, a lower bound for soil moisture is not included. Therefore, the soil moisture output from HIRES can be smaller than wilting level, which is unrealistic. In this study, we have added a lower and an upper bound to the model calculation of soil moisture.

In the original version of HIRES, the precipitation used to calculate soil moisture in the Force–Restore method is the quantity acquired from the specification and solution of the Helmholtz equations. In this study, the quantity used is the sum of the convective and large-scale model forecast precipitation. For convenience, hereafter this revised method is specially referred to as Precipitation Revision.

In the original version of HIRES, each subsequent nesting starts with soil moisture that was specified using surface humidity from the first nesting. In this study, the second and the third nesting uses the initial value of soil moisture from the previous nesting.

In order to compare the Force–Restore method used in the original version of HIRES with the Richards method, the latter is coupled to HIRES. The Richards method has previously been incorporated into an advanced multiple soil-layer soil moisture simulation which established the ALSIS model (Irannejad & Shao, 1998).

Theoretically, a two-layer soil scheme such as the Force–Restore model is not sufficient to fully represent processes involving soil moisture and soil temperature. The Richards model, with additional soil layers, more appropriately represents soil moisture profiles and can better describe the physical process of water flow. The flexible number of soil layers also allows a more effective treatment of root activity and hence a more representative evaporation scheme. A limitation in using additional model soil layers is that it will increase the computation time.

DATA

Unlike many other variables in HIRES, the amplitude of the soil moisture variation is small. A sufficiently long modelling period is needed to assess any differences in the two schemes used. HIRES was initially developed for short-term weather forecasting, with LAPS providing initial state and boundary conditions out to 48 h and GASP providing initial state and boundary conditions out to 168 h. In order to run HIRES over 40 days (from 1 November to 11 December 1997), which covers a sufficiently long period prior to, and following, a large bushfire in the catchment, improved data input to HIRES is needed.

By comparing the forecast results based on LAPS (75 km horiz. res.) and GASP (*c.* 200 km horiz. res.) data for 48 h with observed data of screen temperature, wind, relative humidity and sea level pressure at the closest meteorological station (not shown here), it is shown that results based on LAPS capture the upper and lower bound of sea level pressure much better than the results based on GASP. Therefore, although GASP provides archived boundary conditions out to 168 h, we use the boundary conditions covering the shorter time period out to just 48 h by LAPS, owing to its better verification against observations.

The archived LAPS data consist of two sets of 48-h forecasts per day data with analysis based either on 11 UTC observed data or 23 UTC observed data. Comparing the results based on 11 UTC and 23 UTC analysis data, there are no obvious differences (not shown here). Errors in a model forecast will grow with time due to the chaotic nature of the atmosphere and the uncertainty in specifying the initial conditions (Ibbitt *et al.*, 2001). These errors will pass through the model system and contribute to errors in the forecast. In order to remove this error and to allow us to determine the accuracy of the HIRES results, only LAPS analyses at 11 UTC and 23 UTC (not 6 hourly) are used as initial and boundary conditions. The model was run for a total period of 960 h (40 days). Currently, the maximum possible period that HIRES can run is 999 hours.

The first nesting is at a horizontal resolution of 50 km within the Australian region analysis and forecast domain. The second nested domain covers the eastern half of Australia at 20 km horizontal resolution. Finally, the third nesting covers the area containing the Goulburn River catchment at 5 km resolution. A more detailed description of the Goulburn River catchment and the 1997 bushfire can be found in Liu *et al.* (2003).

RESULTS AND DISCUSSION

The difference in the results with and without the lower and upper bound of soil moisture is readily apparent. The difference in the results with each subsequent nesting starting with soil moisture that was specified using surface humidity from the first nesting and the results with the second and the third nesting using the initial value of soil moisture from the previous nesting, shows little difference in the 48-h forecast results. The logic, however, for this revision is more reasonable. Therefore, we now highlight the results before and after Precipitation Revision, the differences between the Force–Restore (FR) scheme and the Richards (R) scheme, and the differences between the results at different spatial resolutions (50 km, 20 km and 5 km).

From the results it is obvious that before the Precipitation Revision, neither the Force–Restore scheme nor the Richards scheme could capture the expected soil moisture pattern. An example is shown for Bylong station (32°21′57″S, 150°06′52″E) (Fig. 1) at 5-km resolution, where there is too much noise in soil moisture output with the Force–Restore scheme (Fig. 1(a)) and it shows almost no change with the Richards Scheme (Fig. 1(b)), even though there are precipitation peaks; nor in the Force–Restore scheme does the noise correspond to the precipitation peak.

After Precipitation Revision, there is good correspondence between peaks of simulated precipitation and simulated soil moisture (Fig. 1(c) and (d)). There is also a recessing limb of soil moisture after the precipitation peak, which is a common occurrence in soil moisture variation following rain. The CPU time using the Richards

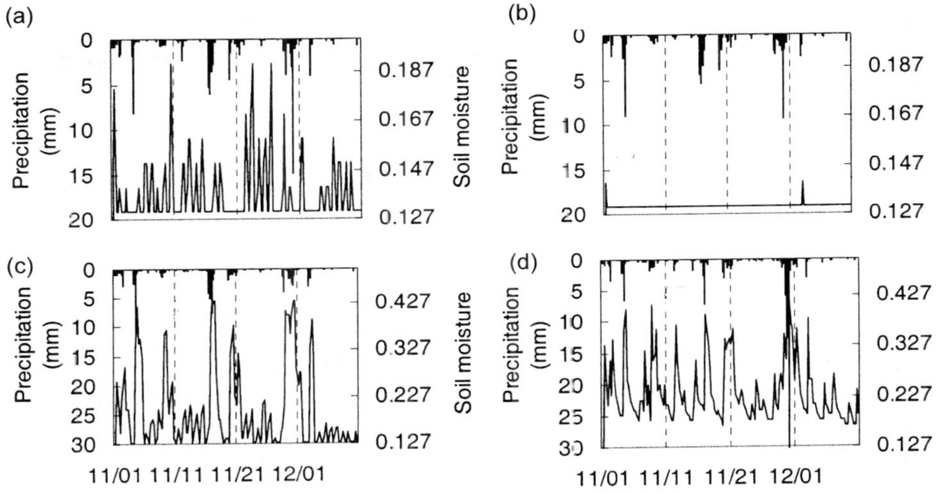

Fig. 1 The simulated precipitation (upper part) and simulated soil moisture (lower part) at the Bylong Station with the resolution of 5 km using (a) the Force–Restore scheme before Precipitation Revision; (b) the Richards scheme before Precipitation Revision; (c) the Force–Restore scheme after Precipitation Revision; and (d) the Richards scheme after Precipitation Revision.

scheme is about twice that for the Force–Restore method owing to the incorporation of nonlinear calculations.

Comparison of the results between the Richards and Force–Restore schemes reveals that changes mostly occurred in precipitation and soil moisture. All the other variables such as temperature, wind speed and sea level pressure reveal insignificant changes as expected. The reason for this is that, although there is a feedback between soil moisture and atmospheric variables in HIRES via surface humidity, the link is still weak. Steps are already underway to strengthen the links between these feedback processes.

Overall, precipitation is still, as expected, the most important factor in determining the temporal and spatial distribution of soil moisture before introducing any advanced soil moisture simulation scheme. It is also true that based on the correct precipitation input, the Richards scheme can provide more realistic soil moisture values. This conclusion is based on the assumption that within the study area the soil property remains the same. The effects of changing the soil properties will be assessed in future. These results will then be compared with the results of a study by Shao *et al.* (1997) with reference to the Australian continent.

Acknowledgements This work is sponsored by a Strategic Partnership with Industry Research and Training Grant (SPIRT) between The University of New South Wales, the NSW Rural Fire Service and the Bureau of Meteorology, Sydney. The first author acknowledges the NSFC project 90211007 and Knowledge Innovation Project of the Chinese Academy of Sciences KZCX2-310. Thanks to Mr G. Mackay and Mr Siu Wong at the Bureau of Meteorology, Sydney, Susana Realica, Allan Raine, Suzanne Bridgman at the Department of Land and Water Conservation, NSW, Australia and Pam O'Neill at National Parks and Wildlife Services, NSW, Australia in providing hydrological, meteorological and bushfire data for the Goulburn River catchment and Dr L. Qi for the help in running the HIRES model.

REFERENCES

Bourke, W., Hart, T., Steinle, P., Seaman, R., Embery, G., Naughton, M. & Rikus L. (1995) Evolution of the Bureau of Meteorology's Global Assimilation and Prediction system. Part 2: resolution enhancements and case studies. *Aust. Met. Mag.* **44,** 19–40.

Cassardo, C., Balsamo, G. P., Cacciamani, C., Cesari, D. & Paccagenella, T. (2002) Impact of soil surface moisture initialization on rainfall in a limited area model: a case study of the 1995 South Ticino flash flood. *Hydrol. Processes* **16**, 1301–1317.

Deardorff, J. W. (1977) A parameterization of ground surface moisture content for use in atmosphere prediction model. *J. Appl. Met.* **16**, 1182–1185.

Eltahir, E. A. B. (1998) A soil moisture rainfall feedback mechanism: 1. Theory and observations. *Water Resour. Res.* **34**(4), 765–776.

Ibbitt, R. P., Herderson, R. D., Copeland, J. & Wratt, D. S. (2001) Simulating mountain runoff with meso-scale weather model rainfall estimates: a New Zealand experience. *J. Hydrol*. **239**, 19–32.

Irannejad, P. & Shao, Y. (1998) Description and validation of the Atmosphere-Land-Surface Interaction Scheme (ALSIS) with HAPEX and Cabauw data. *Global Planet. Change* **19**, 87–114.

Kain, J. & Fritsch, J. M. (1990) A one-dimensional entraining/detraining plume model and its application in convective parameterization. *J. Atmos. Sci.* **47**, 2784–2802.

Leslie, L. M., Mills, G. A., Logan, L. W., Gauntlett, D. J., Kelly, G. A., Manton, M. J., McGregor, J. L. & Sardie, J. M. (1985) A high resolution primitive equations NWP model for operations and research. *Aust. Met. Mag*. **33**, 11–35.

Lin, C. A., Wen, L., Beland, M. & Chaumont, D. (2002) A coupled atmospheric–hydrological modeling study of the 1996 Ha! Ha! River basin flash flood in Quebec, Canada. *Geophys. Res. Letters* **29**(2), 13,1–4

Liu, S., Leslie, L. M., Speer, M., Bunker, R & Mo, X. (2003) The effects of bushfires on hydrological processes using a paired-catchment analysis. *Met. Atmos. Phys*. (in press)

Noilhan, J. & Planton, S. (1989) A simple parameterization of land surface processes for meteorological models. *Mon. Weath. Rev*. **117**, 536–549.

Puri, K., Dietachmayer, G., Mills, G., Davidson, N., Bowen, R., Logan, L. & Leslie, L. (1998) The new BMRC regional assimilation and prediction system. *Aust. Met. Mag.* **47**(3), 203–223.

Seaman, R., Bourke, W., Steinle, P., Hart, T., Embery, G., Naughton, M. & Rikus, L. (1995) Evolution of the Bureau of Meteorology's Global Assimilation and Prediction system. Part 1: analysis and initialisation. *Aust. Met. Mag.* **44**, 1–19.

Shao, Y., Leslie, L. M., Munro, R. K., Iranneijad, P., Lyons, W. F., Morison, R., Short, D. & Wood, M. S. (1997) Soil moisture prediction over the Australian Continent. *Meteorol. Atmos. Phys*. **63**, 195–215.

Wang, Y. (1999) A triply-nested movable mesh tropical cyclone model with explicit cloud microphysics (TCM3). BMRC Research Report no. 74, Bureau of Meteorology Research Centre, Melbourne, Victoria, Australia.

Key word index